Tissue Engineering and Developmental Biology

Tissue Engineering and Developmental Biology

Edited by **Shay Fisher**

SYRAWOOD
PUBLISHING HOUSE
New York

Published by Syrawood Publishing House,
750 Third Avenue, 9th Floor,
New York, NY 10017, USA
www.syrawoodpublishinghouse.com

Tissue Engineering and Developmental Biology
Edited by Shay Fisher

International Standard Book Number: 978-1-68286-167-7 (Hardback)

Printed in the United States of America.

Contents

Preface

Tissue engineering has revolutionized the field of medical science. The scientific progress in the development of biomaterials and stem cells has further accelerated the growth of this discipline. This book is dedicated to the emerging trends in different aspects of tissue engineering and developmental biology such as synthesis, spectroscopy, etc. It also focuses on the tools and techniques for regeneration of germlings and seedlings along with autografts. This book aims to serve as a valuable source of reference for students, academicians and researchers.

This book unites the global concepts and researches in an organized manner for a comprehensive understanding of the subject. It is a ripe text for all researchers, students, scientists or anyone else who is interested in acquiring a better knowledge of this dynamic field.

I extend my sincere thanks to the contributors for such eloquent research chapters. Finally, I thank my family for being a source of support and help.

Editor

A study on the formation and development of *Panax bipinnatifidus* Seem. adventitious root

Bui Dinh Thach, Lenguyen Tu Linh, Nguyen ThiThuy Van, Trinh Thi Ben, Mai Truong and Nguyen Huu Ho

Institute of Tropical Biology, Vietnam Academy of Science and Technology, Vietnam.

This study aimed to generate adventitious roots from root and stem-derived callus of *Panax bipinnatifidus* Seem. Callus formation was observed best (83.3%) in MS media supplemented with 3.0 mg/L 2,4-dichlorophenoxyacetic acid (2,4-D) and 0.1 mg/L indolebutyric acid (IBA). Callus generated adventitious roots in MS media supplemented with 2.0 mg/L IBA at high frequency (66.7%). The optimal culture condition for the development of adventitious roots was liquid MS media supplemented with 40 mg/L sucrose and maintained at pH 6.0.

Key words: Adventitious root, stem, basic MS media.

INTRODUCTION

Panax bipinnatifidus SEEM. (Araliaceae) is a hygrophilous and shade-enduring plant, preferring cool and wet climate conditions with an average temperature of about 12-15°C. In nature, this plant is relatively rare and mostly found in the high mountainous region of Hoang Lien Son in the northwest of Vietnam. The root of *P. bipinnatifidus* has been used as a valuable tonic to increase mental and physical performance, improve thinking and memory, reduce cancer risk, and lower blood sugar in diabetics in the Vietnamese traditional medicine (Tung et al., 2011). Nowadays, bioactive compounds in *P. bipinnatifidus* Seem. are characterized, including compounds in oleanolic triterpenoid saponin group: Bifinoside A-C (1-3), narcissiflorine methyl ester (4), chikusetsu saponin IVa (5), pseudoginsenoside RP1 methyl ester (6), stipuleanoside R1 (7), pseudoginsenoside RT1 methyl ester (8), momordinIle (9) and stipuleanoside R2 methyl ester (10) (Tung et al., 2011). *P. bipinnatifidus* roots contain saponins triterpens (chikusetsu saponin IV, zingibrosid R1, ginseninosid Ro, Rb, Rd, Re, Rg1, Rg2,...), and phytosterol, reducing sugars, oils, uronic acids and fatty acids. Extracts from *P. bipinnatifidus* has metabolic activity (Dua et al., 1989), anti-inflammatory activity (Matsuda et al., 1991), anti-oxidation (Wen-Ta Chiu, 1999), cardiovascular benefits (Chen et al., 2008; Yang et al., 2005).

Adventitious roots are non-transgenic, proliferating through plant growth regulators supplemented in the media. Adventitious roots were rapidly proliferated and accumulated high concentrations of secondary metabolites (Hahn et al., 2003; Yu et al., 2005). This method could be used to increase the accumulation of secondary

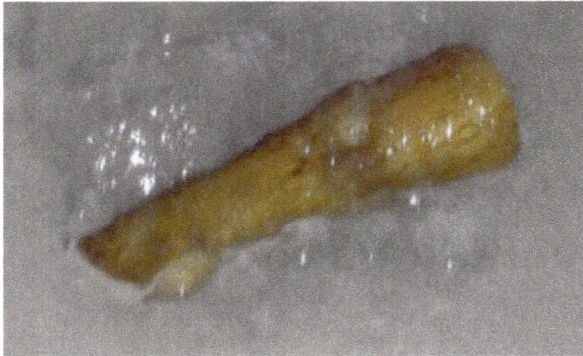

Figure 1. Stem segment.

Figure 2. Root segment.

metabolites in plants. In this research, we have generated adventitious roots from callus of *P. bipinnatifidus* Seem. and investigated effects culture conditions on development of adventitious roots.

MATERIALS AND METHODS

Induction of adventitious roots

The roots and stems of *P. bipinnatifidus* were collected in Sapa, Laocai, Vietnam and were taxonomically identified by botanist Ngo Van Trai (Institute of Medicinal Materials, Hanoi, Vietnam). Explants were washed under running water and soap, then washed again with sterilized distilled water. Explants were then treated with Hypochlorite Na 50% (v/v) solution for 25 min and washed with sterilized distilled water. Explants were then soaked in cefotaxim (250 ppm) for 24 h, and washed with sterilized distilled water. Explants were cut into pieces and cultured on MS media (Murashige and Skoog, 1962). Explants were cut into 0.1-0.3 cm length and cultured on MS medium suplemented with 30 g/L sucrose, 8 g/L agar, pH 5.8-5.9 and 2,4-dichlorophenoxyacetic acid (2,4-D) (2 and 3 mg/L), indolebutyric acid (IBA) (0.1, 0.3, 0.5 mg/L), NAA (0.1, 0.3 and 0.5 mg/L). The callus induction frequency and percentage of callus were recorded after 8 weeks of culture. Callus

were isolated and subcultured on MS media supplemented with 30 g/l sucrose, 8 g/l agar and IBA (0.5, 1.0, 1.5, 2.5 mg/L), NAA (0.5, 1, 1.5, 2, 3 mg/L), with pH adjusted to 5.8. Observations were made after 8 weeks of culture by percentage of explants producing adventitious roots. Cultures were maintained at 25±2°C in the dark.

Effects of culture conditions on the adventitious root

Adventitious roots were harvested from callus cultures and cut into 1 g and was cultured on different media including MS, Chu (N6) Medium and Gamborg's B5; supplemented with 30 g/l sucrose, 8 g/l agar, pH 5.8-5.9 and 2.0 mg/L IBA. Adventitious roots were culture selected media at three status: liquid (0 g/l agar, shaking at 80 rpm, semi-solid (4 g/L agar) and solid (8 g/L agar). Fresh weight and dry weight of cultured roots were recorded after four to eight weeks culture. Cultures were maintained at 25± 2°C in the dark.

Effects of pH, sugar content and mineral proportions on the adventitious root

Adventitious roots was cultured on media with different pH (5.5; 6.0; 6.5; 7.0 and 7.5), sucrose content (20, 30, 40, 50, 60 and 70 g/l), mineral proportions (25, 50, 75 and 100% when compared with the basic media). Fresh weight and dry weight of cultured roots were recorded after 4-8 weeks culture. The experiments were carried out in dark condition, at 25± 2°C, shaking at 80 rpm for liquid media.

Statistics

Experiments were repeated three times. Experimental data were processed by Duncan test (p<0.05, for experiment ≥ 6 treatments) or t test (for experiment with less than 6 treatments) by SAS program (ver. 6.12).

RESULTS AND DISCUSSION

Callus induction

After sterilization, explants were cut into small segments and cultured on MS media without plant growth regulators. The percentage of successfully sterilized explants was 65.2% after 14 days of culture (Figures 1 and 2).

Sterilized *P. bipinnatifidus* explants were cut into 0.1-0.3 cm segments and cultured on media for callus induction. Results are shown in Table 1, Figures 3 and 4.

Results from ANOVA table show that there is no significant difference between the effects of IBA and NAA on the callus induction efficiency (P>0.001); there was a significant difference of the D treatment (2,4-D) (F=127.6 với p<0,001), indicating that 2,4 D played a main role in callus induction capacity of *P. bipinnatifidus*, as in *Panax ginseng* roots (Thành and Yoeup, 2008).

Results indicate that the treatment C4 (2,4-D 3.0 and IBA 0.1 mg/l) induced the highest number of 83.3%.

Adventitious root induction from callus

The results showed (Figure 5 and Table 2), RI$_4$ and RI5 adventitious rooting sample rates are the same, RI$_5$ treatment had higher adventitious rooting average number

Table 1. Effects of plant growth regulators on callus induction.

Treatment	2,4-D (mg/L)	IBA (mg/L)	NAA (mg/L)	Callus induction (%)
C_1	2.0	0.1	0.0	22.3^e
C_2	2.0	0.3	0.0	33.3^e
C_3	2.0	0.5	0.0	33.3_e
C_4	3.0	0.1	0.0	83.3^a
C_5	3.0	0.3	0.0	64.0^{bc}
C_6	3.0	0.5	0.0	55.3^{bc}
C_7	2.0	0.0	0.1	30.3^e
C_8	2.0	0.0	0.3	36.3^{de}
C_9	2.0	0.0	0.5	49.7^{cd}
C_{10}	3.0	0.0	0.1	69.0^b
C_{11}	3.0	0.0	0.3	62.7^{bc}
C_{12}	3.0	0.0	0.5	55.0^{bc}

Source	DF	Type III SS	Mean Square	F Value	Pr> F
D	1	8464.000000	8464.000000	127.60	<.0001
IBA	2	216.777778	108.388889	1.63	0.2161
D*IBA	2	1258.111111	629.055556	9.48	0.0009
NAA	2	30.333333	15.166667	0.23	0.7973
D*NAA	2	852.111111	426.055556	6.42	0.0058

The average score with different letters are significantly different at $p = 0.01$ level. Letters a, b, c, d, e and f in the same column represent the differences among treatments by Duncan test.

Figure 3. Callus induced on C_4 and C_{10} medium.

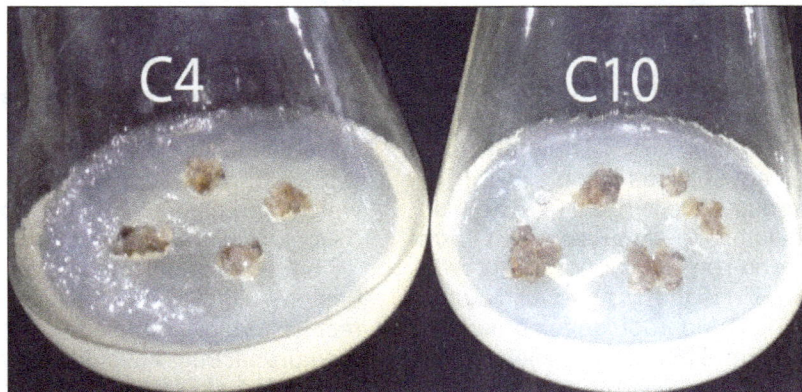

Figure 4. Callus induced on C_4 and C_{10} medium after 12 weeks culture.

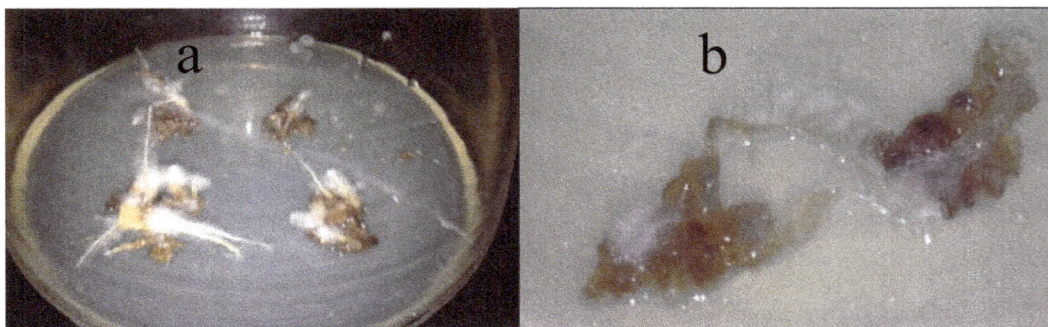

Figure 5. *P. bipinnatifidus*callus-induced adventitious roots after 12 weeks of culture.

Table 2. Effects of plant growth regulators on callus induction.

Treatment	NAA (mg/l)	IBA (mg/l)	Roots that induced callus (%)	Number of roots/explant	Root length (cm)
RN_1	0.5	0.0	17.9[d]	1.23[f]	0.47[d]
RN_2	1.0	0.0	22.2[dc]	2.47[ef]	0.800[bcd]
RN_3	1.5	0.0	36.1[bcd]	3.80[cde]	1.23[ab]
RN_4	2.0	0.0	44.5[abc]	3.87[cde]	1.00[abc]
RN_5	3.0	0.0	58.3[ab]	4.87[bc]	1.10[abc]
RI_1	0.0	0.5	15.1[d]	1.23[f]	0.70[bcd]
RI_2	0.0	1.0	27.8[cd]	3.10[de]	0.77[bcd]
RI_3	0.0	1.5	47.2[abc]	4.37[bcd]	1.07[abc]
RI_4	0.0	2.0	66.7[a]	5.67[b]	1.47[a]
RI_5	0.0	3.0	66.8[a]	7.33[a]	1.03[abc]

The average score with different letters are significantly different at p = 0.01 level. Letters a, b, c, d, e and f in the same column represent the differences among treatments by Duncan test.

Table 3. Effects of different media on adventitious root development.

Media	MS	N6	B5
Average fresh weight (g/75 ml)	1.99[a]	1.70[b]	1.83[ab]
Average dry weight (g/75 ml)	0.206[a]	0.182[b]	0.190[ab]

The average score with different letters are significantly different at p = 0.01 level. Letters a, b, c, d, e and f in the same column represent the differences among treatments by t Tests (LSD).

than RI_4, treatment RI_4 had adventitious rooting longer than RI_5. However, the actual culture, RI_4 created adventitious biomass better than RI_5. Therefore, we choose RI_4 to breeding adventitious biomass of *P. bipinnatifidus*. Callus are group of cells been localized, so the addition of growth regulators to stimulate callus form model differentiation depend on functional of growth regulator. Auxin (NAA and IBA) is the group of stimulants that starts rooting in cuttings and root development in tissue culture (Correa and Fett-Neto, 2004; Fletcher et al., 1965).

The result is simikar to that of Kim et al. (2002), the for-

mation and development roots of *P. giseng* C.A. Meyer had optimum efficiency when environmental IBA (24.6 mM) or NAA (9.8 mM) are added. The ability to stimulate formation of adventitious rooting from callus of *P. giseng* have shown that IBA is suitable for the formation and growth of adventitious rooting, number of roots uncertainty also formed on each additional IBA environment is more when compared with NAA (Thanh and Paek, 2008).

Effects of different media on *P. bipinnatifidus* on adventitious root development

Adventitious root was cultured on three different media (MS, N6, B5) supplemented with 2 mg/L IBA. Fresh weight and dry weight of root cultures were recorded after 6 weeks culture (Table 3 and Figure 6).

The experiments indicated that MS media gave the best results for adventitious root development. Similar results were obtained when culturing *Astragalus membranaceus* adventitious roots (Thwe et al., 2012), *Vernonia amygdalina* (Khalafalla et al., 2009) and *Bupleurum falcatum* (Kusakari et al., 2000). According to

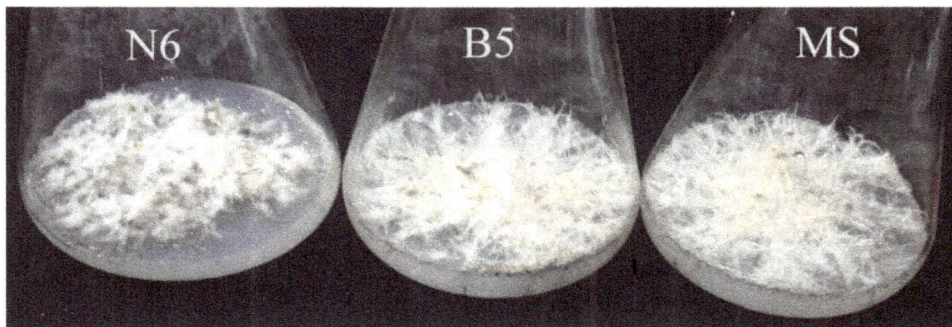

Figure 6. The adventitious root development of *P. Bipinnatifidus* on different media.

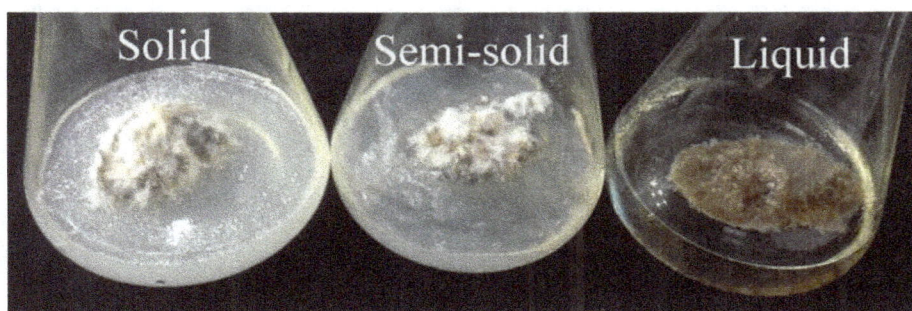

Figure 7. Effects of the physical states of media on *P. Bipinnatifidus* adventitious root development.

Table 4. Effects of physical states of media on *P. bipinnatifidus* adventitious root development.

Physical states of media	Solid	Semi-solid	Liquid
Fresh weight (g/75 ml)	1.90[b]	1.75[b]	2.13[a]
Dry weight (g/75 ml)	0.190[ab]	0.176[b]	0.201[a]

The average score with different letters are significantly different at p = 0.01 level. z: Letters a, b, c, d, e and f in the same column represent the differences among treatments by t-Tests (LSD).

Table 5. Effects of pH.sucrose on *P. bipinnatifidus* adventitious root development.

pH	5.5	6.0	6.5	7.0	7.5
Fresh weight(g/75 ml)	1.917[c]	2.237[a]	2.117[ab]	2.047[bc]	1.950[bc]
Dry weight(g/75 ml)	0.193[b]	0.231[a]	0.206[ab]	0.198[b]	0.192[b]

The average score with different letters are significantly different at p = 0.01 level. Letters a, b, c, d, e and f in the same column represents the differences among treatments by t tests (LSD).

Amzallag et al. (1992), the mineral content affected root development through the metabolism of plant growth regulators.

Effects of the physiscal states of the media on *P. bipinnatifidus* adventitious root development

Adventitious root was cultured on MS media at different physical states: solid (8 g/L agar), semi-solid (4 g/L agar) and liquid (0 g/L agar; shaked at 80 rpm), supplemented with 2 mg/L IBA. Fresh weight and dry weight of cultured roots were recorded after six weeks of culture (Table 4 and Figure 7).

Results show there was an effect from selected media on the adventitious root development of *P. bipinnatifidus*, particularly adventitious roots cultured on media without agar performed better as compared to those cultured on media supplemented with agar according to its fresh and dry weight (Table 4). The good contact of roots in the liquid medium could be the reason.

Effects of pH on *P. bipinnatifidus* adventitious root development

Adventitious roots (1 g/75 ml media) were cultured on liquid MS with pH 5.5, 6.0, 6.5, 7.0 or 7.5, supplemened with 2 mg/l IBA, in dark condition, 25±2°C, shaked at 80 rpm. Fresh weight and dry weight were recorded after 6 weeks of culture (Table 5 and Figure 8).

Results from Table 5 indicate that *P. bipinnatifidus* adventitious root developed best at pH 6.0. pH affected

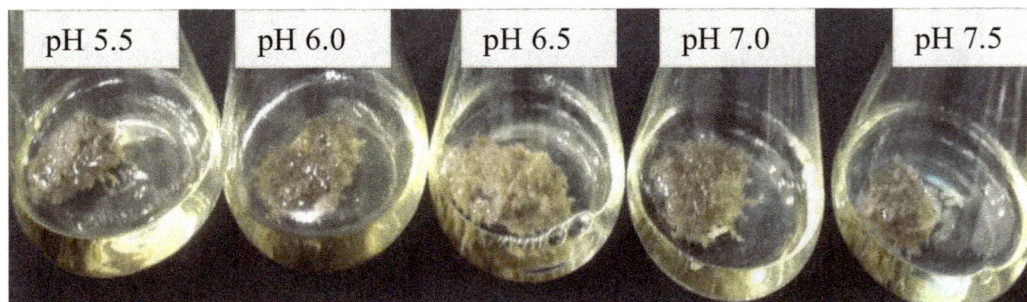

Figure 8. Effects of pH on the development of *P. bipinnatifidus* adventitious roots.

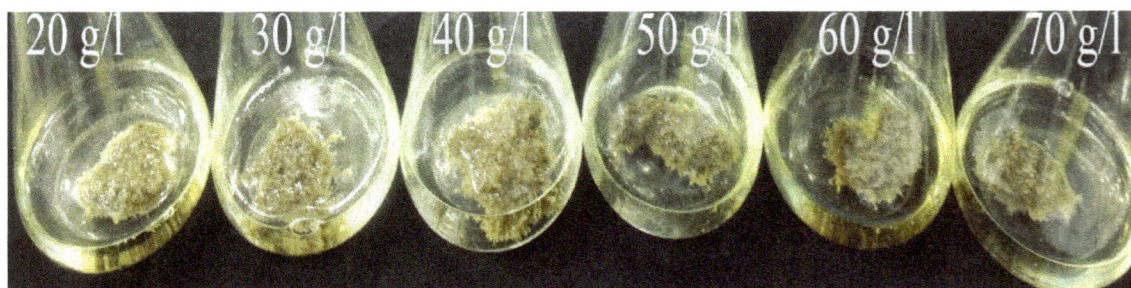

Figure 9. Effects of the sugar content on the root development of *P. bipinnatifidus*.

Table 6. Effects of the sugar content on the adventitious roots development of *P. bipinnatifidus*.

Sucrose content (g/L)	20	30	40	50	60	70
Fresh weight (g/75 ml)	1.989[c]	2.168[b]	2.318[a]	2.198[ab]	1.987[c]	1.931[c]
Dry weight (g/75 ml)	0.199[ab]	0.215[a]	0.220[a]	0.211[ab]	0.199[ab]	0.193[b]

The average score with different letters are significantly different at p = 0.01 level. Letters a, b, c, d, e and f in the same column represent the differences among treatments by t tests (LSD).

the root development through an ion exchange process through the membrane. According to "developing hypotheses about the acid", the size of cells were manipulated by its environment's pH, where it decreases with decreased pH (Cosgrove, 1999). Winch and Pritchard (1999), Evans (1976) and Edwards and Scott (1974) concluded that environemtal pH induced root elongation. The effects of pH on the development of adventitious roots were proved by Ling et al. (2009) while culturing *Orthosiphon stamineus* roots. pH 6.0 was the best among pH 4.0; 5.0; 5.8 and 7.0 and pH 6.0 was also appropriate for the adventitious roots development and saponin accumulation in *P. ginseng* (Kim et al., 2005).

Effects of sugar content on *P. bipinnatifidus* adventitious root development

Adventitious roots was cultured on liquid MS media, with different sucrose concentration (20, 30, 40, 50, 60 and 70

g/L), supplemented with 2 mg/L IBA at pH 6.0 in dark condition, 25±2°C and shaked at 80 rpm. Fresh weight and dry weight of root cultures were recorded after 6 weeks of culture (Table 6 and Figure 9).

The results (Table 6) show that culture media supplemented with 40 g/L sucrose were optimal for *P. bipinnatifidus* adventitious roots development. Carbohydrate was essential for plant metabolism *in vitro,* therefore sugar content in the media had a crucial impact on the root induction and development. Sugar plays an important role in the regulation and expression of the transcription of photosynthesis genes (Sheen, 1990) and also in signals of abscisic acid and ethylene (Leon and Sheen, 2003). Cheng et al. (1992) reported sucrose concentration at 2-3% positively affected the root induction of *Eucalyptus sideroxylon*. Moreover, sugar concentration was reported to have effects on adventitious roots induction of *O. stamineus* and *Scopoliaparviflora* at concentration of 30 and 50%, respectively (Ling et al., 2009; Min, 2007). *Echinacea angustifolia* adventitious roots were cultured on bioreactor on MS media supplemented with 2 mg/l IBA and 50 g/L sucrose. The initial sugar concentration (50 g/L) gave the best result in the mass development and saponin content of *P. ginseng* adventitious roots (Thành and Yoeup, 2008).

Conclusion

This study showed the possibilities of *P. bipinnatifidus* adventitious root development from callus. Results indicated variable effects on root development, forming the basis for further investigations of culturing *P. bipinnatifidus* adventitious roots to collect valuable metabolites.

Conflict of Interests

The author(s) have not declared any conflict of interests.

ACKNOWLEDGEMENT

This research received financial support from National Key Laboratory of Plant Cell Technology, Institute of Tropical Biology.

REFERENCES

Amzallag G, Lerner N, Poljakoff-Mayber HRM (1992). Interaction between mineral nutrients, cytokinin and gibberillic acid during growth of sorghum at high NaCl salinity. J Exp Bot 43:81-87.

Cheng B, CM Peterson, RJ Mitchell (1992). The role of sucrose, auxin and explant source on *in vitro* rooting of seedling explants of *Eucalyptus sideroxylon*. Plant Sci. 87:207-214.

Correa L, Fett-Neto A (2004). Effect of temperature on adventitious root development in microcuttings of *Eucalyptus Saligna* Smith and *Eucalyptus globules* Labill. J. Thermal Biol. 29:315-324.

Cosgrove DJ (1999). Enzymes and other agents that enhance cell wall extensibility. Ann. Review Plant Physiol. Plant Mol. Biol. 50:391-417

Dua PR, Shanker G, Srimal RC, Saxena KC, Saxena RP, Puri A, Dhawan BN, Shukla YN, Thakur RS, Husain A (1989). Adaptogenic activity of Indian *Panax pseudoginseng*. Indian J. Exp. Biol. 27(7):631-4.

Edwards KL, Scott TK (1974). Rapid growth responses of corn root segments: effect of pH on elongation. Planta. 119:27-37.

Evans ML (1976). A new sensitive root auxanometer. Plant Physiol, 58:599-601.

Fletcher R, Peterson R, Zalik S (1965) Effect of Light Quality on Elongation, Adventitious Root Production and the Relation of Cell Number and Cell Size to Bean Seedling Elongation. Plant Physiol. 40:541-548

Hahn EJ, Kim YS, Yu KW, Jeong CS, Paek KY. (2003) Adventitious root cultures of *Panax ginseng* C. A. Meyer and ginsenoside production through large scale bioreactor systems. J. Plant Biotechnol. 5:1-6.

Jung HK, Eun JC, Hoon O (2005). Saponin production in submerged adventitious root culture of *Panax ginseng* as affected by culture condition and elicitors. Asia Pasific J. Mol. Biol. Biotechnol. 12(2):87-91.

Khalafalla MM, Daffalla HM, El-Shemy HA, Abdellatef E (2009) Establishment of *in vitro* fast-growing normal root culture of *Vernonia amygdalina*- a potent African medicinal plant. African J. Biotechnol. 8(21):5952-5957.

Konoshima T, Takasaki M, Tokuda H (1999). Anti-carcinogenic activity of the roots of *Panaxnotoginseng*. II, Biol. Pharm. Bull. 22:1150-1152.

Kusakari K, Yokoyama M, Inomata S (2000). Enhanced Production of saikosaponin in cultured roots of *Bupleurum falcatum* L. using two-step control of sugar concentration Plant Cell Rep. 19:1115-1120.

Leon P, J Sheen (2003). Sugar and hormones connections trend. Plant Sci. 8(3):110-116.

Matsuda H, Samukawa K, Kubo M. (1991). Anti-hepatitic activity of ginsenoside.Planta Med. 57(6):523-526.

Min JY, Jung HY, Kang SM, Kim YD, Kang YM, Park DJ, Prasad DT, Choi MS. (2007). Production of tropane alkaloids by small-scale bubble column bioreactor cultures of *Scopoliaparviflora*adventitious roots. Bioresour. Technol. 98:1748-1753.

Sheen J (1990). Metabolic repression of transcription in higher plants. Plant Cell 2(10):1027-1038.

Wen-Ta Chiu, Paul Chan, Shue-Sen Liao, Jyh-RenLiou, Juei-Tang Cheng (1999). Effect of trilinolein on the activity and gene expression of superoxide dismutase in cultured rat brain astrocytes. Neuroscience Letters 269:17-20.

Winch S, Pritchard J (1999). Acid-induced wall loosening is confined to the accelerating region of the root growing zone. J. Exp. Bot. 50:1481-1487.

Yu KW, Hahn EJ, Paek KY (2005). Production of adventitious roots using bioreactors. Korean J Plant Tissue Cult. 27:309-315.

2

Exogenous NGF favors initiation of lizard tail regeneration while EGF and TGF-β truncate regenerative growth and commit to precocious muscle and cartilage differentiation

Anju Kurup and A.V. Ramachandran*

Division of Developmental Biology, Department of Zoology, Faculty of Science, The M S University of Baroda, Vadodara, Gujarat-390002, India.

Since growth factors exert control over various cellular events during development and as information in this regard on regenerative growth are scant, the present study has evaluated the effect of exogenous *in loco* administration of EGF, TGF-β and NGF during the first few days post-autotomy on the course of tail regeneration in *Hemidactylus flaviviridis*. Further evaluated is the efficacy of NGF to rectify the retardatory influence of hypothyroidism on regeneration. This study reveals that, EGF and TGF-β inhibit tail regeneration while, NGF stimulates it. Histologically, the EGF treated lizards showed greater collagen formation and precocious epithelial and myogenic differentiation. TGF-β had inhibitory influence on dedifferentiation while favoring preponderant chondrogenic differentiation. Overall, the present study on the differential effects of growth factors on various aspects of cell proliferation and differentiation, suggest the need for various growth factors on a precisely synchronized temporal and spatial order for a normal regenerative growth.

Key words: *Hemidactylus flaviviridis*, regeneration, nerve growth factor, fibroblast growth factor, transforming growth factor-β.

INTRODUCTION

Restoration of lost or damaged tissues or organs due to natural causes, disease or accident is the central theme of regenerative medicine. Though tissue engineering and stem cell biology are very much part of the approaches for the same (Stocum, 2004), understanding of ontogenic developmental events and tissue turn over in adult animals can provide great inputs to approaches in regenerative medicine (Rodtke and Clevers, 2005). An alternative approach involves applying the principles of inherent regenerative capacity of non-mammalian models, essentially the molecular events that permit tissue regeneration. Many of the sub mammalian vertebrates from fishes to reptiles demonstrate exemplary ability to replace lost body parts (Tsonis, 2000). An

incidental interest in regeneration is the possibility of finding some clues for the causative mechanisms of cancer as the early phases of regeneration marked by dedifferentiation and proliferation of cells bear close resemblance to oncogenesis. Apart from being a fascinating biological phenomenon, regeneration has also attracted biomedical interest for its potential to replace old or damaged tissues with new ones. Most of the studies on regenerative biology aimed at biomedical applications have focused on stem cells *in vitro* but understanding of regeneration requires *in vivo* studies, as complex interactions and communications within and among the different cell types characterize the process. Model organisms are essential for such *in vivo* studies, which can provide us with necessary information needed for eventual manipulation and control of regenerative properties. The importance of studying the process in lizards in the above context needs no elaboration when we consider the fact that, reptiles represent the closest

*Corresponding author. E-mail: mailtoavrcn@yahoo.co.in.

ancestral stock from which the homeotherms have evolved. Nevertheless, reptilian regeneration is a neglected field and apart from us, only Alibardi (2010) has taken up studies on this aspect in recent times.

Though neglected in comparison to regeneration studies in fishes and amphibians, such studies using lizard tail as a model have been in vogue in this laboratory for nearly four decades. Epimorphic regeneration as exemplified by the lizard tail comprises of post-autotomy preparation phase and a redevelopment phase (regressive and progressive phases) characterized by a non-scarring type of wound healing, blastema formation (akin to a tail bud), re-differentiation and growth phases (Ramachandran, 1996). Other studies have established local histological as well as adaptive biochemical and metabolic alterations in the geckonid lizard, *Hemidactylus flaviviridis* and the scincid lizard, *Mabuya carinata*. Even adaptive systemic responses as well as hemodynamic alterations have also been documented (Shah et al., 1977, 1980, 1982; Ramachandran, 1996; Ramachandran et al., 1979, 1985). Since the establishment of regeneration blastema is crucial for regenerative growth, immediate post autotomy periods deserve attention as a phase of molecular intricacies setting up an appropriate environment for the initiation of regeneration. A number of synchronized inter-related molecular events involving cytokines, growth factors, hormones and modifications of extracellular matrix (ECM) are likely to trigger adaptive changes forming the core of regressive phase changes.

In this context, two of our recent studies have evaluated the participation of signaling molecules and ECM remodeling in the immediate periods subsequent to caudal autotomy (Deshmukh et al., 2008; Nambiar et al., 2008). The above studies showed subtle temporal alterations in signaling molecules, second messengers and enzymes of ECM remodeling. Regeneration of a complex heterogeneous organ like the lizard tail, involving non-scarring wound healing, multiple cellular events, ependymal outgrowth and epithelio-mesenchymal interaction leading to progressive differentiation of heterogeneous tissues in a proximo-distal order, is likely to involve controlling/ regulating molecules like neurotrophic factors, and growth factors in the molecular ecology of the regenerating tail. As of now, understanding of the involvement of such factors in the reptilian regenerating system is poor. Local application is one of the simplest modes of study to understand the role of growth factors.

Hence the present study evaluates the influence of exogenous administration of nerve growth Factor (NGF), epidermal growth factor (EGF) and transforming growth factor-β1 (TGF-β1) on the course of tail regeneration in terms of wound healing, blastema formation and differentiation. Since both thyroxine and NGF are important for the development of nervous system and as thyroxine reportedly increases NGF content (Walker et

al., 1979), the study also assesses the influence of NGF given to hypothyroid lizards to decipher their inter-relationship if any.

MATERIALS AND METHODS

Chemicals

NGF was purchased from Sigma Aldrich, USA, while TGF-β was a gift from Department of Endocrinology, PGIBMS, Chennai and EGF, a gift from Division of Biotechnology, Department of Microbiology, MS University of Baroda, India.

Experimental animals

Adult *Hemidactylus flaviviridis* (10±2 g) of both sexes with snout-vent length of 70 to 80 mm were used for the experiments. The cages housing the animals measured 18 × 15 × 10 with one side of transparent glass and ventilated on three sides. Each cage housing six lizards was balanced for size. The animals were maintained under a normal light-dark photoperiodic schedule and were fed on nymphs and water *ad libitum*. Autotomy was performed by pinching off the tail three segments from the vent

Experimental schedules

Set I: Evaluation of influence of NGF, EGF and TGF-β

Sixty lizards were divided into 6 groups of 10 each.

Group 1(Control): The tail of these lizards was autotomised three segments distal to the vent and injected with 0.6% saline at the cut end of the tail for 15 days.

Group 2 (NGF treated): The tail of these lizards was autotomised three segments distal to the vent and injected with 10 ng of NGF *in loco* for three consecutive days post autotomy.

Groups 3 and 4 (EGF treated): The tail of these lizards was autotomised three segments distal to the vent and injected with 5 or 10 ng of EGF respectively for ten consecutive days post autotomy.

Groups 5 and 6 (TGF-β treated): The tail of these lizards was autotomised three segments distal to the vent and injected with 5 ng TGF-β1 for ten consecutive days post autotomy (Group.5). Group 6 animals were treated with 5 ng TGF-β1 after the formation of blastema (10 to 12 days) for five consecutive days.

Set II: Evaluation of the influence of hypothyroidism, thyroxine replacement or replacement with NGF

40 lizards were divided into 4 groups of 10 animals each.

Group1 (Control): This group of lizards served as euthyroidic control. They were force fed with 0.6% saline on every alternate day. At the end of the 5th dose of saline, the tail was autotomised. Saline feeding was continued until the end of the experimental period.

Group 2 {(Hypothyroid (HT)}: These lizards were rendered hypothyroidic by force-feeding 0.1% 6-propyl, 2-thiouracil (PTU). The tail of these lizards was autotomized at the end of 5th dose of PTU and PTU feeding was continued every alternate day until the end of the experiment.

Table 1. Number of days taken to attain various arbitrary stages in control and experimental lizards.

Manipulation	WH	PB	B	IG
Control	7	9	10	11
EGF	6	11	13	17
TGF-β	7	13	17	-
NGF	5	6	7	8

WH; Wound Healing, PB;Preblastema, B; Blastema, IG; Initiation of growth, EGF; Epidermal growth factor, TGF-β; Transforming growth factor, NGF;Nerve growth factor, T4-Thyroxine.

Group 3 (Thyoxine replacement {TR}): These lizards were also rendered hypothyroid and their tail was autotomized at the end of 5th dose of PTU. Subsequent to autotomy, PTU feeding was continued every alternate day. Apart from this, they were also injected with 10μg of thyroxine *in loco* every alternate day starting from 3 days post autotomy and continued until the end of the experiment.

Group 4 (NGF replacement): These lizards, as those of Group 2, were also rendered hypothyroid and their tail was autotomized at the end of the 5th dose of PTU. Five injections of 10 ng NGF were given *in loco* for 5 consecutive days post autotomy. PTU feeding was continued until the end of the experiment.

Parameters evaluated

Subsequent to treatment with NGF, EGF and TGF-β, the number of days taken to reach the various arbitrary stages of regeneration like, wound healing (WH), preblastema (PB), blastema (B) and initiation of growth (IG) was noted. Since there was inhibitory influence of EGF and TGF-β, the tails of these lizards were cut and processed for histological observations. Prior to dehydration and embedding in paraffin, the tails were decalcified and sections of 3 to 5 micron and 8 to 10 micron thickness were cut and stained with Casson's or Masson's trichome. Some sections were also stained with PAS staining. Since NGF treatment has a stimulatory influence on regeneration, the length of tail regenerated in control, NGF, HT, TR, and HT+NGF groups of lizards was measured every alternate day. The growth rate per day and total replacement of tail were also calculated.

Statistical analysis

The values are expressed as Mean±S.E.M for n = 10. The data was subjected to student's t test and to Duncan's multiple range test.

RESULTS

EGF and TGF-β1 treatment

As shown in Table 1, treatment with both EGF and TGF-β1 resulted in inhibition of regeneration. Though EGF promoted earlier wound healing, the attainment of preblastema and blastema stages was delayed (Figure1).

Figure 1. Photomicrographs of control (a) 1 ng EGF treated (b) 5 ng EGF treated, and (c) lizards, showing the dose dependent inhibitory influence of EGF at the end of 15 days.

In contrast, though TGF-β1 did not show any effect on WH, the formation of PB and B stages was significantly delayed and a poor blastema was formed. However, lizards injected with TGF-β1 after the formation of a blastema did not show any effect and they regenerated their tail at the same rate as the controls.

Histological observations

Control lizards in the wound healing stage showed a single layered wound epithelium with no collagen material and the pre-blastemic and blastemic stages were characterized by a thickened blastemic epithelium with no collagen material and compactly packed mesenchymal cells. Differentiation stage was marked by central cartilaginous neural tube with ependyma inside and differentiating muscle bundles on the lateral sides and differentiating integument (Figure 2).

Histologically, the EGF treated lizards showed heavily deposited collagen below the epithelium (Figure 1). In those cases where regeneration proceeded up to pre-blastema stage, showed thickened epithelium with tendency for keratinization and scale formation. In many cases, specific staining indicated formation of collagen below the blastemic epithelium and even in the mesenchymal mass. Precocious and premature differentiation of muscle elements could be clearly seen even extending up to the distal tip close to the apical epithelium (Figure 3). Lizards given only 1 ng EGF or even those lizards which were given 5 ng EGF, where some growth occurred after the stoppage of EGF treatment, tended to show a hooked stumpy growth and, in these cases, PAS staining indicated less GAG

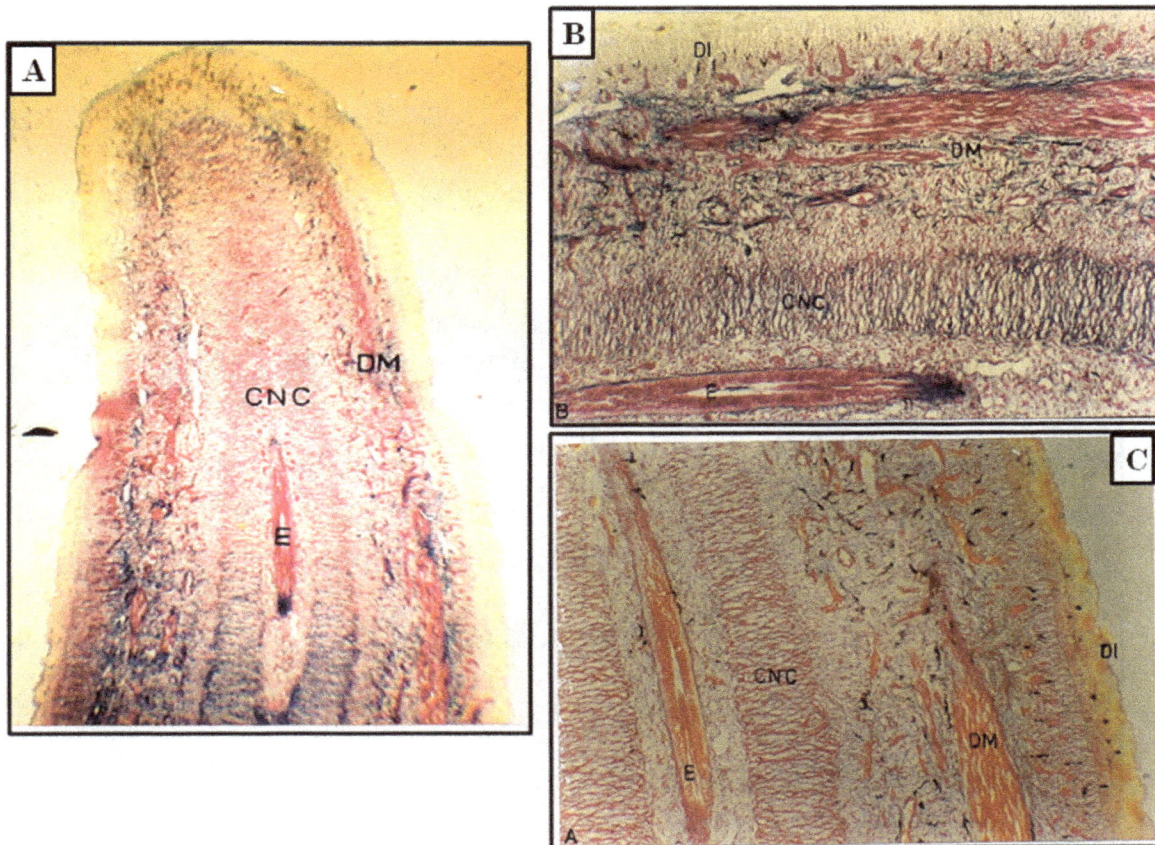

Figure 2. Photomicrographs of a section of regenerating tail of control lizard in the differentiation stage stained with Casson's stain showing various differentiating elements (A), enlarged version of Casson's stained section showing the lateral half with differentiating muscle (DM) and ependyma (E). Integument appears yellow, muscle and ependyma appears orange and cartilaginous neural canal appears pink (B) and central part of section of differentiating tail stained with Mallory's triple stain showing differentiating integument (DI), differentiating muscle (DM), cartilaginous neural canal (CNC) and ependyma (E). Differentiating muscles and ependyma appears red and collagen appears green (C).

substances in the differentiating cartilaginous tube with extensive muscle differentiation all over the tip (Figure 4). Erythropoeitic activity was greatly stimulated in the adipose tissue of the stump tail and large clumps of blood cells could be seen accumulated in large numbers at the cut end below the epithelium (Figure 5). Differential staining revealed increased GAG deposition and precocious and preponderant chondrogenic differentiation (Figure 6). Muscle differentiation was totally inhibited.

NGF treatment

NGF treatment resulted in faster wound healing and early blastema formation and initiation of growth. There was sustained faster rate of growth that resulted in 20% replacement of lost tail within 25 days, as compared to 14% in the control lizards (Table 1).

PTU treatment

PTU induced hypothyroidic lizards showed delayed WH, BL formation and IG. The total length of tail regenerated was significantly less compared to euthyroidic controls and the total replacement at the end of 25 days was only 5%.

T4 and NGF replacement

All the retardatory influence of hypothyroidism was completely nullified by either T4 or NGF replacement. WH, BL formation and IG occurred at periods very much comparable with those noted for the control lizards. The total length of tail regenerated and the total percentage replacement at the end of 25 days were much closer to those of the controls though slightly lesser. On a comparative basis, NGF replacement seems to be a shade

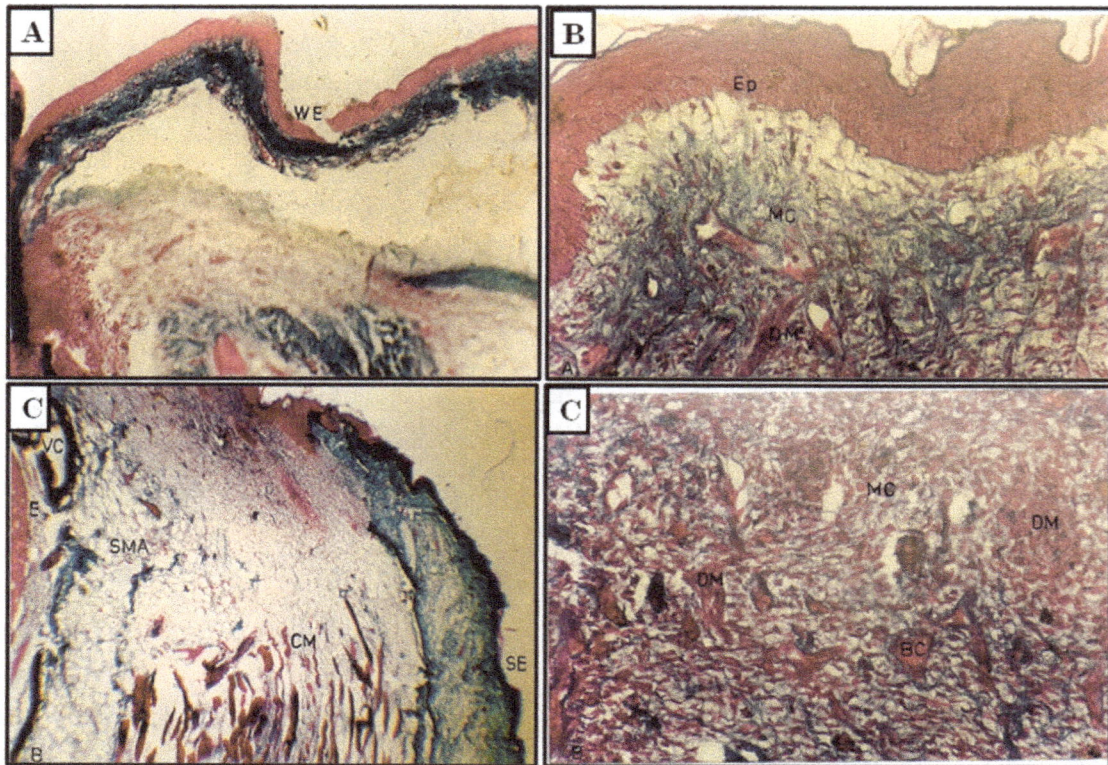

Figure 3. Photomicrographs of the (A) Mallory's triple stained sections of tail of experimental lizards treated with EGF in the wound healing stage showing deposition of thick layer of collagen (green) below the thickened epithelium (WE), (B) enlarged view of the pre-blastema showing presence of collagen (greenish blue) below the thickened epithelium (EP), mesenchymal cells (MC) and precocious differentiating muscle cells (DM), (C) lateral half of the above tail section showing the presence of thick collagen (green) below the stump epidermis (SE), sub muscular adipose tissue (SMA), vertebral column (VC) and part of ependyma (EP) and, (D) enlarged version of the mesenchymal mass showing loose mesenchymal cells (MC), blood cells (BC)-red, differentiating muscle mass (DM)- red and collagen material (green).

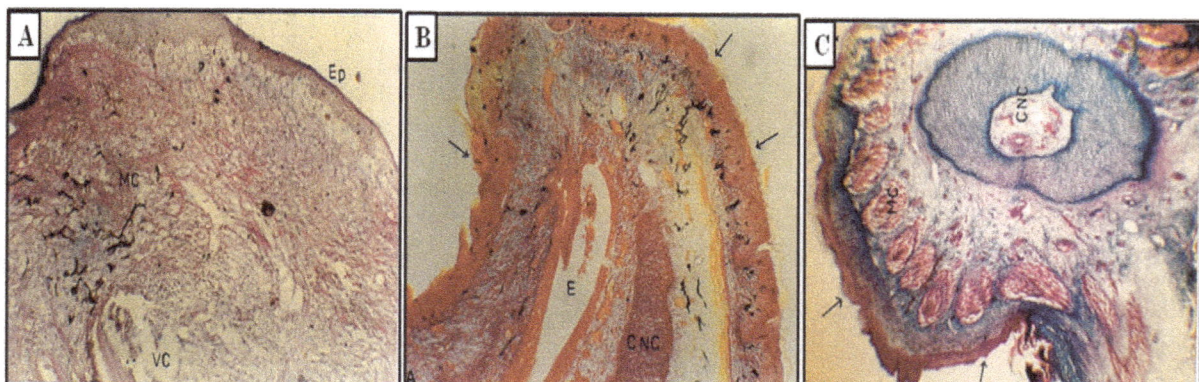

Figure 4. photomicrographs of (A) PAS stained section of a regenerating tail arrested at the pre-blastema stage from a lizard treated with EGF showing loose mesenchymal cells (MC), thick epithelium (Ep) and some GAG material (purplish in color) only around the vertebral column of the tail stump but not in the differentiating cartilage area, (B) hooked tip of the regenerating tail of EGF treated lizard showing precocious differentiation and keratinization of the epidermis (arrow), cartilaginous neural canal (CNC) and ependyma (E) in Mallory's triple stained sections and (C) the tip of a hooked tail of EGF treated lizard showing precocious epidermal differentiation arrows and muscles (MC), cartilaginous neural canal (CNC) and a heavy deposition of collagen (greenish-blue) below the epidermis and around the muscle bundles as well as in the cartilaginous neural canal in Masson's trichome stained sections.

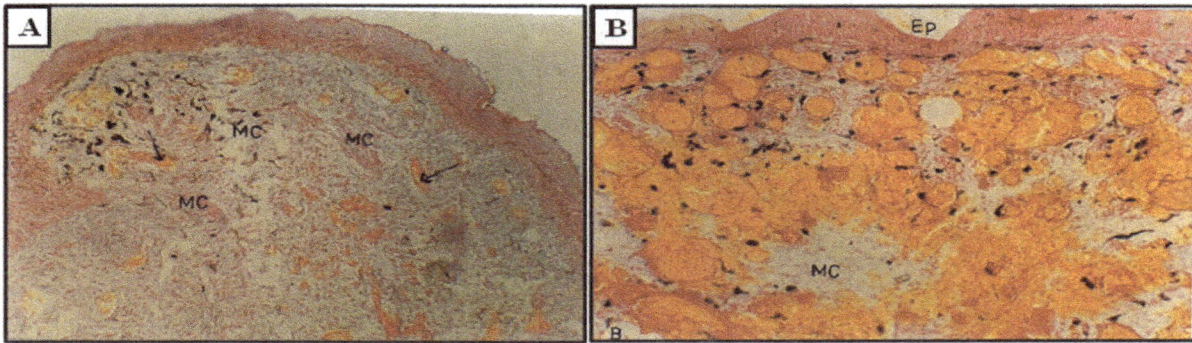

Figure 5. Photomicrographs of TGF-β treated lizards tip of the blastema (A) showing loose mesenchymal cells (MC), scattered accumulation of blood cells (arrow, yellow color), some undifferentiated muscle fibres released from the cut end of the stump (MC) and cartilage condensation in the centre, (B) the tip of the pre-blastema showing epithelial (Ep) and loose mesenchymal mass (MC) studded with clumps of erythrocytes (yellow color).

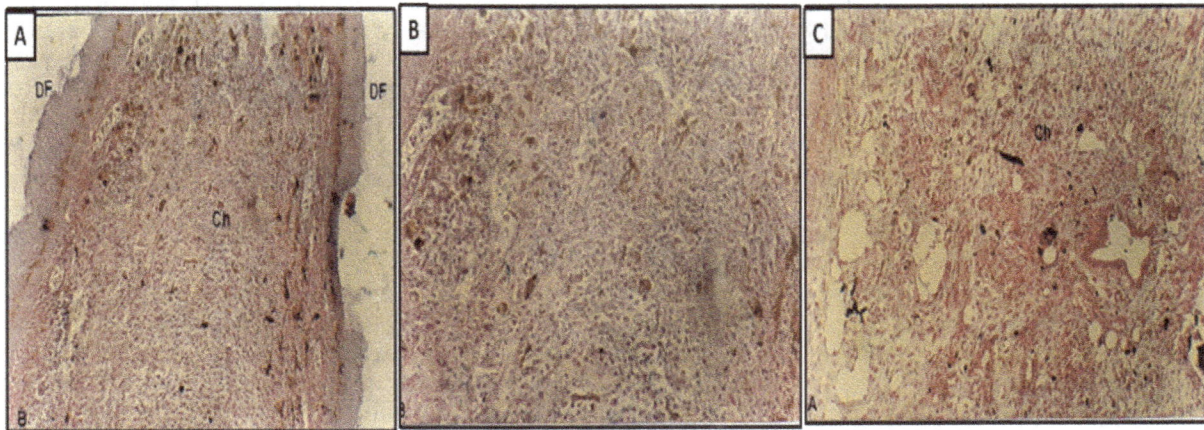

Figure 6. Photomicrographs of sections of regenerating tail of a lizard treated with (A) TGF-β showing only chondrogenic differentiation (Ch) in the mesenchymal tissue and GAG material (purple color) in Casson's stain, (B) section of a regenerating tail of lizard treated with TGF-β1 showing extensive chondrogenesis with no other tissue differentiation. The purplish background represents GAG material.

Table 2. Number of days taken to attain various arbitrary stages in control and experimental lizards.

Manipulation	WH	PB	B	IG
Control	7	9	10	11
NGF	5	6	7	8
PTU	10	11	13	14
PTU+T4	9	10	11	12
PTU+NGF	8	9	10	11

WH; Wound Healing, PB; Preblastema, B; Blastema, IG; Initiation of growth, NGF; Nerve growth factor, PTU;6-Propyl-thiouracil, T4-Thyroxine.

more effective than T4 replacement (Table 2).

DISCUSSION AND CONCLUSION

Of the three growth factors tested, EGF and TGF-β1, show inhibitory effects on regeneration while, NGF depicts significant favorable response. The only study on growth factor in lizard tail regeneration is of Alibardi and Lovicu (2010) who have immunolocalized FGF1 and FGF2 in the regenerating tail of *Lampropholis guinchenoli*. This study conducted in post blastemic regenerating tail suggests some role for FGFs in tissue differentiation especially of muscle and skin. They suggested a possible neurotrophic role for FGF2 in the light of its localization at the growing ends of spinal cord and nerves. Other studies on regenerating systems of fishes and amphibians suggest roles for FGFs in proliferation of neural progenitors and the ependymal cells of the spinal cord (Zhang et al., 2000) and in epithelio-mesenchymal interactions (Bouzaffour et al., 2009). The interactions of FGF2-FG1 seem akin to FGF8-FGF10 interactions in the developing limb bud (Martin, 2001) and, FGFs seem to be good evocators of wound healing and regeneration.

EGF is a small single chain polypeptide that stimulates the proliferation of epidermal and epithelial cells in whole animals and a variety of cell types in culture and controls many other functions (Carpenter, 1993; Wong and Gillaud, 2004). In keeping with its role in epithelial cell proliferation, EGF treatment in the present study promoted early wound healing and helped form a thickened multilayered blastemic epithelium, with delayed blastema formation and stunted growth of the regenerate thereafter. This post-blastemic inhibition on regenerative growth appears related with the heavy deposition of collagen below the epithelium. It is well established that during regenerative wound healing, the deposition of sub-epidermal collagen and/or scar tissue formation do not occur unlike in non-regenerative wound healing (Shah et al., 1980; Harty et al., 2003; Nambiar et al., 2008). Though the deposition of connective tissue material occurs on the lateral sides of regenerate below the differentiating epidermis, it never occurs under the apical wound or blastemic epithelium thereby permitting the continued elongation of the regenerate. The deposition of dermal connective tissue on the lateral sides of regenerate in fact promotes differentiation of integument marked by scale formation and keratinization (Alibardi, 2010). Apparently, precocious and premature epithelial differentiation into scales and keratinization visible even at the apical region in EGF treated lizards finds correlation with the deposition of dermal substance all around the blastemic epithelium. The present observation of EGF induced collagen formation and deposition under the wound epithelium, finds support in the many reports of EGF's role in collagen synthesis under both *in vivo* and *in vitro* conditions (Laato et al., 1987; Babul et al., 2004; Berlanga-Acosta et al., 2009). Another prominent influence of EGF seen in the present study is the precocious and preponderant myogenic differentiation resulting in increased muscle bundles below the incipiently formed blastemic apical epithelium. This myogenic effect of EGF finds support from the observations of Lim and Hauschka (1984), and Olwin and Hauschka (1984) of proliferation of myoblasts in presence of EGF and the loss of receptors for EGF with the fusion of myoblasts and their differentiation into myotubes. Further, Yamane et al. (1997) have shown involvement of EGF in myogenesis of tongue epithelium and induction of myogenic genes (myf5) and myogenic transcription factors. Apparently, as in mammals, EGF is able to promote muscle differentiation in reptiles as well. Another observation of some merit is the decreased amount of glycosaminoglycan (GAG) material in the matrix of differentiating cartilaginous tube and the hooked

nature of the regenerate with stunted growth.

TGF-β is a member of the family of polypeptides, which seems to regulate cellular activity in various metazoan organisms (Herpin et al., 2004). A very important action of this factor is on extracellular matrix synthesis and maintenance and, depending on the cell type, it may exert stimulatory, inhibitory, biphasic or no effect on cell proliferation (Moses et al., 1987). Though TGF-β is capable of stimulating the growth of fibroblasts and osteoclasts, it inhibits the proliferation of many epithelial cells and inhibits growth of developing mammary gland (Pfeilschifter et al., 1990). Another function accredited to this factor is in the various facets of wound healing, which led to its consideration as a potential therapeutic agent for wound healing (Mustoe et al., 1987). Since TGF-β3 is potent in epithelial proliferation and wound healing, TGF-β1 used in the present study did not hasten post-caudal autotomy wound healing. In fact, TGF-β3 is involved in scar less wound healing (Kohama et al., 2002). The most conspicuous effect of TGF-β1 treatment in the present study is the formation of a poor regeneration blastema after a pronounced delay. Further, differential staining techniques revealed increased GAG deposition in mesenchymal mass. This observation is in agreement with the reported effect of TGF-β in stimulating the synthesis of all major matrix proteins, such as collagen and fibronectin (Ignotz and Massague, 1986), tenascin (Pearson et al., 1988), elastin (Liu and Davidson, 1988), glycosaminoglycan (Chen et al., 1987) and thrombospondin (Penttinen et al., 1988). The formation of a poor blastema as well as the inhibition of post-blastemic growth could be due to the inhibitory influence of this peptide on differentiation/ recruitment of cells, as has been observed presently, and on proliferation of the dedifferentiated cells. The purported inhibition of proliferation of dedifferentiated cells is well neigh possible as TGF-β is reportedly a potent inhibitor of growth of a wide variety of cells (Holley et al., 1980; Proper et al., 1982; Lawrence et al., 1984; Roberts et al., 1985). A related observation is the inhibition of DNA synthesis in hepatocytes under the influence of TGF-β in normal and regenerating liver of rat (Strain, 1990). Other observed anomalies consequent to TGF-β1 treatment are inhibited myogenesis and precocious chondrogenesis. These effects appear to be akin to that reported for mammals, as TGF-β purportedly stimulates chondrogenic differentiation and inhibits adipogenic and myogenic differentiation (Pfeilschifter et al., 1990; Miyazono, 2000; Sekiya et al., 2002; Ramón Ríos et al., 2002; James et al., 2009; Kim et al., 2009). Importance of TGF-β in multiple events during regeneration stands validated by the studies of Ho and Whitman (2008), and Tseng and Levin (2008) wherein they have shown the localization of key components of the TGF-β signaling pathway and blockage of Xenopus tail regeneration at multiple points, on inhibition of TGF-β signaling by the use of a specific inhibitor. Apparently, a generalized inhibition of all TGF-β fractions affects multiple events and, the blockage of regeneration suggests the important role of this growth factor in regenerative mechanics. A novel feature observed in the present study with reference to TGF-β1 action is the stimulated hematopoietic activity, predominantly eryhtropoetic, in the adipose tissue of the tail stump, as marked by congregation of RBCs in the loose mesenchymal mass of the poorly formed blastema. In the wake of the hitherto unreported functional involvement of TGF-β in erythropoeitic activity, the present observations implicate this growth factor in such a functional involvement in lower vertebrates

As against the inhibitory actions of EGF and TGF-β1, NGF showed stimulatory influence on regeneration. Administration of NGF in the tail stump, post caudal autotomy, significantly reduced the latent period (by 3 days) in blastema formation and provided greater momentum to growth in the initial phase. This is borne out by the significantly pronounced growth rate recorded during the first 10 days post autotomy. Presumably, the lizard tail has the potential to respond to exogenous NGF or related peptide, and initiate regeneration. The observation of early wound closure in NGF treated lizards indicates its role in promoting wound healing that stands well supported by other studies (Werne and Grose, 2003). Regenerative growth post autotomy/amputation of lizard and amphibian tail is essentially dependent on ependymal outgrowth from the cut end of spinal cord as a *priori* and its triggering action by way of secretion of MMPs, neurotrophic factors etc, which lead to non-scarring wound healing and formation of a regenerative blastema (Alibardi and Lovicu, 2010). Studies on notochord and tail regeneration have suggested the elaboration of important neurotrophic factors like NGF, BDNF, neurotropins and neurotrophic factors as necessary mediatory messages for triggering the initiation of regeneration (Chernoff, 1996; Ferretti et al., 2003).

A probable clue to the involvement of NGF in the autotomised tail comes from the observation of the ability of exogenous NGF to nullify the delay in blastema formation exhibited by hypothyroidic lizards. Thyroxine stands implicated in the outgrowth of the ependyma "a *priori*" for regeneration to occur as, it is responsible for providing the inductive influence for the initiation of regeneration in lizards (Bellaris and Bryant, 1985). Our previous studies (Ramachandran et al., 1984; Ramachandran and Abraham, 1990; Ramachandran and Kurup, 2006) have indicated hypothyroidism induced delay in outgrowth of ependyma to be responsible for retarded regenerative growth. However, in the present study, exogenous NGF could compensate for the lack of thyroxine and initiate a normal regenerative outgrowth (even better than that shown by T4 replacement). This provides substantial circumstantial evidence for ependyma as a source of NGF in the lizard tail. Two relevant observations that lend credence to this

assumption are the ultra-structural changes suggestive of secretory activity of ependymal tip (Alibardi, 2010) and the ability of thyroxine to increase NGF content (Walker et al., 1979). Further evidence comes from the reported expression of NGF receptors in the central nervous system subsequent to the injury of spinal cord (Brunello et al., 1990). The NGF elaborated by the ependyma may also be implicated in neurite outgrowth and other normal functions associated with NGF such as accumulation of neurofilaments, formation of microtubules and axonogenesis. It is also likely that, NGF may induce an environment conducive for regeneration by ECM remodeling by way of expression of MMPs and favor angiogenesis and mitotic activity in the apical epithelial cap. It can be presumed that, the ependyma, under the influence of thyroxine produces NGF required for neurite outgrowth and the formation and action of the mitogenic factor (s) which would contribute to continued regenerative growth.

Overall, the present study has revealed a positive influence of NGF on tail regeneration and a negative effect of both EGF and TGF-β1 by their multiple cellular and sub cellular effects. It is inferable that, uncontrolled or unscheduled expression of EGF and TGF-β temporally or spatially can lead to aberrant regeneration with singular tissue differentiation, cautioning the application of growth factors in regenerative medicine for regrowing or repairing organs or structures comprised of heterogeneous tissues.

REFERENCES

Alibardi L (2010). Morphological and cellular aspects of tail and limb regeneration in lizards: A model system with implications for tissue regeneration in mammals. Adv. Anat. Embryol. Cell. Biol., 207: 1-109.

Alibardi L, Lovicu FJ (2010). Immunolocalization of FGF1 and FGF2 in the regenerating tail of the lizard *Lampropholis guichenoti*: Implications for FGFs as trophic factors in lizard tail regeneration. Acta. Histochemica, 112: 459-473.

Babül A, Gönül B, Dinçer S, Erdoan D, Özoul C (2004). The effect of EGF application in gel form on histamine content of experimentally induced wound in mice. Amino Acids, 27: 3-4.

Bellairs A, Bryant SV (1985). Autotomy and regeneration in reptiles, pp. 304-410. In: Biology of reptilian, 15: 1-731 (eds). Carl Gans and Frank Billet, John willey and Sons, New York.

Berlanga-Acosta J, Gavilondo-Cowley J, Lo´ pez-Saura P, Gonza´ lez-Lo´ pez T, Castro-Santana MD, Lo´ pez-Mola E, Guille´ n-Nieto G, Herrera-Martinez L (2009). Epidermal growth factor in clinical practice—A review of its biological actions, clinical indications and safety implications. Int. Wound. J., 6: 331–346.

Bouzaffour M, Dufourcq P, Lecaudey V, Haas P, Vriz S (2009). Fgf and Sdf-1 Pathways Interact during Zebrafish Fin Regeneration. PLoS ONE, 4(6): e5824.

Brunello N, Reynolds M, Wrathall JR, Mocchetti T (1990). Increased nerve growth factor receptor mRNA in contused rat spinal cord. Neurosci. Lett., 118: 238-240.

Carpenter G (1993). EGF: New tracks for and old growth factor. Curr. Opin. Cell. Biol., 5: 261-264.

Chen J, Hoshi H, Wallace L (1987). Transforming growth factor type β specifically stimulates synthesis of proteoglycan in human adult arterial smooth muscle cells. Proc. Natl. Acad. Sci. USA., 84: 5287-529.

Chernoff EAG (1996). Spinal cord regeneration: A phenomenon unique to urodeles? Int. J. Dev. Biol., 40: 823-831.

Deshmukh PA, Bhatt IY, Jape RR, Jivani PN, Kavale HR, Nambiar VV, Prakashkar SS, Ramachandran AV (2008). Evaluation of signaling molecules in the post-autotomy initiation phase of tail regeneration in the lizard Hemidactylus flaviviridis. J. Endocrinol, Reprod., 12: 31-38.

Ferretti P, Zhang F, O'Neill P (2003). Changes in Spinal Cord Regenerative Ability through Phylogenesis and Development: Lessons to Be Learnt. Dev. Dynam., 226: 245–256.

Harty M, Neff AW, King MW, Mescher AL (2003). Regeneration or Scarring: An Immunologic Perspective. Dev. Dynam., 226: 268–279.

Herpin A, Lelong C, Favrel P (2004). Transforming growth factor-beta-related proteins: An ancestral and widespread superfamily of cytokines in metazoans. Dev. Comp. Immunol., 28(5): 461–485.

Ho DM, Whitman M (2008). TGF-β signaling is required for multiple processes during Xenopus tail regeneration. Dev. Biol., 315: 203–216.

Holley RW, Bolhen P, Fara R, Baldwin JH, Lkeeman G, Armour R (1980). Purification of kidney epithelia cell growth inhibitors. Proc. Natl. Acad. Sci. USA., 77: 5989-5992.

Ignotz RA, Massague J (1986). Transforming growth factor beta stimulates the expression of fibronectin and collagen and their incorporation into the extracellular matrix. J. Biol Chem., 261: 4337-4345.

James AW, Xu Y, Lee JK, Wang R, Longaker MT (2009). Differential Effects of TGF-[beta]1 and TGF-[beta]3 on Chondrogenesis in Posterofrontal Cranial Suture-Derived Mesenchymal Cells In Vitro. Plastic Reconstr. Surg., 123(1): 31-43.

Kim YJ, Hwang SJ, Bae YC, Jung JS (2009). MiR-21 regulates adipogenic differentiation through the modulation of TGF-beta signaling in mesenchymal stem cells derived from human adipose tissue. Stem Cells, 12: 3093-3102.

Kohama K, Nonaka K, Hosokawa R, Shum L, Ohishi M (2002). TGF-beta-3 Promotes Scarless Repair of Cleft Lip in Mouse Fetuses. J. Dent. Res., 81: 688.

Laato M, Kähäri VM, Niinikoski J, Vuorio E (1987). Epidermal growth factor increases collagen production in granulation tissue by stimulation of fibroblast proliferation and not by activation of procollagen genes. Biochem. J., 247(2): 385–388.

Lawrence DA, Pircher R, Krycève-Martinerie C, Jullien P (1984). Normal embryo fibroblasts release transforming growth factors in a latent form. J. Cell. Physiol., 121: 184-188.

Lim RW, Hauschka SD (1984). A rapid decrease in EGF-binding capacity accompanies a terminal differentiation of mouse myoblasts in vitro. J. Cell. Biol., 98: 739-747.

Liu J, Davidson JM (1988). The elastogenic effect of recombinant transforming growth factors-β on porcine aortic smooth muscle cells. Biochem. Biophys. Res. Commun., 154: 895-901.

Martin G (2001). Making vertebrate limb: New players enter from the wings. Bioassays, 10: 865-868.

Miyazono K (2000). Positive and negative regulation of TGF-beta signalling. J. Cell Sci., 113: 1101–1109.

Moses HL, Coffey RJ, Loef EB, Lyons RM, Kesi-Oza J (1987). Transforming growth factor beta regulation of cell proliferation. J. Cell. Physiol., 5: 1-7.

Mustoe T, Pierce JF, Thompson A, Gramates P, Sporn MB, Deuel TF (1987). Accelerated healing of incision wounds in rat induced by transforming growth factor β. Science, 237: 1333-1336.

Nambiar VV, Bhatt IY, Deshmukh PA, Jape RR, Jivani PN, Kavale HR, Prakashkar SS, Ramachandran AV (2008). Assessment of extracellular matrix remodeling during tail regeneration in the lizard Hemidactylus flaviviridis. J. Endocrinol. Reprod., 2: 67.

Olwin BB, Hauscha SD (1984). Cell surface fibroblast growth factor and epidermal growth factor receptors are permanently lost during skeletal muscle terminal differentiation in culture. J. Cell. Biol., 107: 761-769.

Pearson CA, Pearson D, Shibahara S, Hofsteenge J, Chiquet-Ehrismann R (1988). Tenascin: cDNA cloning and induction by TGF-β. Embo. J., 7: 2977-2981.

Penttinen RP, Kobayashi S, Bornstein P (1988). Transforming growth factor β increases mRNA for matrix proteins both in the presence and in the absence of changes in mRNA stability. Proc. Natl. Acad. Sci. USA., 85: 1105-1108.

Pfeilschifter J, Wolf O, Naumann A, Minne HW, Muindy GR, Ziegler R (1990). Chemotactic response of osteoblast like cells to Transforming Growth Factor β. J. Bone Miner. Res., 5(8): 825–830.

Proper JA, Bjornson CL, Moses HL (1982). Mouse embryos contain polypeptide growth factors capable of inducing a reversible neoplastic phenotype in non-transformed cells in culture. J. Cell. Physiol., 110: 169-174.

Ramachandran AV, Abraham S (1990). Effect of chemical thyroidectomy and adreno-cortical suppression on tail regeneration in the Gekkonid lizard, Hemidactylus flaviviridis.Acta. Embyol. Morphol. Exp., 11: 45-52.

Ramachandran AV, Kinariwala RV, Shah RV (1979). Tail regeneration and lipid metabolism: Changes in the content of the total hepatic cholesterol in the scincid lizard, Mabuya carinata. J. Anim. Morphol. Physiol., 26: 21-28.

Ramachandran AV, Kinariwala RV, Shah RV (1985). Haematopoesis and regeneration: Changes in the liver, spleen, bone marrow and hepatic iron content during tail regeneration in the scincid lizard, Mabuya carinata. Amphibia-Reptilia, 6: 377-386.

Ramachandran AV (1996). Biochemistry and metabolism of lizard tail regeneration. J. Anim. Morphol. Physiol., 43: 1-13.

Ramachandran AV, Kurup A (2006). Thyroid hormone control of tail regeneration: Differential in loco and systemic effects and seasonal variation. J. Endocrinol. Reprod., 10(2): 134-142.

Ramachandran AV, Swamy MS, Shah RV (1984). Tail regeneration in the scincid lizard, Mabuya carinata, related to breeding seasons and thyroid activity. Amphibia-Reptilia, 5: 134-144.

Ramón Ríos, Carneiro I, Arce VM, Devesa J (2002). Myostatin is an inhibitor of myogenic differentiation. Am. J. Physiol. Cell Physiol., 282: C993-C999.

Roberts AB, Anzano MA, Wakefield LL, Roche N, Stern DF, Sporn MB (1985). Type β transforming growth factor: a bifunctional regulator of cellular growth. Proc. Natl. Acad. Sci. USA., 82: 119-123.

Rodtke F, Clevers H (2005). Self-Renewal and Cancer of the Gut: Two Sides of a Coin. Science, 307: 1904-1909.

Sekiya I, Vuoristo JT, Larson BL, Prockop DJ (2002). In vitro cartilage formation by human adult stem cells from bone marrow stroma defines the sequence of cellular and molecular events during chondrogenesis. PNAS, 99: 4397–4402.

Shah RV, Kinariwala RV, Ramachandran AV (1977). The effect of tail regeneration on hepatic glycogen content and blood glucose level in the scincid lizard Mabuya carinata. J. Anim. Morphol. Physiol., 24: 76-85.

Shah RV, Kinariwala RV, Ramachandran AV (1980). Haematopoesis and regeneration: Changes in the cellular elements of blood and haemoglobin during tail regeneration in the scincid lizard, Mabuya carinata. Monit. Zool. Ital., 14: 137-150.

Shah RV, Kinariwala RV, Ramachandran AV (1982). Changes in the visceral fat bodies associated with haematopoesis and lipid metabolism, in relation to tail regeneration in the scincid lizard, Mabuya carinata: A histomorphological analysis. Anat, Anz Jena, 151: 137-143.

Stocum DC (2004). Tissue restoration through regenerative biology and medicine. Adv. Anat. Embryol. Cell. Biol., 176: 1-10.

Strain AJ (1990). Epidermal growth factor enhances and transforming growth factor: β inhibits DNA synthesis with similar potency in hepatocytes from normal and from 24 and 48 hr regenerating rat liver. Biochem. Assoc. Trans., 18: 1-7.

Tseng AS, Levin M (2008). Tail Regeneration in Xenopus laevis as a Model for Understanding Tissue Repair. J. Dent. Res., 87: 806.

Tsonis P A (2000). Regeneration in vertebrates. Dev. Biol., 221: 273-284.

Walker P, Moton E, Weichsel Jr., Fisher DA, Guo MS, Fisher DA (1979). Thyroxine increase nerve growth factor concentration in adult mouse brain. Science, 204: 427-429.

Werne S, Grose R (2003). Regulation of Wound Healing by Growth Factors and Cytokines. Physiol. Rev., 83: 835-870.

Wong RWC, Guilland L (2004). The role of epidermal growth factor and its receptors in mammalian CNS. Cytokine Growth Factor Rev., 15: 147-156.

Yamane A, Mayo MI, Bringas P, Chen L, Huynh M, Thai K, Shum I, Slavkin HC (1997). TGF-alpha, EGF, and Their Cognate EGF Receptor are Co-Expressed with Desmin during Embryonic, Fetal, and Neonatal Myogenesis in Mouse Tongue Development.

Zhang F, Clarke JDW, Ferretti P (2000). FGF-2 Up-regulation and Proliferation of Neural Progenitors in the Regenerating Amphibian Spinal Cord *in vivo.* Dev. Biol., 225: 381–391.

3

Synthesis, spectroscopy and biological studies of nickel (II) complexes with tetradentate Schiff bases having N$_2$O$_2$ donor group

Anant Prakash[1]*, Mukesh Pal Gangwar[2] and K. K. Singh[3]

[1]Inorganic Bioinorganic Research Laboratory, Department of Chemistry, Government Post Graduate College, Ranikhet-263645 (Uttarakhand) India.
[2]Department of chemistry, Mahatama Jyotiba Phule Rohilkhand University, Bareilly-243001(Uttar Pradesh) India.
[3]Shri Ram Murti Smarak College of Engineering and Technology (S. R. M. S. CET) Bareilly-243001 (Uttar Pradesh) India.

Complexes of nickel(II) of N,N'-disalicylidene-3,4-diaminotoluene(H2L1), N,N'-bis(3,5-di-tert-butylsalicylidene)-1,3-diaminopropane(H2L2),tetrathiafulvalene-N,N'-phenylene bis (salicylideneimine)(H2L3), o-hydroxybenzaldehyde, o-hydroxyacetophenone ethylene diamine (H2L4) and 1-phenylbutane-1,3-dionemono-S-methylisothio- semicarbazone with 5-phenylazo- o-hydroxybenzaldehyde (H2L5) have been synthesized and characterized by elemental analysis, electronic, (Infra-Red) IR, 1H NMR, magnetic susceptibility measurement, molar conductance and thermal studies. The complexes are found to be non- ionic in nature. The analytical studies show tetrahedral, square planar and octahedral geometries of the complexes. The complexes have been found to posses 1:1 (M:L) stoichiometry. The bioefficacy of the ligands and their complexes have been examined against the growth of bacteria *in vitro* to evaluate their anti-microbial potential.

Key words: Tetradentate schiff base, Nickel (II) ions, N$_2$O$_2$ group, spectra.

INTRODUCTION

Schiff base complexes have remained an important and popular area of research due to their simple synthesis, versatility and diverse range of applications (Taylor and Relinski, 2004; Yamada, 1999). Tetradentate Schiff base complexes of nickel afford two main differences relative to macrocyclic ligands, easier access to mixed donor environments and an open equatorial ring, the hole size of which can in principle accommodate more easily the expected changes in metal size upon oxidation/reduction (Cristin and de Castro, 1988). The coordination chemistry of nickel metal complexes with salen-type ligands had achieved a special status (Klein et al., 2000; Isse et al., 1992), because of their very interesting O$_2$-binding reactivity, redox chemistry, unusual magnetic and structural properties, as well as their usage as catalysts for the oxidation and epoxidation reactions (Veli et al.,

2005; Zhang et al., 1990). Salen complexes have also been recently used as catalytically active materials to develop surface-modified electrodes for sensoring applications and as sources of planar supramolecular building blocks (Aubert et al., 1999; Mao et al., 2000; Jager et al., 1997; Chiehak et al., 2002). In the area of bioinorganic chemistry, interest in Schiff bases complexes had centered on synthetic applications (Khaddar et al., 2005), whereas, unsymmetrical Schiff base ligands have clearly offered many advantages over their symmetrical counterparts in the elucidation of the composition and geometry of the metal ion binding sites in the metal-proteins and enzymes and selectivity of natural systems with synthetic materials (Daneshvar et al., 2003; Bu et al., 1997). Salen-type Schiff base complexes exhibiting potentially large nonlinear optical (NLO) responses have attracted attention in last decades (Bella, 2000, 2002; Chiang et al., 1996). In such compounds having generally a planar or a pseudo planar structure, the metal atom s strategically placed at the center of the charge-transfer system, allowing the d electrons of the metal to take part in the conjugation scheme of the organic ligands. As a result enhanced optical nonlinearities are observed after complexation.

EXPERIMENTAL

Materials and methodology

All the chemicals were used of analytical grade and used as procured. Solvents used were of analytical grade and were purified by standard procedures. The stoichiometric analyses (C, H and N) of the complexes were performed using Elementar vario EL III (Germany) model. Metal contents were estimated on an AA-640-13 Shimadzu flame atomic absorption spectrophotometer in solutions prepared by decomposing the complex in hot concentrated HNO_3. The molar conductances at 10^{-3} molar dilution were measured by Elico-Conductometer Bridge.

The IR spectra were recorded on Perkin-Elmer FTIR spectrophotometer in KBr and polyethylene pellets. The UV-visible spectra were recorded in water on Beckman DU-64 spectrophotometer with quartz cells of 1 cm path length. 1H NMR spectra were recorded in DMSO solvent (solvent peak 3.8 ppm) on a Bruker Advance 400 instrument.

Synthesis of ligands

Synthesis of H2L1

It was prepared by dissolving 3.05 gm (25 mmol) of 3, 4-diaminotoluene in 100 ml of ethanol and stirred for 3 h. After that 50 mmol (6.10 ml) of salicylaldehyde was mixed in 150 ml of ethanol. The 3, 4-diaminotoluene solution was added to the salicylaldehyde solution using an overhead stir for complete mixing. The crude product obtained was recrystallised from dichloromethane.MP 120°C.

Synthesis of H2L2

It was prepared by refluxing 20 mmol of 3, 5-di-tertbutylsalicylaldehyde and 10 mmol of the 1, 3-diaminopropane in 60 ml of ethanol for 1 h. The product was recrystallised from ethanol and dried (melting point, 144°C).

Synthesis of H2L3

A solution of 5,6-diamino-2-(4,5-bis (propylthio)-1,3-dithio-2-ylidene)-benzo 1,3-dithiole (5 mmol) and salicylaldehyde (10 mmol) in ethanol was stirred for 6 h. The resulting precipitate was filtered, washed with ethanol and dried in vacuum. The analytical pure ligand was obtained as an orange powder (melting point (MP) 185°C).

Synthesis of H2L4

Hot ethanolic solution of ethylene diamine (0.61 g, 10 mmol) was added drop-wisely with continuous stirring to a hot ethanolic solution of o-hydroxy acetophenone (10 mmol) followed by o-hydroxybenzaldehyde (10 mmol). The mixture had been refluxed on water bath for about 1 h then allowed to cool at room temperature. The solid product was filtered off and recrystallised from ethanol (MP 170°C).

Synthesis of complexes (general method)

One millimole of $NiAc_2.4H_2O$ was dissolved in ethanol and stirred for 2 h and 1 mmol of requisite ligand was suspended in hot ethanol and stirred for 3 h. Ethanolic solutions of the Schiff base were added to ethanolic Nickel (II) acetate tetra hydrate solutions and the resulting mixtures refluxed; after cooling, solids were filtered off, washed with ethanol and diethyl ether and dried under vacuum over P4O10.

Synthesis of ligand (H2L5) and complex (NiL5)

A warm solution of $NiAc_2.4H_2O$ (1.0 mmol) in methanol (10 cm^3) was added to a solution containing phenylbutane-1, 3-dionemono-S-methylisothiosemicarbazone (1.0 mmol), 5-phenylazo-salicylaldehyde (0.2 cm^3, 1.5 mmol) and triethylamine (1 cm^3). Precipitation of complex occurred immediately. The precipitate was washed with methanol and dried (Abe et al., 2004).

RESULT AND DISCUSSION

The synthesized compounds are crystalline colored, non-hygroscopic, insoluble in water, partially soluble in ethanol but soluble in chloroform, acetone, Dimethylformamide DMF and Dimethyl sulfoxide (DMSO). They were obtained with excellent yield because of the intramolecular hydrogen bond between the fairly acidic phenolic hydrogen and the azomethine nitrogen atom in salicylaldehyde or its derivatives, which catalyzes the condensation reactions. The coordination of the Ni metal of the quadridentate (ONNO) ligands is realized by means of nitrogen atoms N1 and N4 of the ligands and two oxygen of the ligands. Composition and identity of the assembled system were deduced from elemental analysis and IR, UV-vis, NMR, TGA and molar conductance. The analytical data of the complexes indicated 1:1 metal to ligand stoichiometry. The complexes were decomposed in the range of 200 to 270°C. Possible compositions of the complexes were calculated and compared with the experimental values as presented in Table 1.

Molar conductance

The complexes were dissolved in DMF and the molar conductivities of 10^{-3} M of their solutions at room temperature were measured. Table 2 shows the molar conductance values of the complexes. It is concluded from the results that complexes are found to have molar conductance values in the range of 2 to 14 ohm^{-1} mol^{-1} cm^2 indicating the non-electrolytic nature of these complexes (Figure 1) (Prakash et al., 2010; Rabie et al., 2008).

I.R. spectra

The IR spectra of the complexes were interpreted by

Table 1. Composition, colours, melting point and elemental analyses.

Complexes composition	Melting point	Colour	Yield	Elemental analyses (found|calcd.)			
				C	H	N	Ni
NiL[1] $C_{21}H_{14}N_2O_2Ni$	240	Red	86	64.45 (64.40)	3.70 (3.76)	7.54 (7.51)	15.72 (15.74)
NiL[2] $C_{33}H_{48}N_2O_2Ni$	205	Green	83	71.43 (70.33)	9.35 (8.58)	5.55 (4.97)	1 8.98 (18.19)
NiL[3] $C_{30}H_{26}N_2O_2S_6Ni$	215	Orange	87	51.85 (51.65)	3.75 (3.76)	4.08 (4.02)	9.12 (8.42)
NiL[4] $C_{17}H_{20}N_2O_2Ni$	200	Yellow	84	54.66 (54.44)	5.32 (5.37)	7.75 (7.47)	16.00 (15.66)
NiL[5] $C_{25}H_{21}N_5O_2SNi$	270	Brown	83	58.70 (58.39)	4.25 (4.12)	13.48 (13.63)	11.87 (11.42)

Table 2. I. R., Electronic spectra, Magnetic moment, Molar conductance and Geometry of the complexes.

Complexes	I.R. v (C=N) (cm^{-1})	Electronic (nm)	μ_{eff}. B. M..	Geometry	Molar Conductance $(\Omega^{-1} mol^{-1} cm^2)$
NiL[1]	1600	478,448,378	Diamagnetic	Square planar	14
NiL[2]	1605	270,340,600	Diamagnetic	Square planar	6
NiL[3]	1588	280,310,350	Diamagnetic	Square planar	2
NiL[4]	1578	340,360,410	2.50	Octahedral	8
NiL[5]	1585	290,380,490	2.40	Tetrahedral	10

comparing the spectra with that of the free ligands. The absence of the broad band at 2500 to 3200 cm^{-1} due to v(OH) of the intra molecularly bonded N-H-O in the spectra of the complexes indicated the de-protonation of the salicylaldimine moiety of H_2Lx in the Complexation (Abe et al., 2004; Rigamonti et al., 2006). The shift of the characteristic imines (CH=N) band from 1590 to 1620 to 1578 to 1605 cm^{-1} indicated coordination of the azomethine nitrogen's to the nickel atom (Gradinaru et al., 2007). Further coordination of azomethine is confirmed with the presence of new bands at 430 to 480 cm^{-1} region assignable to v (Ni-N) for these complexes. A new band in the 400 to 450 cm^{-1} region in the spectra of the complexes is assignable to v (Ni-O) (Garg and Kumar., 2003). A very broad band at about 3300 to 3446 cm^{-1} is present in the spectra of NiL[4] complex. The presence of this broad band is associated with coordinated water molecules. The presence of coordinated water in the complex had been inferred on the basis of a medium intensity at 728 to 777 cm^{-1} (OH- rocking) Figure 2 (El-Wahab and.El-Sarrag, 2004).

Magnetic susceptibility measurement and ^{1}H NMR spectroscopy

The complexes NiL[1], NiL[2], NiL[3] are diamagnetic at RT indicates their square planar geometry around Ni(II) ion while the complexes NiL[4] and NiL[5] having magnetic tetrahedral geometry, respectively. The TGA study of the NiL[4] complex shows the presence of two coordinated

water molecules which is in the support of its octahedral geometry.

All tetradentate Schiff bases showed a narrow inter se singlet in the region of δ 13.00 to 13.99 ppm assigned to hydrogen bonded salicylic proton. The presence of a doublet signal in the region of δ 7.42 to 8.40 ppm indicated the presence of two azomethine groups, as two signals are recorded for the azomethine protons. The ^{1}H NMR spectra of all the complexes showed a down-field shift in the frequency of azomethine protons confirming coordination of the metal ion. In all the complexes, no signal was recorded for phenolic hydrogen in the 12.5 to 14.00 ppm region, as in the case of the Schiff base, indicating deprotonation of the orthohydroxyl group (¯as et al., 2006) and confirmed coordination through phenolic oxygen. Protons of the bridging methylene groups attached to a nitrogen atom, N-CH2-, resonances in the region of δ 3.57 to 3.91 ppm as a triplet pattern and doublet pattern in H_2L^2 and H_2L^4, respectively (Figure 3).

Electronic absorption spectra

The UV-vis spectra of the complexes 15 in chloroform are reported in Table 2. They show essentially three sets of common bands, falling in the range 270 to 600 nm. Thevery intense bands at low wavelengths have been assigned to charge transfer transition, for complexes with aromatic bridges these bands occur at longer wavelengths, as expected from the higher aromaticity of the ligands which eases delocalization of electron density.

Synthesis, spectroscopy and biological studies of nickel (II) complexes with tetradentate Schiff bases having N2O2 donor group

21

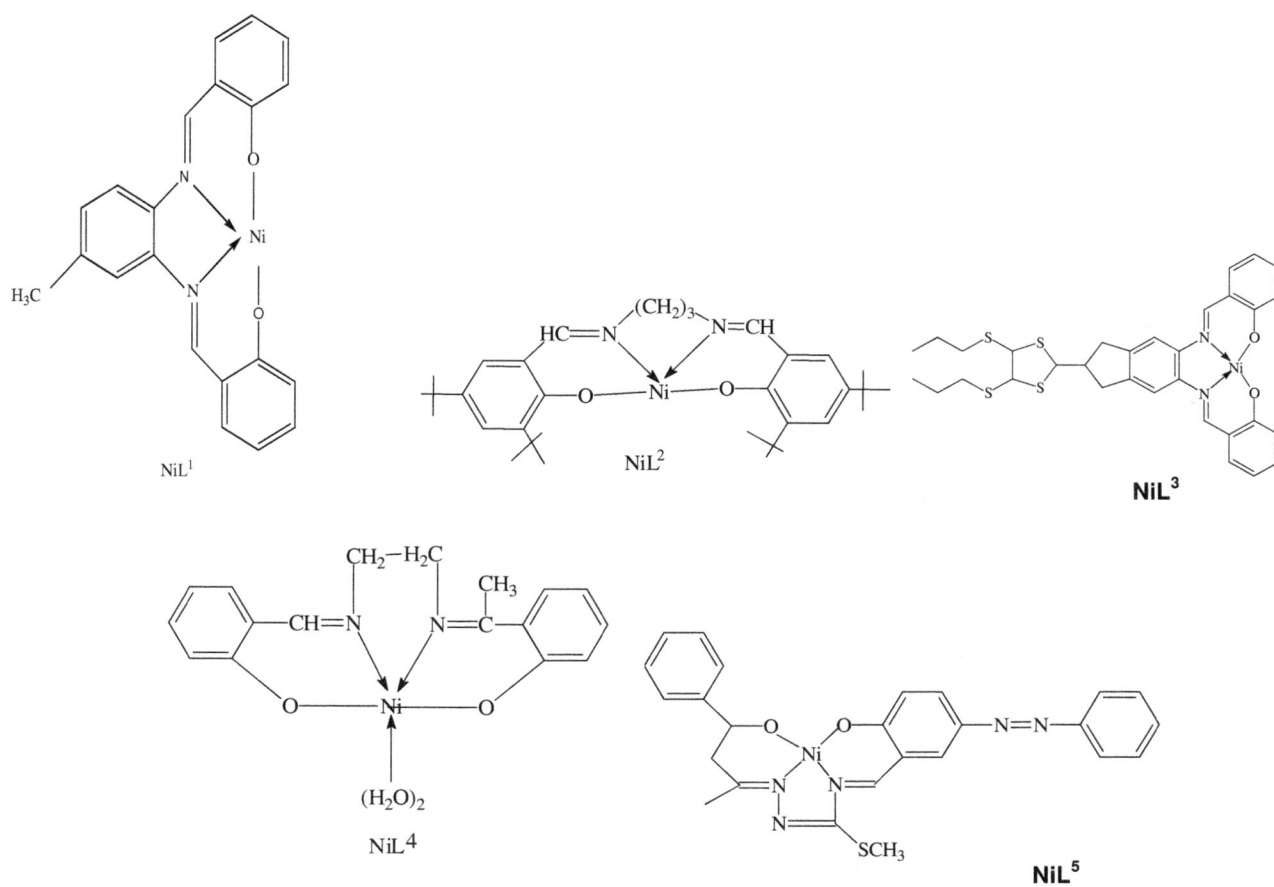

Figure 1. Structures of the complexes.

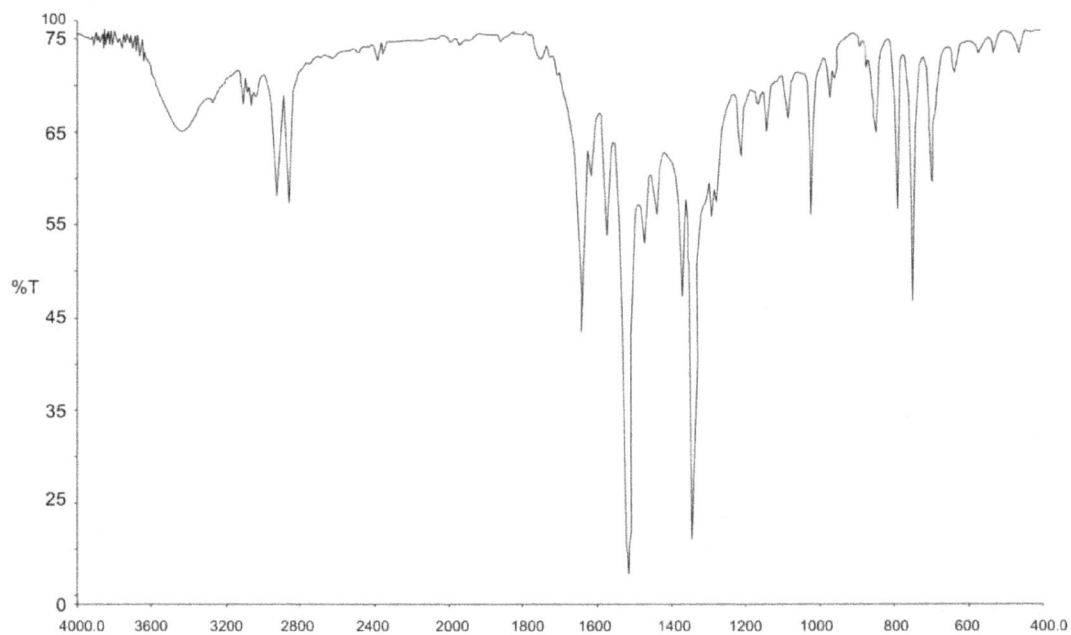

Figure 2. I.R. spectrum of complex NiL^4.

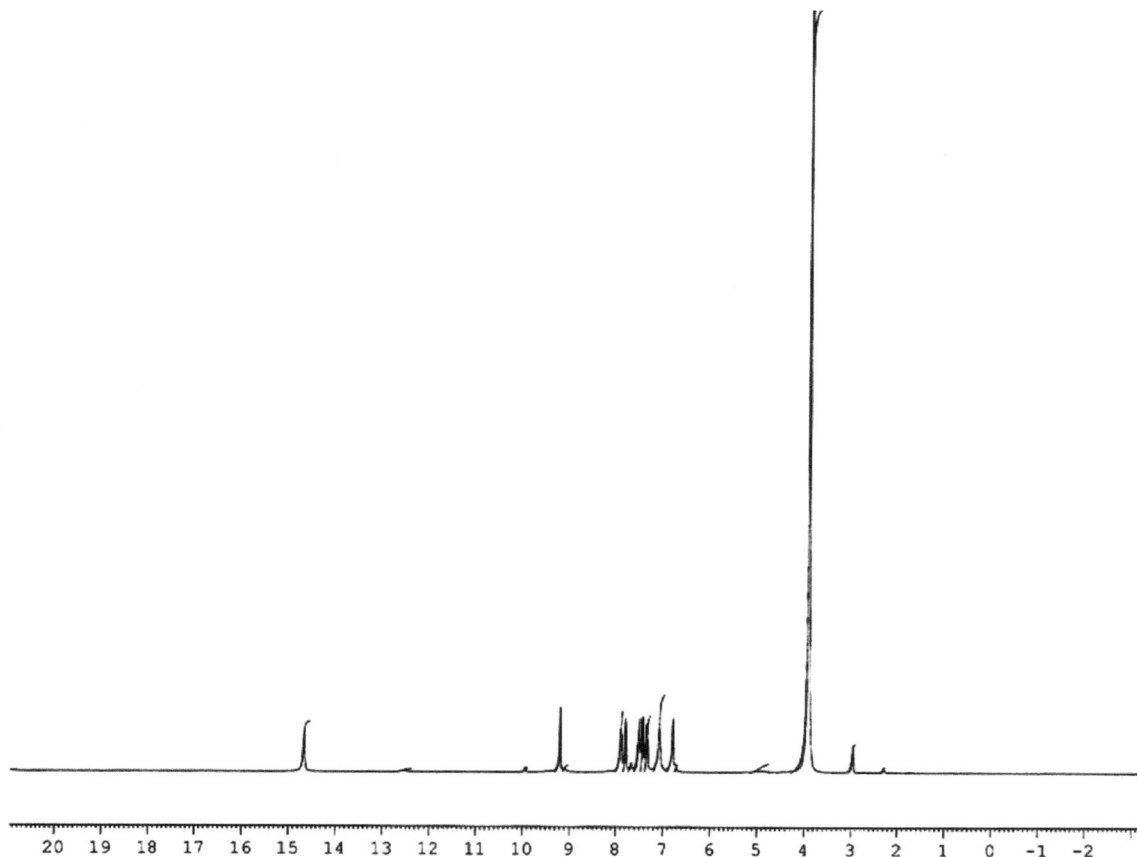

Figure 3. ^{1}H NMR (300 MHz) spectra NiL^1 in DMSO-d6.

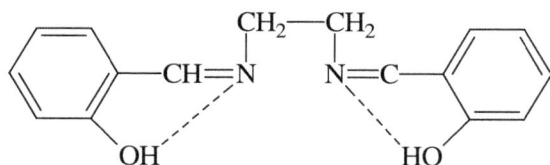

Figure 4. Intramolecular H-bonding H2L4.

On the other hand, the observed new shoulder around 450 nm in the spectra of the complexes solutions can be likely ascribed to an intermolecular transition from the ligands molecules to the vacant orbitals localized on the coordinated metal ions, that is, L→MCT (Figure 4).

The weaker band in the region 520 to 600 nm in the spectra of complexes with aliphatic imines is assigned to unresolved transitions from the four low-lying d-orbitals to the empty dxy orbital (Hoyt and Everett, 1969). This band could not be observed for complexes with aromatic imines bridges since it is masked by the high-intensity charge transfer transitions (Figure 5).

Electronic absorption spectra of NiL^1 to NiL^3 are characterized by a broad band in the range of 270 to 600 nm. This behavior can be assigned to $^1A_{1g}$→$^1A_{2g}$, $^1A_{1g}$→$^1B_{1g}$ and $^1A_{1g}$→$^1E_{1g}$ transitions confirmed square planar geometry of these complexes (Garg and Kumar., 2003). The electronic spectra of NiL^4 are characterized by a broad band covering the long wavelength region 390 to 435 nm. This behavior can be assigned to an octahedral $^3A_{2g}$→$^3T_{1g}(P)$ transitions. In NiL^5, the band at 290 to 490 nm is assigned to $^3T_1$→$^3T_1(P)$ transitions which are also in support of a tetrahedral geometry (Lever, 1984).

Thermal analysis

TG and DTA studies were carried out on the ligands and their complexes in the temperature range of 30 to 650°C. The thermal analyses show that there are two endothermic peaks in the DTA curve of the ligands. The first is the melting point of the ligands, because no loss of weight was observed in the TG curve and second corresponds to decomposition of the ligands. TG studies of NiL^1, NiL^2, NiL^3 and NiL^5 complexes showed no weight loss upto 170°C, indicating the absence of coordinated

Table 3. Antibacterial activity of the synthesized compounds.

S/N	Compound	Diameter of zone of inhibition in mm for *E. coli*		Diameter of zone of inhibition in mm for *S. aureus*	
		50 ppm	100 ppm	50 ppm	100 ppm
1	H_2L^1	7	6	2	4
2	NiL^1	10	11	8	7
3	H_2L^2	2	3	2	2
4	NiL^2	4	6	4	4
5	H_2L^3	5	6	2	4
6	NiL^3	7	10	8	12
7	H_2L^4	6	5	5	6
8	NiL^4	6	7	10	11
9	H_2L^5	4	5	5	4
10	NiL^5	7	6	4	10
11	Chloramphenicol (Reference)	11	22	10	20

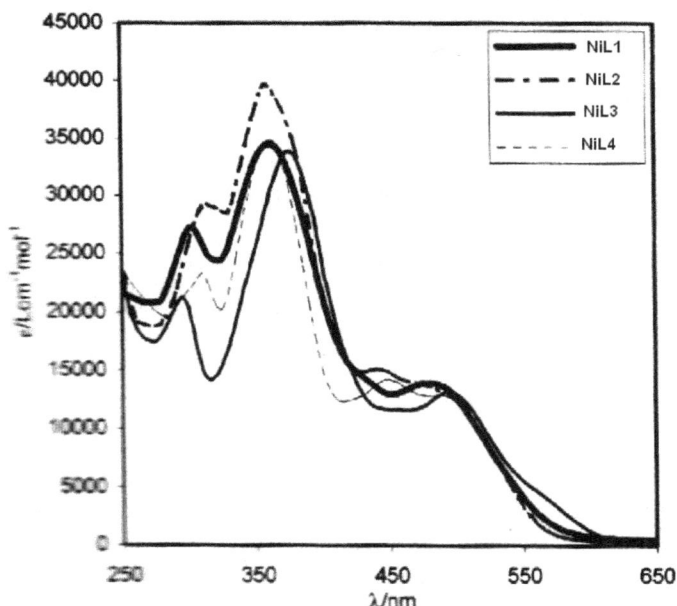

Figure 5. Experimental UV-vis spectra of complexes.

water molecules in the complexes. The complex NiL^4 showed weight loss at about 170°C, corresponding to two water molecules, this suggests the presence of two coordinated water molecules in this complex.

Biological studies

The free ligands and their Ni(II) complexes were screened against *Escherchia coli* and *Staphylococcus aureus* bacteria to assess their potential antimicrobial activity. The results are quite promising. The bacterial

screening results (Table 3) reveal that the free ligand (H_2L^1) and complex I showed maximum activity against *E. coli* bacteria, but the complex III and IV show maximum activity against *S. aureus* whereas ligand (H_2L^3) and complex III showed the better activity against *E. coli* bacteria and the complex III shows better agent for *S. aureus* bacteria. The antimicrobial data reveal that the complexes are more bioactive than the free ligands. The enhanced activity of the metal complexes may be ascribed to the increased lipophilic nature of the complexes arising due to chelation. It is probably due to faster diffusion of the chelates as a whole through the cell membrane or due to the chelation theory.

Conclusion

Metal complexes are found to be monomer and involved coordination of metal ion through azomethine nitrogen atom, phenolic oxygen atom and water molecules of the ligand molecules and forms different types of geometry. Kinetic decomposition studies reveals the first order kinetics and proceeds in two/ three step decomposition. The ligands and their metal complexes exhibit noble anti microbial activity against the reported bacterial species.

ACKNOWLEDGEMENT

The author (Anant Prakash) is thankful to CSIR, New Delhi, India for financial assistance. The authors also wish to thank Dr Shamim Ahmad, Retd. Reader, Bareilly College Bareilly (U.P.) for helpful discussion.

REFERENCES

Aubert PH, Neudeck A, Dunsch I, Audebert P, Capdevielle P, Maumy

M (1999). Electrochemical synthesis and structural studies of copolymers based on the electrooxidation of pyrrole and some salen compounds. J. electroanal. Chem., 470; 77.

Bu XR, Jackson CR, Derveer DV, You XZ, Meng QJ, Wang RX, (1997). New copper(II) complexes incorporating unsymmetrical tetradentate ligands with cis-N2O2 chromophores: synthesis, molecular structure, substituent effect and thermal stability. Polyhedron, 16: 2991.

Chiang W, Vancngen D, Thompson ME (1996). Second-order non-linear optical properties of Fe(SALEN) complexes. Polyhedron, 15: 2369.

Chiehak K, Jacquenmard U, Brand NR (2002). Highly Efficient Synthesis and Solid-State Characterization of 1,2,4,5-Tetrakis(alkyl- and arylamino)benzenes and Cyclization to Their Respective Benzobis(imidazolium) Salts. Eur. J. Inorg. Chem., pp. 357.

Cristin AF, de Castro B (1988). Reductive electrochemical study of Ni(II) complexes with N2O2 Schiff base complexes and spectroscopic characterisation of the reduced species. Reactivity towards CO. J. Chem. Soc. Dalton Trans., pp1491.

Daneshvar N, Entezami AA, Khandar AA, Saghtaforoush LA (2003). NEW NICKLE (II) complexes INCORPORATING DISSYMMETRIC tetradentate schiff base ligands derived from aminother pyridine with N2os chromophores: synthesis, spectroscopic characterization and crystal structures of [Ni(pytAzosal) ClO4.H20. Polyhedron, 22: 1437.

Di Bella S, Fragela I (2000). Organometallic Complexes for Nonlinear Optics. 24. Reversible Electrochemical Switching of Nonlinear Absorption. Synth, Met. 115: 191.

Di Bella S, Fragela I (2002). Two-dimensional characteristics of the second-order nonlinear optical response in dipolar donor–acceptor coordination complexes. I.New J. Chem., 26: 285.

El-Wahab ZH, El-Sarrag MR (2004). Derivatives of phosphate Schiff base transition metal complexes: synthesis, studies and biological activity. Spectrochimica Acta Part A, 60: 271.

Garg BS, Kumar ND (2003). Spectral studies of complexes of nickel(II) with tetradentate schiff bases having N2O2 donor groups. Spectrochimica Acta Part A, 59: 229.

Gradinaru J, Forni A, Druta V, Tessore F, Zecchin S, Quici S, Garbalau N (2007). Structural, Spectral, Electric-Field-Induced Second Harmonic, and Theoretical Study of Ni(II), Cu(II), Zn(II), and VO(II) Complexes with [N2O2] Unsymmetrical Schiff Bases of S-Methylisothiosemicarbazide Derivatives†. Inorganic Chem., 46: 884.

Hoyt WC, Everett GW (1969). J. Inorg. Chem., 8: 2013.

Isse AA, Cenno A, Vianello E (1992). Electrochemical reduction of Schiff base ligands H2salen and H2salophen. Electrochim, Acta . 47113.

Jager EG, Schunmann K, Gorls H (1997). Inorg. Chim. Acta, 255: 295.

Khaddar A, Hosseni-yazdi SA, Zarei SA (2005). Synthesis, characterization and X-ray crystal structures of copper(II) and nickel(II) complexes with potentially hexadentate Schiff base ligands Inorg. Chem. Acta, 358: 3211.

Klein LT, Alleman KS, Peters DG, Karty JA, Reilly JP (2000). Two-Dimensional Cysteine and Cystine Cluster Networks on Au(1¯1) Disclosed by Voltammetry and in Situ Scanning Tunneling Microscopy. J. Electro Anal. Chem., 481: 24.

Lever ABP, (1984). The interaction of the VO2+ cation with oxidized glutathione. Inorganic Electronic Spectroscopy, 2nd Ed. Elsevier. Amsterdam, New York,

Mao L, Yamamato K, Zhou W, Jin L (2000). Electrochemical Nitric Oxide Sensors Based on Electropolymerized Film of M(salen) with Central Ions of Fe, Co, Cu, and Mn. Electroanalysis, 12: 72.

Prakash A, Singh BK, Bhojak N, Adhikari D (2010). Synthesis and characterization of bioactive zinc(II) and cadmium(II) complexes with new Schiff bases derived from 4-nitrobenzaldehyde and acetophenone with ethylenediamine. Spectrochimica Acta Part A, 76: 356.

Rabie UM, Assram ASA, Abou-El-Wafa MHM (2008). Unsymmetrical Schiff bases functionalize as bibasic tetradentate (ONNO) and monobasic tridentate (NNO) ligands on complexation with some transition metal ions. J. Mole. Structure, 872; 113.

Rigamonti L, Demartin F, Forni A, Righetto S, Pasini A (2006). Copper(II) Complexes of salen Analogues with Two Differently Substituted (Push–Pull) Salicylaldehyde Moieties. A Study on the Modulation of Electronic Asymmetry and Nonlinear Optical Properties. Inorg. Chem., 45: 10976.

Tas E, Aslanoglu M, Kilic A, Kaplan O, Temel H (2006). Synthesis and spectral characterization of macrocyclic NiII complexes derived from various diamines, NiII perchlorate and 1,4-bis(2-carboxyaldehydephenoxy)butane. J. Chem. Res. (S), 4: 242.

Taylor MK, Relinski J, Wallace D, (2004). Coordination geometry of tetradentate Schiff's base nickel complexes: the effects of donors,

Veli TK, Seniz OY, (2005). Synthesis, spectroscopy and electrochemical behaviors of nickel(II) complexes with tetradentate shiff bases derived from 3,5—salicylaldehyde. Esref Tas, Spectrochimica Acta part A 62:716.

Yamada S (1999). Advancement in stereochemical aspects of Schiff base metal complexes. Coordin, chem. Rev., 192; 537.

Zhang W, Loebach JL, Wilson SR, Jacobsen EN (1990). Enantioselective epoxidation of unfunctionalized olefins catalyzed by salen manganese complexes. J. Am. Chem. Soc., 112: 2801.

4

Photosynthetic properties of the protoplasts from *Bryopsis hypnoides* Lamouroux. during the early regeneration process

Nai Hao Ye[1]*, Hong Xia Wang[2], Zheng Quan Gao[3] and Guangce Wang[2]

[1]Yellow Sea Fisheries Research Institute, Chinese Academy of Fishery Sciences, Qingdao, 266071, China.
[2]Key Laboratory of Experimental Marine Biology, Institute of Oceanology, Chinese Academy of Sciences, Qingdao 266071, China.
[3]School of Life Sciences, Shandong University of Technology, Zibo, 255049, China.

The scanning electronic microphotograph (SEM), photosynthetic oxygen evolution rate and excitation spectra at liquid nitrogen temperature (77 K) of the protoplasts from Bryopsis hypnoides were determined at the early regeneration period (12 h). SEM revealed an existence of the tight aggregation; cell membrane and wall were about 10 min, 6 and 12 h respectively, after the culture. Oxygen evolution rate decreased slowly during the whole process and a minus value appeared at 6 h. Peaks originated from photosystem II (PS II) and photosystem I (PS I) apparently shifted to shorter wavelength during the regeneration process in the excitation spectra. No peak from PS I was determined until in the spectrum of 12 h, which was excited by the wavelength of 436 nm; while the peak from PS I appeared in the spectrum of extruded protoplast excited by the wavelength of 660 nm and strengthened during the rest of the regeneration process.

Key words: Photosynthetic oxygen evolution rate, absorbance spectra, excitation spectra.

INTRODUCTION

Isolated protoplasts from single cells show their attractiveness by acting as a valuable system for plant breeding and for various fundamental studies, such as photosynthetic property query (Quick and Horton, 1986; Pier and Berkowitz, 1989; Matsue et al., 1993; Peng et al., 1996; Goh et al., 1997; Kao et al., 2002; Shiku et al., 2005).

Bryopsis is a coenocytic unicell giant alga whose protoplasts can swirl into sub-protoplasts and regenerate into new lives in seawater (Kim et al., 2001; Ye et al., 2005). Ye et al. (2005) suggested that both the total soluble proteins and the ratio of chlorophyll a to chlorophyll b between the regenerated B. hypnoides and the wild type were quite different. Shiku et al. (2005) investigated the respiration and photosynthesis activities as a function of the size of the protoplast from Bryopsis plumosa and their

results indicated that the respiration rate was linear to the cube of the sample radius, while the photosynthesis rate was linear to the square of the sample radius. It is suggested, therefore, the former was controlled by the volume of the protoplast, and the latter was controlled by the surface area of the protoplast.

The information about the photosynthetic properties of the protoplasts from Bryopsis is still limited. In this paper, the microphotograph, photosynthetic oxygen evolution rate and excitation spectra of the protoplasts from B. hypnoides were employed to evaluate the photosynthetic characteristics during their regeneration process.

MATERIALS AND METHODS

Plant material

Gametophytic specimens of Bryopsis hypnoides were collected from the intertidal zone (35.35°N, 119.30°E, 20 - 50 cm depth) of Zhanqiao Wharf, Qingdao, China. Seawater was samped using a pump placed 2 m deep under the water surface, and filtered with

*Corresponding author. E-mail: yenh@ysfri.ac.cn

nested plankton nets (20 µm). Collected water was autoclaved and made up into enriched seawater (ES) (McLachlan, 1979). In the laboratory, the thalli were examined and those that were intact were isolated, washed several times with sterile seawater, disinfested with 1% sodium hypochlorite (NaClO) for 2 min, then rinsed with autoclaved seawater. The prepared material was placed into a sterile aquarium (diameter = 40 cm, height = 30 cm) containing ES enriched seawater, and maintained at 20°C under 20 µmol photons m-2s-1 provided 12 h.d-1 by cool-white fluorescent tubes.

Preparation and culture of protoplasts and the thallus segments

For the regeneration investigations, Bryopsis hypnoides thalli were sterilized again with 1% sodium hypochlorite for 2 min and rinsed with autoclaved seawater, and the surface moisture was removed at once with 3 layers of absorbent paper. The clean algae were cut into segments (about 4 mm), and protoplasts were squeezed out with 8 layers of muslin into 10 ml Eppendorf tubes. The initially extruded protoplasts were cultured in 10 cm Petri dishes containing autoclaved ES enriched seawater at 20°C under 20 µmol photons m-2s-1. Some of the protoplasts were used at intervals for the spectral analysis; some of them were used in the oxygen evolution determination; some of them were cultured in Petri dishes and fixed at intervals for SEM.

Longer gametophytic filaments were selected and cut into 0.5 mm long segments using a sapphire knife and cultured in 6 cm Petri dishes containing enriched ES seawater. The segments were cultured two days for healing before being used for spectra analysis. The culture conditions were as follows: 20°C, 20 µmol photons m-2 s-1, 12 h.d-1. Scanning electron microscope (SEM)

Some of the initially extruded protoplasts were cultured and fixed at intervals for SEM. Cultures were mixed for 1 h at 4°C with an equal volume of a fixative containing 5% glutaraldehyde (all the chemicals used were products of Sagon, shanghai, China) and 2% osmium tetroxide dissolved in culture medium. After fixation, samples were placed on poly-L-lysin-coated glass plates, rinsed, dehydrated with ethanol, and then substituted with isoamyl acetate before critical point drying with a Hitachi HCP-2 (Hitachi Koki Co. Ltd., Tokyo, Japan). Imaging was carried out on a KYKY 2800B scanning electron microscope (KYKY Technology Development Ltd., Beijing, China) at an acceleration voltage of 25 kV. The chamber was kept in low vacuum mode under a pressure of 59.6 Pa at an ambient temperature (21°C).

Oxygen evolution of the protoplast/sub-protoplasts

Photosynthetic O2 evolution in the light was monitored at 20°C polarographically using a Clark type O2 electrode (Control Box 980321, Hansatech, King's Lynn, UK). Thermostated water of 20°C was circulated through the outer jacket of the reaction chamber (TB-85, Shimadzu, Japan). Prior to the measurement, samples were kept under the executive light condition for 5 min for adaptation, then 10 µl of samples were introduced into the chamber with the stirrer bar at the level of 3. The reaction seawater of 1 ml for the assay of protoplast photosynthesis contained 1 mM NaHCO3 for optimal CO2 at pH 8.0 and the protoplasts equivalent to 1 µg chlorophyll (Chl). Illumination of 300 µmol m-2 s-1 was provided by a 35 mm slide projector (halogen lamp, 224 V/500W), which was adjusted by the distance.

Spectral properties determination

Fluorescence excitation spectra (77K) of samples at different stages were taken and measured with a spectrofluorometer

(Hitachi).

RESULTS

The regeneration process of the protoplasts

The extruded protoplasts swirled into aggregation soon after being placed into autoclaved seawater (Figure 1 - 1). As light microscope revealed, protoplasts turned into compact aggregations 10 min later with the newly formed primary envelope, however, no primary envelope was found hereon because of the destructive treatment of the sample preparation for SEM (Figure 1 and 2). Protoplasts in microphotographs of 6 h (Figure 1, C 3) and 12 h (Figure. 1D) were all covered by certain substances and there was no exposed organelle could be found any more.

Oxygen evolution of the protoplasts

Figure 2 shows the total oxygen content changes in the chamber with twelve selected oxygen evolution rates during the early regeneration period. The content of the oxygen increased slightly at the very start and gradually reached a maximum value at about 6 h, then began a decreasing all the rest of the time. However, the oxygen evolution rate decreased slower during the whole process and the minus value appeared 6 h later.

Spectral properties of the protoplasts and the thalli segments

Fluorescence spectra excited by the wavelengths of 436 and 660 nm were determined at liquid nitrogen temperature (77K). The result showed that emission peaks in both spectra moved toward shorter wavelength distinctly. Peaks of PS II were at 730 nm in the spectrum of 10 min, while it moved to 694 nm in the spectrum of 12 h (Figure 3a). Peaks attributed to PS I excited by the wavelength of 436 nm could not be found till in the spectrum of 12 h (Figure 3a), while weak peaks of PS I presented in all of the spectra excited by wavelength of 660 nm (Figure 3b).

DISCUSSION

Extruded protoplasts of the coenocytic green alga *Bryopsis hypnoides* aggregate spontaneously in ES enriched seawater and some of the aggregations can grow into new plants (Ye et al., 2005). The regeneration process of the protoplasts from *B. hypnoides* revealed by the SEM (Figure 1) indicates that the new lives are derived from entirely disturbed protoplasts from the giant 'mother' cells and the regeneration of the protoplasts was quickly. It is reported that most the primary organelles and about 15% of the original cell membrane of *B. plumosa* are recycled to make the body of new individuals (Kim et al., 2005).

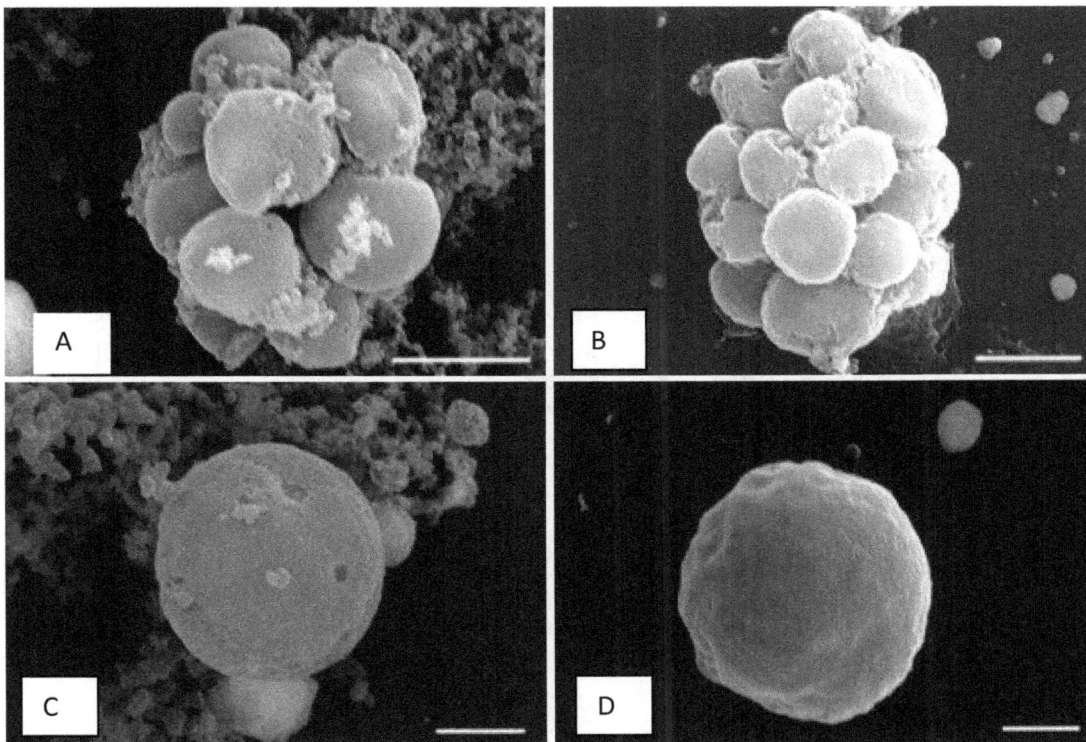

Figure 1. The regeneration process of protoplasts of *Bryopsis hypnoides*. Scale bars, 5 μm. (A) the aggregation of the protoplasts 10 seconds after the culture; (B) the aggregation of the protoplasts 10 min after the culture; and (C and D), aggregations being covered with some substances, which were fixed after 6 and 12 h after the culture respectively.

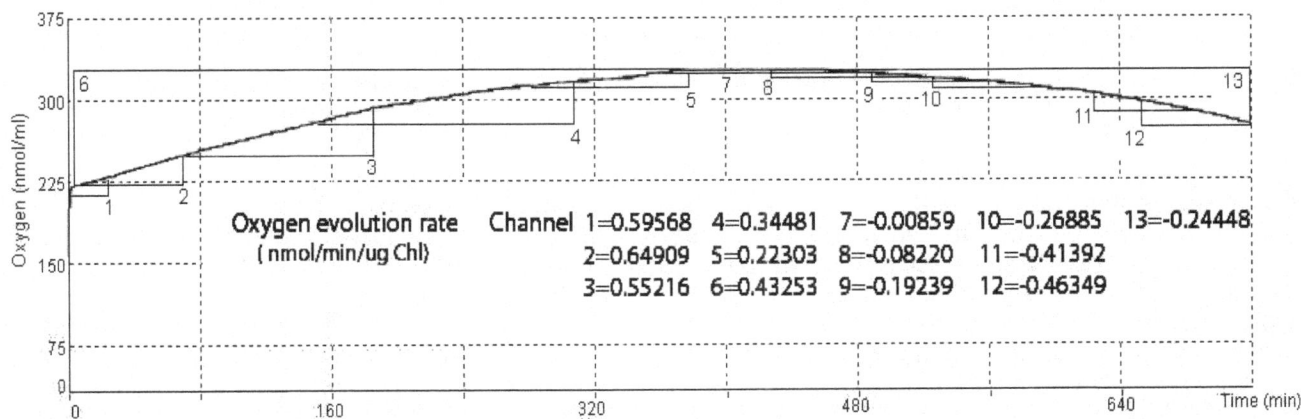

Figure 2. The oxygen content and the evolution rates of the protoplasts during the regeneration period of the experiment. 13 Channels were selected and the oxygen evolution rates were calculated using the software Oxygen Graph V 2.32. Chlorophyll was estimated from the fixed leaves. The method of chlorophyll estimation was that of Arnon (1949). The total chlorophyll was calculated according to the formula given by Arnon (1949).

The protoplasts swirled into tight globes within 10 min and started to form new cell membrane and wall within 12 h, which were consistent with the fluorescence dye results from Kim et al. (2001). According to the regeneration process, three stages of the protoplasts (10 min, 6 h and 12 h after the culture) were selected for the comparison of thalli segments for the photosynthetic property analysis.

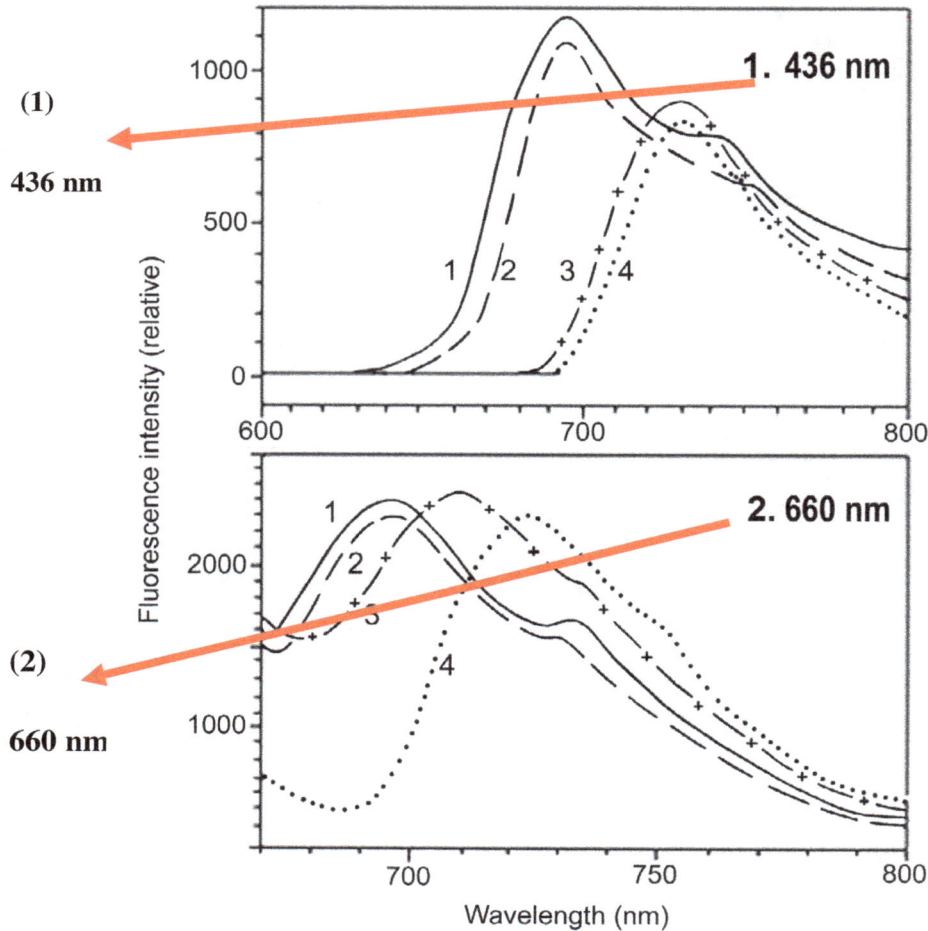

Figure 3. Fluorescence excitation spectra (77 K) of the protoplasts and thalli segments at different regeneration stages. 1, fluorescence excitation spectra excited by the wavelength of 436 nm; 2, fluorescence excitation spectra excited by the wavelength of 660 nm. ——, Spectra of thalli segments; - - -, Spectra of protoplasts 12 h after the culture; — + —, Spectra of protoplasts 6 min after the culture; Spectra of protoplasts 10 min after the culture.

SEM indicates that the regeneration of the protoplast from B. hypnoides is an absolute recombination process of the organelles (Figure 1); the excitation spectra (Figure 3) also suggest that the regeneration of protoplasts from B. hypnoides comprises a recombination or renaissance procedure of the organelles, including some important systems, such as photosystem I (PS I), which may be the main reason leading to the decline of the net oxygen evolution for energy consumption. The changes in the photosynthetic capacity during the cell cycle of *Scenedesmus quadricauda* are explained by the heterogeneity of photosystem II (PS II) (Kaftan et al., 1999). Kaftan et al. (1999) found that the decline in the photosynthetic oxygen evolution during the cell cycle mainly corresponded to an increase in inactive centers of PS II (PS IIx). Some studies also indicate that the energy storage (ES) and the yield of oxygen evolution rapidly increase during the first 3 - 4 h of the cell cycle, but in the second half of

the cycle, ES almost keeps a constant value, whereas the yield of oxygen steadily decreases (Szurkowski et al., 2001).

Previous studies revealed that PS I and PS II of red algae are evolutionarily in different positions: PS I is closer to that of higher plants whereas PS II is more similar to that of cyanobacteria (Szurkowski et al., 2001). Xiong et al (1998) suggested that PS I and PS II were independent to each other in their evolutionary history. It is thought that the emergence of individuality during the unicellular-multicellular transition is based on the evolution of cells that differentiate and specialize in reproductive and survival-enhancing vegetative functions (Michod et al., 2006). Taking the regeneration process of the protoplasts from B. hypnoides as a condensed evolution history, the result of this experiment reflects homologous cases: 1) chlorophyll b and PS I are closer to those of higher plants than chlorophyll a and PS II because the latter take action earlier than the former; 2) chlorophyll a and b, as well as PS I and PS II are independent to each other in the evolutional course because PS II and chlorophyll a can take action independently.

ACKNOWLEDGEMENTS

This work was supported by the National Natural Science Foundation of China (Grants # 40706050, 40706048 and 30700619), the National Science & Technology Pillar Program (Grants # 2006BAD01A13, 2008BAC49B04), Qingdao Municipal Science and Technology plan project (Grants # 08-1-7-6-hy) and the Hi-Tech Research and Development Program (863) of China (Grants # 2006AA10Z414).

REFERENCES

Arnon DZ (1949). Copper enzymes in isolated chloroplast, polyphenoloxidase in *Beta vulgarris,* Plant Physiol. 24: 1-5.

Goh CH, Oku T, Shimazaki KI (1997). Photosynthetic properties of adaxial guard cells from Vicia leaves. Plant Sci. 127: 149-159.

Kaftan D, Meszaros T, Whitmarsh J, Nedbal L (1999). Characterisation of photosystem II activity and heterogeneity during cell cycle of the green alga Scenedesmus quadricauda, Plant Physiol. 120: 433-441.

Kao FJ, Wang YM, Chen JC. Cheng PC, Chen RW, Lin BL (2002). Micro-spectroscopy of chloroplasts in protoplasts from Arabidopsis thaliana under single- and multi-photon excitations. J. Lumin. 98: 107-114.

Kim GH, Klotchkova TA, Kang YM. (2001). Life without a cell membrane: regeneration of protoplasts from disintegrated cells of the marine green alga Bryopsis plumose, J. Cell Sci. 114: 2009-2014.

Matsue T, Koike S, Uchida I (1993). Microamperometric Estimation of Photosynthesis Inhibition in a Single Algal Protoplast. Biochem. Biophys. Res. Comm. 197: 1283-1287.

McLachlan J (1979). Growth media—marine. In: Handbook of Phycological Methods, edited by J.R. Stein, Cambridge University Press, Cambridge pp. 25-51.

Michod RE, Viossat Y, Solari CA, Hurrand M, Nedelcu AM. (2006). Life history evolution and the origin of multicellularity. J. Theor. Biol. 239: 257-272.

Peng GH, Shi DJ, Fei XG., Zeng CK (1996). Isolation and photosynthesis of spheroplasts in *Spirulina platensis.* Act. Bot. Sin. 38: 861-866.

Pier PA, Berkowitz GA (1989). The effects of chloroplast envelope-Mg^{2+}, cation movement, and osmotic stress on photosynthesis. Plant Sci. 64: 45-53.

Quick WP, Horton P (1986). Studies on the induction of chlorophyll fluorescence in barley protoplasts. III. Correlation betweeen changes in the level of glycerate 3-phosphate and the pattern of fluorescence quenching, BBA – Bioenergetics 849: 1-6.

Shiku H, Torisawa Y, Takagi A, Aoyagi S, Abe H, Hoshi H, Yasukawa T, Matsue T (2005). Metabolic and enzymatic activities of individual cells, spheroids and embryos as a function of the sample size. Sens. Actuat. B. 108: 597-602.

Szurkowski J, Bascik-Remisiewicz A, Matusiak K, Tukaj Z (2001). Oxygen evolution and photosynthetic energy storage during the cell cycle of green alga *Scenedesmus armatus* characterised by photoacoustic spectroscopy, J. Plant Physiol. 158: 1061-1067.

Xiong J, Inoue K, Bauer CE. (1998). Tracking molecular evolution of photosynthesis by characterization of major photosynthesis gene cluster from *Heliobacillus mobilis.* Pro. Natl. Acad. Sci. USA. 95: 14851-14856.

Ye N, Wang G, Wang F, Zeng C. (2005). Formation and Growth of Bryopsis hypnoides Lamouroux Regenerated from Its Protoplasts. J. Int. Plant Biol. 47: 856-862.

5

Influence of osteogenic supplements on the osteoclastogenesis of human monocytes

Christiane Heinemann[1]*, Sascha Heinemann[1], Corina Vater[2], Hartmut Worch[1] and Thomas Hanke[1]

[1]Max Bergmann Center of Biomaterials and Institute of Materials Science, Technische Universität Dresden, Budapester Str. 27, D-01069 Dresden, Germany.
[2]Department of Orthopedic Surgery, University Hospital Carl Gustav Carus, Fetscherstr 74, D-01307 Dresden, Germany.

A basic requirement for bone remodeling is the communication between bone forming osteoblasts and bone resorbing osteoclasts – the so-called cross talk. Corresponding *in vitro* co-culture models might be a valuable technique in order to investigate the influence of novel biomaterials on the cross talk. Assuming that both cell types are derived by precursor cells, the role of the common osteogenic supplements concerning the osteoclastogenesis is still little known. Therefore, the approach of the present study was to analyse osteoclast formation in the presence of both, osteoclast differentiation factors as well as osteogenic differentiation factors dexamethasone (Dex), β-glycerophosphate (β-GP) and 1.25-dihydroxy vitamin D3 (VitD3) in typical concentration ranges as used for osteoblastogenesis of bone marrow stromal cells (BMSC). Human monocytes were isolated from buffy coat and separated by using magnetic activated cell sorting (MACS). DNA amount, activity of tartrate-resistant acid phosphatase isoform 5b (TRAP 5b), morphological features of the cells as well as gene expression for TRAP, cathepsin K (CTSK), calcitonin receptor (CALCR) and vitronectin receptor (VTNR) were evaluated. Finally, we are able to suggest conditions which allow both, osteoblastogenesis and osteoclastogenesis of human precursor cells in a combined cultivation medium.

Key words: Human bone marrow stromal cell, human monocytes, osteoclasts, osteoclastogenesis, osteogenic supplement, co-culture.

INTRODUCTION

Understanding communication of osteoblasts and osteoclasts is a pivotal topic in the development of bone substitution materials (Matsuo and Irie, 2008). An approach to the *in vivo* condition is ensured by the use of precursor cells for *in vitro* models. On the one hand, BMSC (also known as mesenchymal stem cells, MSC) can be commonly differentiated into bone forming osteoblasts. Therefore, osteogenic supplements like Dex, β-GP and VitD3 are used to stimulate BMSC to differentiate along the osteoblastic lineage. On the other hand bone-resorbing multinucleated osteoclasts differentiate from monocytes (Mc) which is derived from hematopoietic stem cells. The additional supplementation of the cytokines

macrophage colony stimulating factor (M-CSF) and receptor activator of nuclear factor κB ligand (RANKL) is required to regulate formation, activity and survival of osteoclasts (Heinemann et al., 2010). M-CSF is mostly important for osteoclast precursor cell survival, migration and cytoskeletal reorganisation (Lagasse and Weissman, 1997; Yoshida et al., 1990). RANKL binds to its receptor RANK on the surface of osteoclast precursors and induces the differentiation to osteoclasts (Hadjidakis and Androulakis, 2006). Both cytokines combined induce expression of genes that typify the osteoclast lineage, including those encoding TRAP, CTSK, CALCR and VTNR (Boyle et al., 2003). In an *in vitro* co-culture situation of BMSC/osteoblasts and monocytes/osteoclasts the stimulating supplements may affect differentiation behaviour of the precursors. The scope of the present study was to evaluate the influence of the osteogenic differentiation factors on the osteoclastogenesis

---

*Corresponding author. E-mail: christiane.heinemann@tu-dresden.de.

from monocytes. Therefore, human precursor cells were cultivated in media supplemented with varied concentrations of osteogenic supplements Dex, β-GP and VitD3. Evaluation was performed using biochemical measurement of DNA and TRAP 5b.

Reverse transcriptase-polymerase chain reaction (RT-PCR) was used to detect genes characterizing osteoclasts. Light microscopy revealed the morphological features of the cells. The future aim is to identify a cell culture media regime, which allows both osteoblasto-genesis and osteoclastogenesis simultaneously.

MATERIALS AND METHODS

Isolation of human monocytes

Human monocytes (hMc) were isolated from buffy coats (purchased from the German Red Cross blood donation service, Dresden, Germany) obtained from the blood of healthy anonymous donors. The isolation is based on the OptiPrep density-gradient media technique with some modifications. OptiPrep™ (ProGen Biotechnik GmbH, Heidelberg, Germany) was mixed with the α-modification of minimal essential medium (α-MEM, Biochrom, Berlin, Germany) to obtain the Optiprep™WorkingSolution (WS), a 1.078 g/ml gradient solution and a 1.068 g/ml gradient solution. Buffy coats were centrifugated at 450 g for 20 min and the leukocyte-rich fraction (LRF) at the interface was collected. Optiprep™WS was mixed with the LRF to obtain a density of 1.100 g/ml. In a 50 ml falcon vessel, the OptiPrep™WS/LRF mixture was placed under a layer of 1.078 g/ml lymphocyte-specific gradient solution. A layer of Hepes buffered saline (HBS) was placed on top and centrifuged at 700 g for 20 min. The peripheral blood mononuclear cells (PBMC) fraction was collected, washed with phosphate buffered saline (PBS) containing 2 mM ethylenediaminetetraacetic acid (EDTA) and 0.5% bovine serum albumin (BSA) and centrifuged at 400 g for 10 min.

OptiPrep™WS was mixed with the PBMC fraction to obtain a density of 1.100 g/ml and was covered by layers of 1.078 g/ml gradient solution, 1.068 g/ml gradient solution and HBS. By centri-fugation at 600 g for 25 min the monocyte-enriched PBMC fraction floated into the 1.068 g/ml layer. That fraction was collected, washed with PBS/EDTA/BSA and finally monocytes were purified by negative selection using MACS (Miltenyi, Bergisch Gladbach, Germany).

Osteoclastogenesis

Isolated monocytes were seeded at a density of 4×10^5 cells/cm^2 in 48-well tissue culture dishes and cultivated in α-MEM, supplemen-ted with heat-inactivated 7.5% fetal calf serum (FCS), 7.5% human A/B serum, 2 mM L-glutamine, 100 U/ml penicillin and 100 μg/ml streptomycin. Medium and all supplements were obtained from Biochrom (Berlin, Germany). To induce differentiation, 50 ng/ml M CSF and 50 ng/ml RANKL were supplemented. For testing the effect of osteogenic supplements on osteoclastogenesis concentra-tions of VitD3, β-GP and Dex have been varied in several ranges that are typically used to differentiate hBMSC into osteoblasts (Vater et al., 2010).

Colorimetric measurements

All measurements were performed with cell lysates obtained after 3 and 10 days of cultivation. Cell lysis was achieved with 1% Triton X-100 (Sigma) in PBS. For all colorimetric measurements, a Spectra

Fluor Plus microplate reader (Tecan, Crailsheim, Germany) was used.

DNA assay

Examination of DNA amount was carried out using Quant-iT™ PicoGreen ® dsDNA Reagent.

TRAP 5b activity assay

Osteoclast differentiation was evaluated by the measurement of TRAP 5b activity using naphthol-ASBI phosphate (N-ASBI-P, Sigma) as a substrate according to a slightly modified protocol of Jankila et al. (2001). Cell lysates were added to the TRAP 5b reaction buffer consisting of 2.5 mM N-ASBI-P in 100 mM Na-acetate (sigma) buffer containing 50 mM Na-tartrate (sigma), 2% NP-40 (sigma) and 1% ethylene glycol monomethyl ether (EGME, sigma) adjusted to pH 6.1 and the mixtures were incubated at 37°C for 1 h. The enzymatic reaction was stopped by adding 0.1 M NaOH and fluorescence was measured at an excitation wavelength of 405 nm and an emission wavelength of 535 nm. The relative fluorescence units were correlated to a TRAP 5b standard.

Statistics

All measurements were collected in triplicate and expressed as mean ± standard deviation. ANOVA test was employed to assess the statistical significance of results. P values less than 0.001 were considered highly significant.

Microscopy

Light microscopy of cell-seeded polystyrene (PS) was performed using a Zeiss Axiovert 40 CFL equipped with a digital camera (Canon).

RT-PCR

For RT-PCR, cells were washed twice with PBS and total RNA isolation was performed using the peqGOLD MicroSpin Total RNA Kit (Peqlab, Erlangen, Germany) according to the manufacturer's instructions. Complementary DNA (cDNA) was transcribed from 250 ng of total RNA (measured using a Peqlab Nanodrop ND 1000) in a 20 μL reaction mixture containing 200 U of Superscript II Reverse Transcriptase (invitrogen), 0.5 mM dNTP (invitrogen), 12.5 ng/μL random hexamers (MWG Biotech, Ebersberg, Germany) and 40 U of RNase inhibitor RNase OUT (invitrogen). For cDNA synthesis, the reaction mixtures were incubated for 50 min at 42°C followed by 15 min at 70°C in a Primus 25 Advanced Thermocyler (Peqlab). For PCR experiments, 1 μL of cDNA was used in a 20 μL reaction mixture containing specific primer pairs (MWG Biotech) to detect transcripts of TRAP, CALCR, CTSK, VTNR and the housekeeping gene glyceraldehyde-3-phosphate dehydrogenase (GAPDH). Primer sequences used and annealing temperatures are summarized in Table 1. After the initial activation step at 95°C for 4 min, 25 to 35 PCR cycles were run with each denaturation step at 95°C for 45 s, an annealing step for 45 s and a synthesis step at 72°C for 1 min followed by a final synthesis step at 72°C for 10 min. The same single stranded cDNA was used to investigate the expression of the genes described. The resulting PCR-products were analysed using the FlashGel™ Dock and documentation system (Cambrex Bio Science, East Rutherford, NJ, USA).

Expression of the markers was normalised to the expression of

Table 1. Primers for RT-PCR.

Gene	Forward primer (5'-3')	Reverse primer (5'-3')	T_A (°C)
GAPDH	GGTGAAGGTCGGAGTCAACGG	GGTCATGAGTCCTTCCACGAT	55
TRAP	TTCTACCGCCTGCACTCCAA	AGCTGATCTCCACATAGGCA	57
CALCR	GCAATGCTTTCACTCCTGAGAAAC	CAGTAAACACAGCCACGACAATGAG	57
CTSK	GATACTGGACACCCACTGGGA	CATTCTCAGACACACAATCCAC	57
VTNR	AAGTTGGGAGATTAGACAGAGG	CTTTCTTGTTCTTGAGGTGG	57

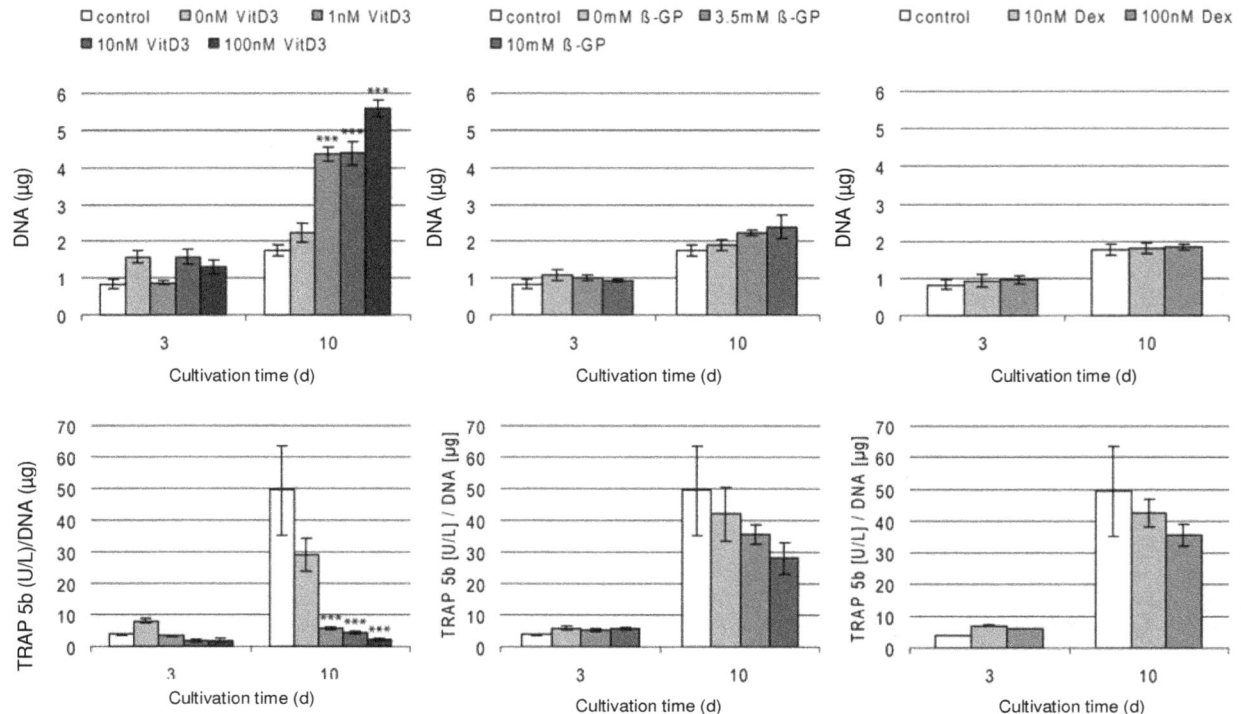

Figure 1. Influence of VitD3, β-GP and Dex concentrations on DNA amount and relative TRAP activity. The concentrations of other supplements were kept constant in each case as follows: 50 ng/ml M-CSF, 50 ng/ml RANKL, 10 nM Dex, and 10 mM β-GP for analysis of VitD3 influence; 50 ng/ml M-CSF, 50 ng/ml RANKL, and 10 nM Dex for analysis of β-GP concentration; 50 ng/ml M-CSF, 50 ng/ml RANKL, and 3.5 mM β-GP for analysis of Dex influence. Stars indicate significance compared to the 0 nM VitD3 sample.

GAPDH by image analysis using the BioImaging System Gene Genius with the acquisition software GeneSnap and the GeneTools software (SynGene, Cambridge, UK).

RESULTS

Influence of vitamin D3

An increase of DNA amount by factors of up to 3 after 10 days was detected for increasing concentrations of VitD3. The cultivation without VitD3 strongly stimulates relative TRAP 5b activity, whereas increasing VitD3 concentrations cause significant decrease in a dose-dependent manner Figure 1. Corresponding light microscopy images are shown in Figure 2h, d, c, b, a, and reveal clear

influence due to supplementation of VitD3. Even though all modifications result in the formation of multinucleated cells (exemplary indicated by white arrows), characteristic network-like accumulations of cells (exemplary indicated by *ac) were formed with increasing concentrations of VitD3. Only few of these cell accumulations are visible for the cultivation with 1 nM VitD3, but they dominate with 100 nM VitD3. Concurrently, the cell number seems to increase, which confirms the results of the DNA analysis. On the other hand the amount of large multinucleated cells is directly associated with the TRAP 5b activity measurements. RT-PCR analysis shows slightly increased expression of TRAP and VTNR at 10 nM VitD3 as well as 100 nM VitD3.

CTSK expression is not influenced. Interestingly the

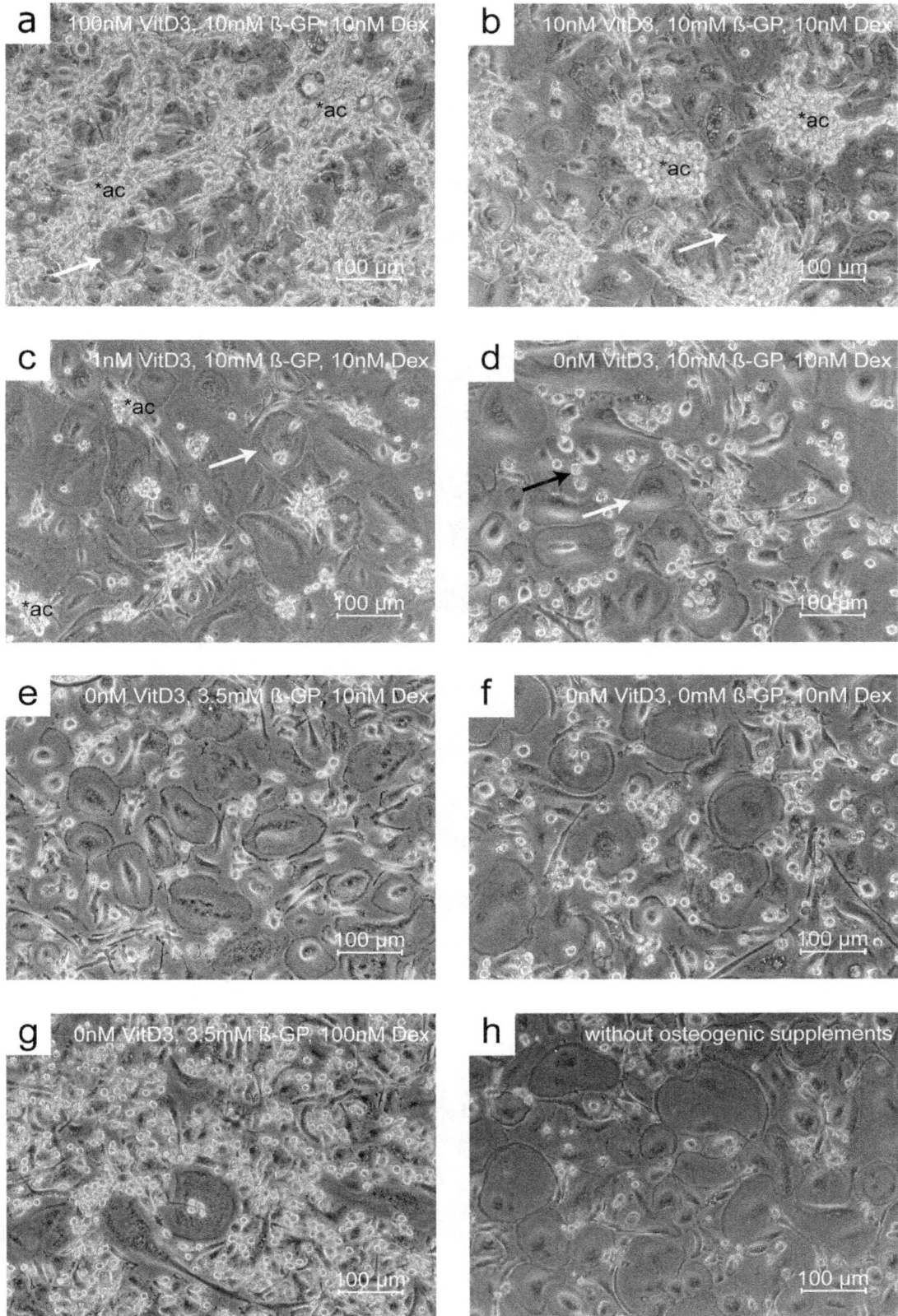

Figure 2. Light microscopy images of cells after 10 days of cultivation in different media compositions. Network-like cell accumulations (*ac), mononucleated cells (black arrows) and multinucleated cells (white arrows) are indicated.

Table 2. Expression of the markers normalised to the expression of the housekeeping gene GAPDH calculated by image analysis of Figure 3. For comparison, highest values were set to 10, respectively. HMc represents the expression profile of human monocytes as isolated. Letters correspond to the media compositions as indicated in Figure 2.

Markers	hMc	a	b	c	d	e	f	g	h
TRAP	5	10	10	7	7	10	6	10	6
CALCR	1	2	1	1	7	6	6	10	7
CTSK	2	8	8	7	8	6	7	10	8
VTNR	5	9	8	6	6	10	6	8	5

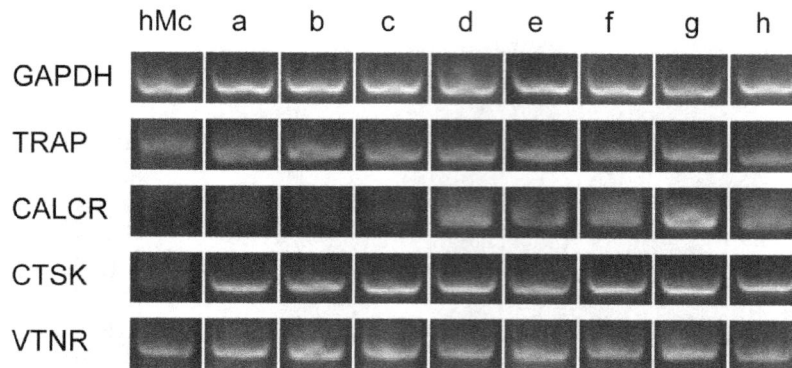

Figure 3. Gene expression of TRAP, CALCR, CTSK and VTNR after 10 days of cultivation in different media compositions. HMc represents the expression profile of human monocytes as isolated. Letters correspond to the media compositions as indicated in Figure 2.

expression of CALCR is strongly reduced even for low concentrations (1 nM) of VitD3 (Figures 3a, b, c and Table 2).

Influence of β-glycerophosphate

At day 10 the addition of up to 10 mM β-GP shows a slightly increasing but not a significant effect on the DNA amount. Relative TRAP 5b activity decreases by a factor of about 0.75 comparing the cultivation with 0 and 10 mM β-GP. Performing light microscopy (Figures 2h, f, e and d) no differences in the cell's morphology were detected for the three β-GP concentrations as well as compared to the control sample. The gene expression profile is marginally influenced by varying the β-GP concentration, with slight advantages for the 3.5 mM β-GP modification (Figures 1, 3h, f, e, d and Table 2).

Influence of dexamethasone

Dex has clearly no influence on the DNA amount during the cultivation and results in negligible decrease of relative TRAP 5b activity. Morphological influence on

osteoclastogenesis is detected by light microscopy (Figures 2h, e, and g). The addition of 10 nM Dex seems not to influence osteoclast formation. However with 100 nM Dex strong effects are visible, while less multinucleated cells are formed and a lot of cells remain mononucleated. In contrast to the microscopic observation gene expression levels of TRAP, VTNR, CALCR, and CTSK are slightly increased at 100 nM Dex compared to the control and 10 nM Dex (Figures 3h, e, g and Table 2).

DISCUSSION

Although monocytes are known to be postmitotic cells (Nijweide et al., 1986), slightly increase of DNA values from day 3 to 10 was recorded for all modifications. On the one hand, this observation can be addressed to further adherence of the cells at day 10. On the other hand, this can be explained by the presence of other proliferative cell types. The isolation and purification of monocytes from buffy coat is done by negative selection, whereby non-monocytes are labelled using antibodies against other cell types (Miltenyi, 2010). Although the isolation and purification of monocytes is carefully performed, low fractions of other cell types in the final suspension cannot be excluded. In the present case flow cytometry using an anti-CD14-FITC antibody detected the purity of monocytes to be about 90%. The significant increase of DNA values detected for high VitD3 concentrations possibly can be addressed to dendritic cells. These cells also derive from monocytes; however, it is still unknown how their lineage commitment is regulated (Miyamoto et al., 2001). The role of VitD3 on osteoclast differentiation and activity is controversially is discussed in the literature. On the one hand VitD3 has been shown to support osteoclastogenesis by increasing the expression of RANKL, and decreasing the expression of its antagonist OPG in a co-culture with stromal cells/osteoblasts (Horwood et al., 1998). Although the precise mechanism of interaction is still unclear, evidence suggests that VitD3 has direct effects on osteoclast precursors by increasing the expression of VTNR (Medhora et al., 1993). This corresponds with our findings, where VTNR is slightly increased by VitD3. In contrast, other reports describe the suppression of osteoclast differentiation due to VitD3 and a decreased resorptive activity (Itonaga et al., 1999; Kogawa et al., 2010; Takasu et al., 2006).

In the present study the TRAP 5b activity is strongly decreased by the addition of VitD3, whereas TRAP gene expression was slightly higher for 10 nM VitD3 and 100 nM VitD3. Measurement of the total TRAP 5 (5a and b) using the method developed by Lau et al. (1987) resulted in equal values for all concentrations. This indicates that cells cultivated with VitD3 start to proliferate and obviously exhibit more TRAP 5a (calculated as a difference of total TRAP and TRAP 5b) at an expense of TRAP 5b activity. Other TRAP 5a positive cells are macrophages and dendritic cells as it was discussed on the basis of DNA analyses (Halleen et al., 2002). In the present study the low TRAP 5b activity detected for high VitD3 concentrations conforms to the low numbers of multinucleated osteoclast-like cells observed using light microscopy. Moreover, expression of CALCR is strongly reduced in case of VitD3 addition. This receptor is considered as a late osteoclast differentiation marker, and its presence is often related to mature bone resorbing osteoclasts (Kurihara et al., 1990). Inhibiting effects of phosphate on osteoclastogenesis using co-culture models have been reported by Takeyama et al. (2001) and Kanatani et al. (2003). For increasing concentration of β-GP a similar trend is also observed in the present study, however, without being significant. No influence was recognized by microscopy. Similar to our experiments, Sivagurunathan et al. (2005) have ascertained that Dex has no significant influence on the number of TRAP-positive cells. Hozumi et al. (2009) described decreasing sizes of the osteoclasts as well as decreasing numbers of their nuclei in case of cultivation with 100 nM Dex in a co-culture with bone marrow adipocytes. Furthermore, Kartsogiannis reported increasing yield of osteoclasts due to the addition of Dex, but it may be

detrimental to osteoclast morphology (Kartsogiannis and Ng, 2004). The reports cited correspond with our findings that high expression levels of TRAP, CALCR and CTSK were recorded for high Dex concentrations being accompanied by a disturbance of the typical osteoclast morphology.

Taking into account, that all media modifications result in the partial formation of cells that show features characterizing osteoclasts, the total fractions should be described as osteoclast-like cells (OLC) – a term which is generally used in the literature. However the varying degree of severity of these typical features as detected in the present study, allows the identification of disadvantageous and advantageous cultivation media supplement concentrations, with the future aim of precursor cell co-cultivation and differentiation.

Conclusion

The osteogenic supplements Dex and β-GP slightly inhibit osteoclast formation but not thus far that osteoclastogenesis does not work at all. Nevertheless a co-cultivation of human BMSC/osteoblasts and hMc/osteoclasts should not use more than 10 nM Dex and can use up to 10 mM β-GP if necessary for osteoblast differentiation. Already in minute quantities the addition of VitD3 results in the reduction of both TRAP 5b activity and expression of the late osteoclastic marker CALCR. Therefore it should not be used for osteogenic differentiation of human BMSC in a co-culture with monocytes.

ACKNOWLEDGEMENT

This study was partly supported by the Bundesministerium für Bildung und Forschung, Grant 03FPB00379 (Forschungsprämie).

REFERENCES

Boyle WJ, Simonet WS, Lacey DL (2003). Osteoclast differentiation and activation. Nature, 423: 337-342.

Hadjidakis DJ, Androulakis, II (2006). Bone remodeling. Ann. N. Y. Acad. Sci., 1092: 385-396.

Halleen JM, Ylipahkala H, Alatalo SL, Janckila AJ, Heikkinen JE, Suominen H, Cheng S, Vaananen HK (2002). Serum tartrate-resistant acid phosphatase 5b, but not 5a, correlates with other markers of bone turnover and bone mineral density. Calcif. Tissue Int., 71: 20-25.

Heinemann C, Heinemann S, Bernhardt A, Lode A, Worch H, Hanke T (2010). In vitro osteoclastogenesis on textile chitosan scaffold. Eur. Cell. Mater., 19: 96-106.

Horwood NJ, Elliott J, Martin TJ, Gillespie MT (1998). Osteotropic agents regulate the expression of osteoclast differentiation factor and osteoprotegerin in osteoblastic stromal cells. Endocrinology, 139: 4743-4746.

Hozumi A, Osaki M, Goto H, Sakamoto K, Inokuchi S, Shindo H (2009). Bone marrow adipocytes support dexamethasone-induced osteoclast differentiation. Biochem. Biophys. Res. Commun., 382: 780-784.

Itonaga I, Sabokbar A, Neale SD, Athanasou NA (1999). 1,25-Dihydroxyvitamin D(3) and prostaglandin E(2) act directly on circulating human osteoclast precursors. Biochem. Biophys. Res. Commun., 264: 590-595.

Janckila AJ, Takahashi K, Sun SZ, Yam LT (2001). Naphthol-ASBI phosphate as a preferred substrate for tartrate-resistant acid phosphatase isoform 5b. J. Bone Miner. Res., 16: 788-793.

Kanatani M, Sugimoto T, Kano J, Kanzawa M, Chihara K (2003). Effect of high phosphate concentration on osteoclast differentiation as well as bone-resorbing activity. J. Cell. Physiol., 196: 180-189.

Kartsogiannis V, Ng KW (2004). Cell lines and primary cell cultures in the study of bone cell biology. Mol. Cell. Endocrinol., 228: 79-102.

Kogawa M, Findlay DM, Anderson PH, Ormsby R, Vincent C, Morris HA, Atkins GJ (2010). Osteoclastic metabolism of 25(OH)-vitamin D3: a potential mechanism for optimization of bone resorption. Endocrinology, 151: 4613-4625.

Kurihara N, Gluck S, Roodman GD (1990). Sequential expression of phenotype markers for osteoclasts during differentiation of precursors for multinucleated cells formed in long-term human marrow cultures. Endocrinology, 127: 3215-3221.

Lagasse E, Weissman IL (1997). Enforced expression of Bcl-2 in monocytes rescues macrophages and partially reverses osteopetrosis in op/op mice. Cell, 89: 1021-1031.

Lau KH, Onishi T, Wergedal JE, Singer FR, Baylink DJ (1987). Characterization and assay of tartrate-resistant acid phosphatase activity in serum: potential use to assess bone resorption. Clin. Chem., 33: 458-462.

Matsuo K, Irie N (2008). Osteoclast-osteoblast communication. Arch. Biochem. Biophys , 473: 201-209.

Medhora MM, Teitelbaum S, Chappel J, Alvarez J, Mimura H, Ross FP, Hruska K (1993). 1 alpha,25-dihydroxyvitamin D3 up-regulates expression of the osteoclast integrin alpha v beta 3. J. Biol. Chem., 268: 1456-1461.

Miltenyi (2010). http://www.miltenyibiotec.com/en/PG_58_47_ Monocyte_Isolation_Kit_II.aspx

Miyamoto T, Ohneda O, Arai F, Iwamoto K, Okada S, Takagi K, Anderson DM, Suda T (2001). Bifurcation of osteoclasts and dendritic cells from common progenitors. Blood, 98: 2544-2554.

Nijweide PJ, Burger EH, Feyen JH (1986). Cells of bone: proliferation, differentiation, and hormonal regulation. Physiol. Rev., 66: 855-886.

Sivagurunathan S, Muir MM, Brennan TC, Seale JP, Mason RS (2005). Influence of glucocorticoids on human osteoclast generation and activity. J. Bone Miner. Res., 20: 390-398.

Takasu H, Sugita A, Uchiyama Y, Katagiri N, Okazaki M, Ogata E, Ikeda K (2006). c-Fos protein as a target of anti-osteoclastogenic action of vitamin D, and synthesis of new analogs. J. Clin. Invest., 116: 528-535.

Takeyama S, Yoshimura Y, Deyama Y, Sugawara Y, Fukuda H, Matsumoto A (2001). Phosphate decreases osteoclastogenesis in coculture of osteoblast and bone marrow. Biochem. Biophys. Res. Commun., 282: 798-802.

Vater C, Kasten P, Stiehler M (2010). Culture media for the differentiation of mesenchymal stromal cells. Acta Biomater., 7: 463-477.

Yoshida H, Hayashi S, Kunisada T, Ogawa M, Nishikawa S, Okamura H, Sudo T, Shultz LD, Nishikawa S (1990). The murine mutation osteopetrosis is in the coding region of the macrophage colony stimulating factor gene. Nature, 345: 442-444.

6

Time trends, and survival of patients with oral and pharyngeal malignancies

Waguih Mohamed Abouzeid[1]*, Samiha Ahmed Mokhtar[2], Nehad Hassan Mahdy[2], Mohamed Sherif Ahmed[3] and Fayek Salah El Kwsky[4]

[1]Research Department, Alexandria Dental Research Center, Alexandria University, Egypt.
[2]High Institute of Public Health, Alexandria University, Egypt.
[3]Faculty of Medicine, Alexandria University, Egypt.
[4]Medical Research Institute, Alexandria University, Egypt.

An accurate assessment of oral and pharyngeal malignancies in cancer treatment trends, and survival of the disease was missing in Egypt. Accordingly, all new cases treated in Alexandria and El Behira governorates during the last decade were studied retrospectively. Data were collected through all accessible archives using a special data collection sheet. The total populations of different governorates were obtained from the "Central Administration of Census and Statistics" reports, by gender, and residential selective distributions. The personal history, socio-demography, staging and site of the tumor, treatment and complications of treatment, response, as well as survival were explored. The mean age of 1254 investigated subjects was 52.02 ± 16.13 years, where 15% were educated. Pharyngeal cases represented 41.5%, while the oral were 58.5%. Those of stage 1 recorded 52%, while stage 4 was 47.7%. Surgery followed by irradiation was the line of treatment for 54.3% of cases. The estimated population for non-censal years was determined as the average value of both the "Arithmetic Progression" and the "Geometric Progression" technique estimates. The annual incidence rates through the period of study were plotted and analyzed using the relevant regression line to test significance (Di Bonito, 1983; Saunders and Trapp, 1990). Tracing trends revealed a decreasing incidence in all situations, except in females of El Behira governorate, which resulted in an increasing trend of El Behira as a whole, as all trends were not statistically significant. The 5-year survival was computed using the actuarial method, and presented graphically using the Kaplain Meier curve (Ederer and Cutler, 1958). The overall 5-year survival probability was 0.54%. Survival for stage 1 was 74.5%, while it was 46.38% for stage 4. Smoking showed an apparent adverse effect on survival. Stepwise logistic regression revealed that, the best predictor for overall survival was gender, as males have 1.74 times the risk compared to females, followed by stage, as stage 4 was the worst. Results of the present study suggest that, the database coded cases were quite important for treatment and follow up. Smoking should be prohibited in a decisive manner. Care is to be given for raising the socioeconomic status, especially for categories living under potentially higher stress. Early referral of cases to oncologists is highly mandatory, and whenever surgery is indicated; safety margin combined with alleviating complications is of great effect on survival.

Key words: Oral, pharyngeal, cancer, incidence, survival, quality of life.

INTRODUCTION

Malignant tumor is defined as a lesion arising from proliferation of cells, which is autonomous and persists after the initiating stimulus has been removed. It is a manifestation of an abnormality of the process involved in the control of all growth. The term "cancer" is a general term, which applies to malignant tumor of any type

*Corresponding author. E-mail: waguihabouzeid@gmail.com.

(Cawson and Odell, 1998)a. The ultimate definition of malignancy is the ability to metastasize. Metastasis is the spread of tumor cells from primary site to one or more separate distant secondary sites (Cawson and Odell, 1998b) In case of an epithelium, the cells are capable of invading through the basement membrane to make transition from *in situ* to invasive carcinoma (Cawson and Odell, 1998b).

Cancer is second to coronary artery disease as being the most common cause of death in the western world (Smyth, 1999). Oral cancer accounts for less than 1% of all cancer deaths among white females in the United States, and over 40% in various parts of India (Schottenfeld et al., 1993). The highest death rates from oral and pharyngeal cancer were in Hong Kong for both sexes followed by France and Puerto-Rico. The lowest mortality figures were reported for El-Salvador, Egypt, and Honduras in both sexes (Schottenfeld et al., 1993). Oral cancer registrations and incidence are increasing through Europe and in the United Kingdom (Schottenfeld et al., 1993). Worldwide 197,000 deaths from cancer of the oral cavity and pharynx occur per year, of which 74% are in developing countries (Pisani et al., 1999) Over 95% are well differentiated or moderately differentiated, arising from the mucosa, therefore classified as squamous cell carcinoma (SCC) (Cawson and Odell, 1998).

MATERIALS AND METHODS

Study setting

Study was conducted in Oncology Department, Faculty of Medicine, Maxillo-facial Department, Faculty of Dentistry, and Statistics Department, Medical Research Institute (Alexandria University), Oncology Department-Gamal Abdel Nasser Health Insurance Hospital and Damanhour Oncology Center (Ministry of Health and Population).

Study design

A retrospective study for 10 years (1991 to 2000) was conducted in oncology centers for all oral and pharyngeal malignancy records. A prospective study for survival and response was done up to the end of March 2002.

Target population

The target population was oral and pharyngeal cancer cases in Alexandria and El Behira region.

Data collection technique

Record review

a) Records of the "Cancer Registry" at the Medical Research Institute were studied for having an initial idea about the size and distribution of cases in the region.
b) All accessible files of oral and pharyngeal malignancies were reviewed in Alexandria University hospitals, Gamal Abdel Nasser Insurance hospital, and in Damanhour Oncology center.
c) Data about Alexandria and El Behira total population were collected from the Central Administration for Census and Statistics reports (Central Directory for Census and Statistics, 1998).

Follow up

Follow up was carried through the records and monthly regular clinic visits of the patients, also through telephone calls, letters, relatives, and home visits. Accessible cases of follow up were 852.

Data collection tool

Available data were collected, while missing data were tried to be completed by the researcher personal communications with patients or their families. Data available were collected using a pre-designed data collection sheet including:

1) Clinic name
2) Personal history: Patient name, phone number, address if available socio-demographic data (age, gender, residence, educational level, occupation, and marital status), smoking habit for number of cigarettes smoked daily, and duration of smoking (ex-smokers are those having a history of smoking before one year of diagnosis or more, and those smoke less than 10 cigarettes were considered light smokers).
3) Data about oral and pharyngeal cancer:

a. Date of diagnosis.
b. Site of cancer according to ICD-10 (World Health Organization, 1996) classification (including major salivary glands).
c. Stage of the tumor (TNM, 1997) according to TNM classification.
d. Grade of the tumor, including three grades according to degree of cell differentiation.
e. Main treatment line, broad lines of treatment were: surgical irradiation, chemotherapy, or their combinations. Also salvage or palliation were recorded as well.

4) Response, complete response, partial, no response, or progressive disease.
5) Data about survival.

Statistical design

Data were revised, coded, as a "Foxpro" database file. The "SPSS"-version 11 was used for data analysis.

Statistical analysis

Incidence and time trends of the disease

Reference population was estimated according to the official numbers of the "Central Directory of Census and Statistics"-Egyptian government (1986 to 1996). Screening showed that 173 cases were not from Alexandria or El Behira residents, so they were not included in the incidence analysis.

Estimation of the population: The estimated population was determined as the average of both figures obtained from the two methods.

1. Mid-year estimated population according to "Arithmetic Progression" method (Swaroops, 1960). This method assumes that, the population increases or decreases by constant value from year

to year between any two census years.

2. Mid-year estimated population according to "Geometric Progression" method (Swaroops, 1960) This method is more precise if the population is large, and it usually gives a higher estimate than the previous method. It assumes that the increase in population occurs at a constant rate throughout the period of estimation.

Incidence rates: Dividing the number of cases of every year by the estimated population, multiplying by 100000, incidence rates were calculated.

Trend analysis: The annual incidence rates throughout the period of study were plotted and analyzed using the simple regression line to test its significance (Di Bonito, 1983; Saunders and Trapp, 1990).

Survival analysis

Calculation of survival using the actuarial method: This method was based on information available for each case namely the date of diagnosis, and the cut off date of follow up. Computation was performed by recording the following for each one year interval follow up:

1. The number alive at the beginning of the interval.
2. The number who died during the interval.
3. The number lost to follow up during the interval.

Construction of life table

Registered cases survival during the 10 years was traced to 31st of March 2001.

Columns of this table were constructed according to the following scheme:

Column 1: Year of observation (x to x+1), time elapsed from date of diagnosis in intervals of one year.
Column 2: Alive at the beginning of interval (lx). The first entry in this column is the number of patients at diagnosis. Then the new entries are achieved as follows:
$l_{x+1} = l_x - (d_x + w_x)$
Column 3: Withdrawn during interval (w_x).
Column 4: Died during interval (d_x).
Column 5: Effective number exposed to risk of dying l'_x. It is assumed that patients lost or withdrawn from observation during an interval were exposed to the risk of dying, on average, for one half of the interval, $l'_x = l_x - \frac{1}{2} w_x$.
Column 6: Proportion dying during the interval (q_x).
Column 7: Proportion surviving during the interval: $p_x = 1-q_x$.
Column 8: Cumulative proportion surviving P_x. This is generally referred to as the cumulative survival rate, and is obtained by cumulatively multiplying the proportion surviving each interval:

$Px = p1 . p2 . p3 ....p_x$

The successive entries in this column give the 1, 2 .....and 6-year cumulative survival rates (Saunders and Trapp, 1990; Ederer and Cutler, 1958).

Kaplain-Meier product limit estimates

This is a method for estimating the probability of survival at a distinct point of time. It is best presented graphically by the Kaplain-Meier curve (Beth, 1999). The significance of difference between survival curves was calculated by Breslow test (Generalized Wilcxon analysis) (Beth, 1999):

$$U = \sum_{i=1}^{K} W_1 (Oi - Ei)$$

where Wi is the weight for the time i (the weights are the number at risk at each time point). The test is based on computing the weighted difference between observed and expected number of death at each of the time points.

Multiple Cox-regression analysis: (Clayton and David, 1992; Christensen, 1987)

This standard statistical technique was performed using the stepwise method. It is used to discover the hazardous attributes for survival, where there are multiple covariates, and the additional complications of censored cases. This model allows the covariates (independent variables) in the regression equation to vary with time. The dependent variable is the years after diagnosis (Cox, 1999)

Cox regression model:

$H (T) = h_o (t) e (B_1X_1+B_2X_2+.......B_pX_p)$

Where: H (T) is the hazard rate of early death at time t.
h_o (t) is the baseline hazard at time t.
e is the well known constant.
B is the Cox regression coefficient which denotes the magnitude of the increase or decrease in the value of the independent variable while holding all other explanatory variables constant.
$B_1.....B_p$:   the respective coefficient for each of the independent variables.

For each subject, two quantities were used to define the outcome survival. Binary model for death (Di Bonito, 1983), or otherwise which is called censoring (0). Also needed is an exposure time which is the length of observation for a patient from diagnosis date till death or last follow up whichever occurs first. The output of Cox Regression analysis is as follows (Ann and Sarabjot, 1998):

B: is the Cox regression coefficient that denoted the magnitude of the increase or decrease in the value of the independent variable.
SE (standard error): It estimates the variability in regression coefficient and can be used to construct confidence interval.
Wald test: is test to show the significance of the relation.
Exponentiation of the coefficient (B) estimates the hazard ratio of the outcome for each unit increase in factor (X).
P value: To test the null hypothesis (no association) between the exposure and survival.
Hazard ratio (HR) = e^B
95% CI= confidence intervals of HR = $e^{Bi \pm 1.96.SE (B)}$
If the confidence interval does not include the value of 1, null hypothesis could be rejected that the variable is not related to survival.

RESULTS

Frequency distribution of oral and pharyngeal cancer patients:

According to age; 59.5% were in the age of 30 to <60 years. Those below 30 represented 9.8%. The mean age was 52.02±16.13 years. Males represented 62%, and

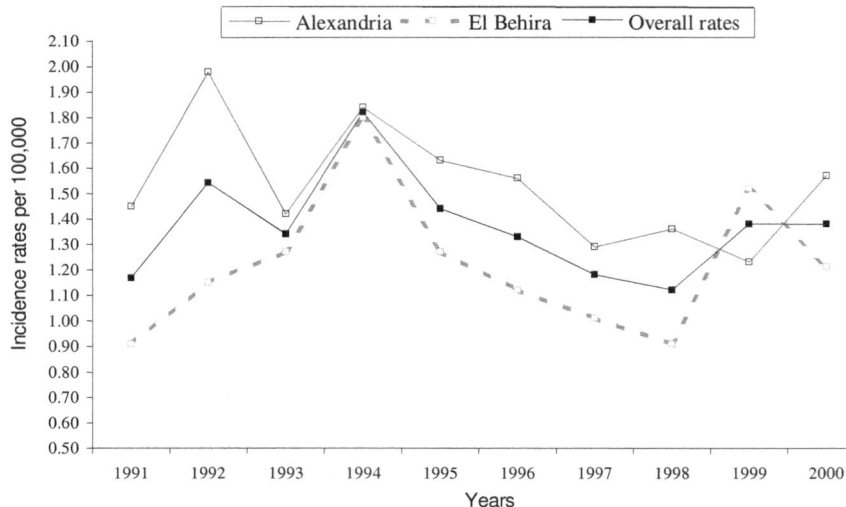

Figure 1. Incidence rates of registered oral and pharyngeal cancer cases "Alexandria and El Behira", 1991-2000.

Table 1. Incidence of registered oral and pharyngeal cancer cases rates per 100,000 persons in Alexandria and El Behira (1991 to 2000).

Year	El Behira	Overall rates
1991	0.91	1.17
1992	1.15	1.54
1993	1.27	1.34
1994	1.80	1.82
1995	1.27	1.44
1996	1.12	1.33
1997	1.01	1.18
1998	0.91	1.12
1999	1.52	1.38
2000	1.21	1.38

52.7% were urban. Married were 78.3%, where uneducated conformed to 85.1%. All types of workers represented 30%, as housewives and retired were nearly equal (26.3 and 26% respectively). Employee and professionals were 8.6%. Oral malignancies represented 58.5% of cases, as pharyngeal were 41.5%. Hard palate with cheek and retro-molar areas were affected in 19.6%, tongue was affected in 12.6%, followed by 9.6% in the major salivary glands. Lip cancer occupied 7.3%. Gum and floor of the mouth were the least (5.6 and 3.8% respectively). Naso-pharynx included 17.3% of the whole cases, and hypo-pharynx included 14%, while only 8.1% where for the oro-pharynx. Stages 1, 2, 3, and 4 were of an ascending trend (5.2, 14.1, 33, 47.7%).

Concerning grade; 296 patients (36.8%) had moderately differentiated tumors, followed by well differentiated (24.5%), then the least were the poorly differentiated and undifferentiated tumors (21.5 and 17.2% respectively). Distribution of patients according to the main line of treatment showed that, 54.3% received irradiation with surgery, and 14.4% were subjected to surgery only. Chemotherapy with irradiation and surgery were given to 19% of cases, as chemotherapy with surgery was the treatment of 5.6% of patients. Chemotherapy was afforded to only 1.2% of cases.

Trends of oral and pharyngeal cancer patients

Table 1 and Figures 1 and 2 display the incidence rate per 100,000 for Alexandria, El Behira, and both. The highest incidence rate in Alexandria and El Behira

$$Y = 29.91 - 0.01X, F = 0.37, p = 0.56$$

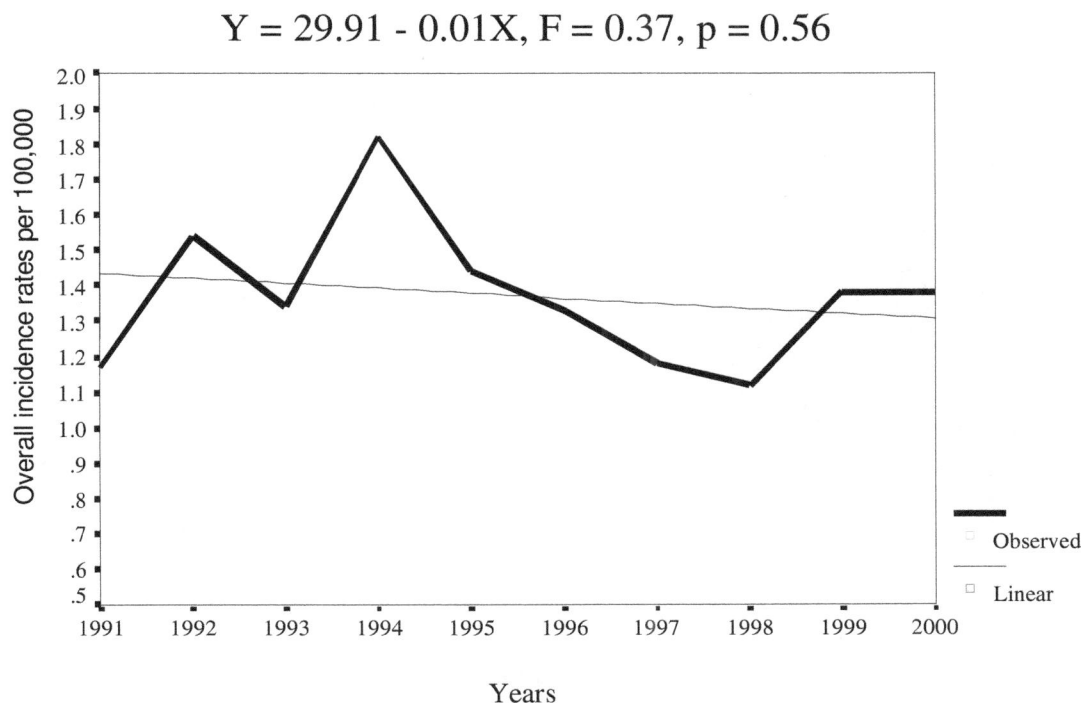

Figure 2. Trend of annual incidence rates of registered oral and pharyngeal cancer cases "Alexandria and El Behira", 1991-2000.

Table 2. Incidence of registered oral and pharyngeal cancer cases rates per 100,000 persons among total, male, and female in Alexandria and El Behira (1991 to 2000).

Years	Males	Females	Overall rates
1991	1.54	0.78	1.17
1992	1.88	1.19	1.54
1993	1.75	0.91	1.34
1994	1.86	1.77	1.82
1995	1.60	1.21	1.44
1996	1.57	1.07	1.33
1997	1.48	0.90	1.18
1998	1.28	0.95	1.12
1999	1.80	0.97	1.38
2000	1.59	1.15	1.38

governorates together lied in the year 1994 (1.82), followed by the year 1992 (1.54), as the lowest incidences were through the years 1998 and 1991 (1.12 and 1.17 respectively). Alexandria revealed its highest peaks of incidence rates in 1992 and 1994 also (1.98, and 1.84 respectively), where the lowest values of incidence rates were seen in years 1990 (1.23) and 1997 (1.29). El Behira, incidence rate was highest relatively also in the year 1994 (1.8), followed by the years 1993 and 1995 where incidence was 1.27 for each. During the years 1991 and 1998, El Behira had the lowest incidence rates (0.91 cases for each).

Table 2 display the incidence rated for the whole males, whole females, besides the overall rates. The highest incidence rates were shown among males in years 1992, 1994, and 1999 (1.88, 1.87, and 1.8 respectively), while the years 1998 and 1997, showed only 1.28 and 1.48 respectively. Females of Alexandria

$$Y = 77.514 - 0.0381X, F\ 0.901 - p = 0.37$$

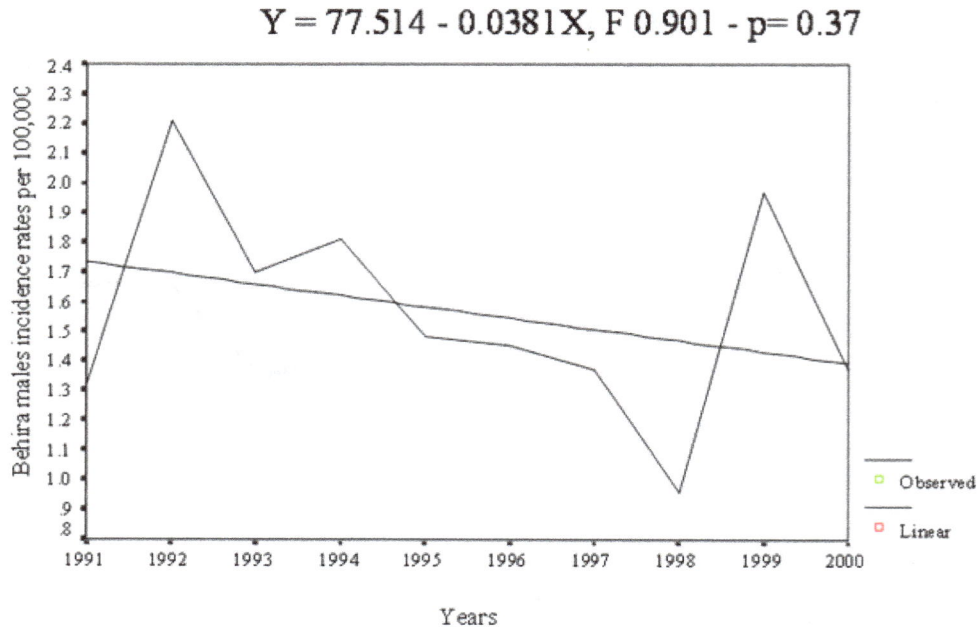

Figure 3. Trend of annual incidence rates of registered male oral and pharyngeal cancer cases " El Behira", 1991-2000.

and El Behira together showed their upper most value in 1994, that was 1.77 cases. The lowest incidence rate for females was shown through the year 1991 (0.78). A non-significant overall decreasing trend was seen. Figures 3 and 4 dismantle the only increasing trend which was in females of El Behira governorate, even though it was not significant.

Five year survival rates for oral and pharyngeal cancer cases

Table 3 summarizes cumulative probability of survival for 5 years of all variables of study. Knowing that the overall 5 year cumulative probability of survival was 53.4%, it was interesting to see the highest 5 year survival for those who received surgery alone (82.2%), followed by cases of stage 1 tumors in general (74.5%). The lowest rates of 5 year survival were seen in heavy smokers, and those received chemotherapy and /or irradiation (37.87 and 39.92% respectively). It may be worth noting that, the educated had 41.07%, workers had 40.24%, and widowed with divorced had 41.29% cumulative 5 year survival probabilities.

Cox-regression analysis

Table 4 shows the results of stepwise Cox's regression analysis of various independent factors on 5 year probability of survival. The Table 4 shows that, out of eight factors studied, only three were significantly associated with the probability of 5 year survival. The first predictor was gender. Males had higher risk of death than females (hazard ratio =1.741, 95% C.I. = 1.22 – 2.48).

Cases of stage 2 had higher risk of death compared to these with stage 1 (hazard ratio = 1.527, 95% C.I. = 0.586 - 3.978). Cases of stage 3 had about two times the risk of the first stage (95% C.I. = 1.215 – 6.759), while the risk in stage 4 was about three times (95% C.I. = 1.215-6.759). Studying the risk of death upon grade; grade 2 risk of dying was about 1.8 times (C.I. = 1.176 – 2.683), while grade 3 revealed only a risk of about 1.3 times the risk of grade 1 (95% C.I. = 0.765 –2.316); Model $\chi2$ = 35.029; p < 0.01.

DISCUSSION

Over the past decade we have seen patients at a much younger age, in the third and fourth decades, suffering from head and neck cancer, especially cancer of the oral cavity and tongue. Around the world, oral age-standardized death rate through 46 countries revealed the least male or female value in Greece and Israel (Murphy et al., 1995). Incidence and mortality rates for oral and pharyngeal cancer have been increasing in several parts of the world, most notably in countries of central and Eastern Europe in the last two or three decades (Franceschi et al., 1999). From the present

Table 3. Summary of 5-year survival rates for registered patients with oral and pharyngeal cancer according to certain factors (Alexandria and El Behira, 1991 - 2000).

	Variable	No. of cases	5-year survival
	Overall	852	0.5342
Gender	Male	516	0.5014
	Female	336	0.585
Residence	Urban	447	0.5504
	Rural	388	0.5011
Age	<30	85	0.5213
	30-<60	495	0.5544
	60-<75	229	0.5209
	>75	43	0.4151
Marital status	Married	503	0.4432
	Single	46	0.5502
	Widowed + divorced	104	0.4129
Education	Uneducated	518	0.4246
	Educated	88	0.4107
Occupation	Working	367	0.4024
	Not Working	338	0.5477
Site	Oral	516	0.5614
	Oro-pharygeal	72	0.5344
	Naso and hypo-pharyngeal	264	0.4823
Stage	1	39	0.745
	2	93	0.6065
	3	212	0.5062
	4	309	0.4638
Grade	1		0.6233
	2	322	0.4683
	3	97	0.605
Treatment	Surgery	60	0.822
	Chemo. and/or irradiation	49	0.3992
	Surgery+ irradiation	457	0.5604
	Surgery+ chemoth.	50	0.4477
	Surgery+ irradia.+ chemo.	177	0.4236
Type of surgery	Excisional	211	0.6602
	Safety margin	86	0.6927
	Radical neck dissection	95	0.5655

pharyngeal malignancy cases represented 58.5 and 41.5% respectively. The oro-pharynx, naso-pharynx, and hypo-pharynx accounted for 8.1, 17.3 and 14% respectively. The highest percent of occurrence was for

Table 4. Stepwise Cox regression analysis of the effect of various independent factors on 5-year survival probability (Alexandria and El Behira, 1991-2000).

Independent variables (covariates)	B	SE	Wald	P	Hazard ratio	95% CI LL - UL
Gender (male)	0.554	0.181	9.353	0.002	1.741	1.22 - 2.483
Stage of tumor			9.507	0.023		
Stage 2	0.423	0.489	0.749	0.387	1.527	0.586 - 3.978
Stage 3	0.795	0.441	3.256	0.071	2.215	0.934 - 5.254
Stage 4	1.053	0.438	5.787	0.016	2.866	1.215 - 6.759
Grade of differentiation			7.965	0.019		
Grade 2	0.574	0.211	7.444	0.006	1.776	1.176 - 2.683
Grade 3	0.286	0.283	1.023	0.312	1.331	0.765 - 2.316

$X_6^2 = 35.029$.

cases with stage 4, while one third had stage 3, and stage 1 represented the least percent. Regards grade; the highest percent was for those with moderately differentiated, but the lowest was for undifferentiated. About one half of all patients were exposed to surgery with irradiation treatment. Nearly one seven were subjected to surgical treatment only, and those received the three lines together were of less percent.

Although there was an ascending trend of incidence in El Behira governorate, but in general there was a decreasing trend of occurrence of oral and pharyngeal cancer in the years between 1990 and 2001, which was not statistically significant. Site and sex distribution of 6789 cases registered in "Cairo Metropolitan Registry" report for cancer (1987), showed that 5.8% of male malignancies was in the oral cavity and pharynx, compared to 3% only among female cases (The profile of cancer in Egypt, 1987). Incidence was quite different according to place, race, and time. As the average total incidence was 1.37 in our data; it is estimated with about 3 cases per 100,000 in U.S.A. in year 2001, where incidence of men was 2.6 times that of women (Silverman, 2001). In Catania (Italy), malignant tumors markedly prevailed in the males with an incidence that was 3 times that observed in females (Sortino and Milici, 1998).

Overall 5 year cumulative survival probability was 0.534, as it was 0.5 in males, and 0.585 in females. Male/female proportion of cases was 1.63, and this was matching that of Alexandria Registry (1.61) most recent reports, (Medical Research Institute, 2001) albeit "Cairo Metropolitan Registry" report was 1.8 male to females (The profile of cancer in Egypt, 1987). While male/female proportion of mean incidence in Alexandria Metropolitan region was 1.576; census male/female proportion in Alexandria and El Behira governorates ranged from 1.03 to 1.058 according to residential distribution. This survival rate was very impressive, as it was shown to be 49.8 in Scotland (1988 to 1992), (Information and Statistics Division, 2002). 52.5% for oral cancer in USA (1983 to 1990) and 33% for pharyngeal cancer (1981 to

1986), (Oral Cancer Background Papers, 2002) where it was 48% in Germany (Prevention of oral cancer, 2002). Finland showed 51% as 5 year cumulative survival rate in year 1978 to 1985 (Survival of Cancer Patients in Europe, 1995). A recent big research in Mumbai (India) showed a range of 20 to 43% as 5 year cumulative survival for oral cancer, and a range of only 8 to 25% 5 year survival in pharyngeal cancer (Rao et al., 1998).

The highest 5 years survival rate was found in patients age 30 to < 60, followed by those below 30 years, and 60 to <75 age groups. Urban cumulative survival probability was 55% in the 5th year, where it was 50.1/100 person in the rural areas. There was a significant statistical difference between 75+ age group and the other intervals. Educated patients recorded higher 5 year cumulative survival rate compared to uneducated. Those who did not marry had higher survival; and the same was seen with retired. Smoking revealed high statistical significance between any smoking rate and being a non smoker. The best cumulative survival rate was for oral sites, followed by oro-pharyngeal sites, as the worst survival was found in the other pharyngeal sites, and the difference was statistically different. Apparently, decrease of cumulative survival was there with the advance of age. Grade was not consistent with survival. The best cumulative survival rates existed in surgical treatment category, followed by surgery and irradiation, then surgery with chemotherapy, and this was statistically significant. Gender, stage, and grade were the most important predictors of survival. It is quite clear that stages 3 and 4 accounted for more than 80% of cases. TNM staging system, the most accurate prognostic variable in patients with oral carcinoma, (Noguchi et al., 1999) indicated that 80 to 90% of cases were presented with stage 3 or 4 malignancies in a recent research on a high-risk population of India (Rao et al., 1998).

RECOMMENDATIONS

1) Registration of malignancy cases must have a code

for every patient, which is to be used through a net covering all the oncology centers. This is a very beneficial regarding treatment of cases wherever they are, and provide an excellent database and a registry for such a disease. Planning and saving data would be available then, and also keeping the medico-legal dimensions. Stressing on the importance of meticulous and careful registration of patients data including the cause of death especially in the Ministry of Health is of prime importance.

2) Being a problem of non-controversial impact, a decisive fight against smoking must be launched. Intellectuals, politicians, and economists are to take their role in management of stopping such a bad habit.

3) Care is to be given for raising the socio-economic status, including the level of civilization and the infrastructure particularly in the rural areas (stressing on avoiding crowding and pollution). Life style and working circumstances seem to have an apparent impact on survival, hence care about occupational risk, and behavior therapy is mandatory.

4) Health education should be directed as the early referral to oncologists, as the earlier the surgery and /or stage were of significant higher survival feedback, thence it was the condition treatment modality.

5) Categories living potentially higher stress; are apt to develop malignancy; but have also lower survival rates comparatively. Those categories lie specifically within out-doors, uneducated, and males in addition to divorced or widow/widower females or males.

6) Whenever surgery is indicated, and patient is operable, safety margin is of great effect on survival.

ACKNOWLEDGEMENT

The authors of this paper would like to express their gratitude for the invaluable continuous excel help of Professor Dr. Mohamed Hussein Mohamed– the founder of Biostatistics Department, High institute of Public Health, Alexandria University.

REFERENCES

Ann ES, Sarabjot SA (1998). Cox regression to give an individual's point prediction and patient survival with multiple attributes. JR Stat. Soc., 34: 187-220.

Beth D (1999). Kaplain Meier product limit estimates. In: Basic and Clinical Biostatistics. Robert G. (ed.). Appelton and Lang,, pp. 192-194.

Cawson RA, Odell EW (1998a). Oral cancer-in-Essentials of Oral Pathology and Oral Medicine. Churchill Livingstone, Edinburgh, UK, Chapter 17, pp. 228-257.

Cawson RA, Odell EW (1998b). Neoplasia-in-Essentials of Oral Pathology and Oral Medicine. Churchill Livingstone, Edinburgh, UK, Chapter 4, pp. 95-113.

Central Directory for Census and Statistics (1998). Census report No: 1102-M/TH, Egypt; Alexandria sub-directory, Foaad street.

Chilvers ER, Hunter JAA, Boom NA (1999). Churchill, Livingstone. London, Chapter 16, pp. 1049-1094.

Christensen E (1987). Multivariate Survival Analysis Using Cox's Regression Model. Hepatology, 7(6): 1346-1358.

Clayton H, David C (1992). Modelling with logistic regression. In: Advanced Statistical Models in Epidemiology. New York, Oxford, Oxford University Press, pp. 1-10.

Cox DR (1999). Regression models and Life Tables. JR. Stat. Soc, 34: 187-220.

Di Bonito L (1983). Cancer of endometrium in the province of triesta: Epidemiological considerations. Eurp. J. Gynecol. Oncol., 41(1): 44-46.

Ederer F, Cutler SJ (1958). Maximum utilization of the life table method in analyzing survival. J. Ch. Dis., 8(6): 699-712.

Franceschi S, Levi F, La Vecchia C (1999). Comparison of the effect of smoking and alcohol drinking between oral and pharyngeal cancer. Int. J. Cancer., 83:1-4.

Information and Statistics Division (ISD). Trinity Park House, South Trinity Road. Edinburh. Scotland. EH5 35 Q. WWW.show.scot.nhs.uk/isd/Last revised 05/2002.

Levi F, Pasche C, La Vecchia C, Lucchini F, Franceschi S (1998). Food groups and risk of oral and pharyngeal cancer. Int .J. Cancer, 77: 705-709.

Medical Research Institute (2001). Cancer Registry Reports. Statistics Department, Alexandria University, Egypt.

Murphy GP, Lawrence W, Lenhard RA (1995). American Cancer Society of Clinical Oncology. 2nd. ed. USA, pp. 1-76.

Noguchi M, Kido y, Kubota H, Kinjo H, Kohama G (1999). Prognostic factors and relative risk for survival in N1-3 oral squamous cell carcinoma: A multivariate analysis using Cox's Hazard model. Brit. J. Oral Maxillofac. Surg., 37: 433-437.

Oral Cancer Background Papers (2000). Prepared for the National Strategic Planning Conference, for the: Prevention and Control of Oral and Pharyngeal Cancer. (Chapter one). Chicago, Illinois. Last revised 15/02/2002.WWW.ncbi.nlm. nih.gov.

Pisani P, Parkin DM, Bray F, Ferlay J (1999). Estimates of the World Mortality from 25 Cancers in 1990. Int. J. Cancer, 83: 18-29.

Rao DN, Shroff PD, Chattopadhyay G, Dinshaw KA (1998). Survival analysis of 5595 head and neck cancers-results of conventional treatment in a high-risk population. Brit. J. Cancer, 77(9): 1514-1518.

Saunders BD, Trapp RG (1990). Methods for Analysing Survival Data. In: Basic and Clinical Biostatistics. Prentice-Hall International Inc.. Chapter 11, pp. 186-206.

Schottenfeld D, Fraumeni JF, Mahboubi E, Sayed GM (1993). Oral cavity and pharynx. In: Cancer Epidemiology and Prevention. USA, Philadelphia. W. B. Saunders Company, Chapter 33, pp. 583-593.

Smyth JF (1999). Principles of Oncological and Palliative Care. In: Davidson's Principles and Practice of Medicine. Eighteenth ed. Haslett C.

SOL Silverman S (2001). Jr. Demographics and occurrence of oral and pharyngeal cancers. The outcomes, the trends, the challenge. J. Am. Dent. Assoc., 132 Suppl: 7S-11S.

Sortino F, Milici A (1998). Epidemiology of oral cavity tumors. Minerva Stomatol,, 47(5): 197-202.

Survival of Cancer Patients in Europe (1995). Lyon, International Agency for Research on Cancer, IARC, Scientific Publication No. 132.

Swaroops (1960). Analysis of Demographic Data. In: Introduction to Health Statistics. E & S Livingstone, Ltd., Edinburgh and London, Chapter 15, pp. 154-156.

The Profile of Cancer in Egypt (1987).The Cancer Registery for the Metropolitan Cairo Area. (CRMCA) – Cairo University, pp. 49-90.

TNM (1997). International Union Against Cancer, 5th. ed., Springer Verlag Geneva.

World Health Organization (1996). International Classification of Diseases. Injuries and Causes of Death. 1998 Version 10. (ICD-10), WHO: Geneva.

7

Regeneration of germlings and seedlings development from cauline leaves of *Sargassum thunbergii*

Feng Li[1], Shoutuan Yu[2], Yuze Mao[1] and Naihao Ye[1]*

[1]Yellow Sea Fisheries Research Institute, Chinese Academy of Fishery Sciences, Qingdao 266071, China.
[2]Hongdao Street Office of QingdaoChengyang District, China.

A method of producing artificial *Sargassum thunbergii* seedlings is urgently desired to meet the increasing demand of raw materials used for aquaculture and for the sake of environmental protection. This is the first report of attempt to study the vegetative propagation of this species using cauline leaves under laboratory conditions. On average, 45.75% of the excised leaves survived and one leaf could produce several new individuals. Adventitious burgeons grew into branches about 2 mm in length after 3 months of culture. The new individuals were cutoff and could be used as seedlings for raft culture and seabed restoration after a further culture to elongate enough for hand-planting.

Key words: *Sargassum thunbergii,* artificial seedling, brown seaweed, cauline leaves.

INTRODUCTION

Sargassum spp. have drawn lot of attentions for its active substances, such as antitumor polysaccharide (Fujihara et al., 1984; Zhuang et al., 1995; Jo et al., 2005), anticoagulative polysaccharide (Athukorala et al., 2006), anthelmintic components (Lee and Min, 1970), as well as peroxynitrite-scavenging and reactive oxygen scavenging constituents (Seo et al., 2004; Park et al., 2005). *Sargassum thunbergii* (Mert.) O. Kuntze, named rat-tail alga in China after its morphological characteristics, is becoming increasingly valuable as a resource of diet for aquaculture, especially with regards to feeding sea cucumber (Liu et al., 2004). During the past ten years, wild stocks of the species have been subject to intense harvesting by local inhabitants. With increasingly high levels of harvesting, some places of natural stands of *S. thunbergii* have collapsed (Zou et al., 2005). Recently, sexual breeding of *S. thunbergii* have been achieved under laboratory conditions (Wang et al., 2006). However, availability of a production technique for artificial seedling breeding is needed and it would be advantageous over the current method for large-scale breeding.

Knowledge regarding the biology of *S. thunbergii* is exceedingly limited and this precludes the development of seedling cultivation techniques aimed at alleviating the negative impact of indiscriminate harvesting currently affecting the wild stocks. Under natural conditions, vegetative propagation from holdfasts is an important method sustaining the population (Yoshida et al., 1999). Practices have proved that seedlings with holdfasts grow faster and have lower morality than those without, and this is the key reason makes *S. thunbergii* cultivation different from other commercial algae, such as *Gracilaria lemaneiformis*, which can be successfully cultured using any parts of its vegetative filaments (Hurtado-Ponce et al., 1992; Nelson et al., 2001; Ryder et al., 2004). Studies reveal that zygotic embryos of *Sargassum* for making seedlings are only formed seasonally (Tokuda et al, 1987; Nanba and Okuda, 1992; Sun et al., 2007). Tissue culture of several other species of *Sargassum* has been conducted. Spontaneous formation and development of adventitious embryos from the cauline leaves of *S. macrocarpum* are observed in laboratory culture. Swellings from the single leaf of *S. macrocarpum* became cylindrical protuberances and grew into 'daughter' thalli, which detach from their 'mother' thalli and develop into individual thalli exhibiting the same morphological processes as zygotic embryos (Yoshida et al., 1999) However, factors induce adventitious embryos in *S. macrocarpum* are still uncertain and the process is complicated and time consuming (Yoshida et al., 1999). In the present study, we report the regeneration of germlings developed from cauline leaves of *S. thunbergii* and their

*Corresponding author. E-mail: yenh@ysfri.ac.cn.

Figure 1. Seedling regeneration of *S. thunbergii from the* cauline leaves under laboratory conditions: (A) cauline branches obtained from the wild, (B) leaves used for the regeneration, (C) a newly developed germlings from the cauline leaves, 3 months after the culture, (D) the developing germlings on the leave, 4 months after the culture, (E) an individual attaching to the stone after 1 month of field culture; (F) a *S. thunbergii* plant with several newly-formed cauline branches after 2 months of field culture, (G) a mature *S. thunbergii* plant after 3 months of field culture (Bar, 1 cm).

availability as artificial seedlings.

weekly changed.

MATERIALS AND METHODS

Samples were collected between August and November 2005 from the intertidal zone (35.35°N, 119.30°E) of the Second Bathing Beach, Qingdao, China using a pump placed 2 m depth under the water surface, and filtered with plankton nets (200 μm net and 20 μm net nested inside), then, autoclaved and prepared into enriched seawater with ES (enriched seawater) (Mclachlan, 1979). In the laboratory, intact plants were isolated, washed several times with sterile seawater, sterilized with 1% sodium hypochlorite for 2 min, then rinsed with autoclaved seawater and placed into a sterile aquarium (d=40 cm, h=30 cm) containing enriched seawater, and maintained at 15°C, with illumination of 50 photons m-2s-1 provided by cool-white fluorescent tubes and with a photonperiod of 12:12 h. The branches of *S. thunbergii* fronds were cut (2 to 3 cm in length) and rinsed with antibiotic enriched seawater solution made of ampicillin, penicillin, rifampicin, nystatin (0.2mg ml^{-1} each) andof GeO (0.1g ml^{-1}) for about 6 h. The leaves were cutoff using a surgeon knife (sigma) and cultured in flasks (5 L) containing autoclaved enriched seawater (1:80, w/w), which were bubbled 24 h in the incubator (Ningbo Jiangnan, GXZ-430ABC, Ningbo, China). Ten flasks were included in the treatment and culture media were

RESULTS AND DISCUSSION

The percentage of the survived excised leaves from *S. thunbergii* plant was 45.75% (n=201) and regeneration of adventitious embryos was observed (Figure 1). Longer cauline branches (Figure 1A) were selected and cut into segments for sterilization, then glossy and intact leaves were cutoff using a surgeon knife (Figure 1B). After 3 months of culture, semispherical swellings, about 2 mm in diameter, arose from the petioles end of the cauline leaves and became cylindrical protuberances (Figure 1C). Each protuberance developed into a new germling on the mother plant leaf (Figure 1D). New germlings were easily detached from the mother plant by scraping with forceps. The morphological characteristics of these germlings were quite similar to those developed from zygotic embryos with a few small cauline leaves (Figure 1D). Within several weeks of detachment, rhizoids emerged at the basal part of the germlings and they attached to   the culture substances (Figure 1E). The cauline leaves, which began to produce adventitious embryos, were gradually covered with one or several of newly developed germlings. The growth of each germling was very slow. It took 3 to 5 months to grow big enough for the convenience of manual operation. The germlings could be stored in culture vessels for about half a year under the culture conditions described previously with the medium being renewed weekly. After several months in laboratory culture, germlings were taken out of the vessels and cultured outdoor with the substances, such as stones (Figure 1E). Rhizoids developed fast at the basal part of the germlings and they attached to the culture substances and leaves were produced consecutively at the top of the axis of each plant (Figure 1F). After another 3 months of culture, the length of the largest cauline branch was more than 10 cm (Figure1G). The primary branches showed rapid elongation in the field.

Their daily growth rate was about 0.5 cm/day from early March to late June. At that time, the germlings had three to thirteen cauline branches (Figure 1G).

All plants began to mature after 4 months of field culture. They were all, males or females, the same as their matrixes indicating that all new germlings were clones. Vegetative propagation of the genus *Chondrus* (Chen and Taylor, 1978), *Gracilaria* (Goldstein, 1973; Santelices and Varela, 1995) and *Eucheuma* (Doty, 1987) through thallus fragmentation is well known and it has allowed a rapid expansion in farming of these species all over the world, suggesting that vegetative propagation is a good method of generating propagules for large scale cultivation. However, there is an exceedingly limited amount of information regarding the regulatory effect of physical factors such as temperature, light, etc on the regeneration processes of *S. thunbergii*. In this study, incubation at 15°C under 50 photons·m^{-2} · s^{-1} and 12:12 h LD illumination were suitable conditions used in the seedling breeding as they showed high burgeon ratio and fast growth rate of the germlings. Considering the sustenance of the resource, it is becoming increasingly apparent that the present status imposes a severe pressure on the wild stocks of *S. thunbergii*, and the available information indicates that some stands have already been abandoned by fishermen as a result of over-exploitation. From the current observations, it is clear that the leaf-origin germlings of *S. thunbergii* can grow and mature the same as the cauline branches in the wild and, therefore, can be useful for artificial seedling production. Also, the results indicate that the produced seedlings could be stored for a long term in the laboratory without loss of growth activity and vegetative regeneration capability.

ACKNOWLEDGEMENTS

This work was supported by the National Natural Science Foundation of China (40706050, 40706048 and 30700619), the National Science and Technology Pillar Program (2008BAC49B04), National special fund for transgenic project (2009ZX08009-019B), Natural Science Foundation of Shandong Province (2009ZRA02075), Qingdao Municipal Science and Technology plan project (09-2-5-8-hy) and the Hi-TechResearch and Development Program(863) of China (2009AA10Z106).

REFERENCES

Athukorala Y, Jung WK, Vasanthan T, Jeon YJ (2006). An anticoagulative polysaccharide from anenzymatic hydrolysate of *Ecklonia cava*. Carbohydr. Polym., 66: 184-191.
Chen LC, Taylor AR (1978). Medullary tissue culture of the red alga *Chondrus crispus*. Can. J. Bot., 56: 883-886.
Doty MS (1987). The production and use of Eucheuma. In: Doty MS, Caddy JF and Santelices B (eds), Case studies of seven commercial seaweed resources. FAO Fisheries Tech. Pap. 281. FAO, Rome, 123-161.
Fujihara M, Iizima N, Yamamoto I, Nagumo T (1984). Purification and chemical and physical characterization of an antitumour polysaccharide from the brown seaweed *Sargassum fulvellum*. Carbohydr. Res., 125: 97-106.
Goldstein M (1973). Regeneration and vegetative propagation of the agarophyte *Gracilaria debilis* (Forsskal) Boerg. (Rhodophyceae). Bot. Mar., 16: 226-228.
Hurtado-Ponce AQ, Samonte GPB, Luhan MR, Guanzon JN (1992). *Gracilaria* (Rhodophyta) farming in Panay, Western Visayas, Philippines. Aquaculture, 105: 233-240.
Jo EH, Cho SD, Ahn NS, Jung JW, Yang SR, Park JS, Hwang JW, Lee SH, Park JR, Kim SJ, Park HK, Lee YS, Kang KS (2005). Inhibition of human breast carcinoma by BLC (*Sargassum fulvellum*) and BLC/HEN egg *in vitro* and *in vivo*. Korean J. Vet. Res., 45: 85-91.
Lee WH, Min KN (1970). Detection of anthelmintic components of *thunbergii Kuntze*. Korean J. Pharmacognol., 1: 19-22.
Liu X, Zhu G, Zhao Q, Wang L, Gu B (2004). Studies on hatchery techniques of the sea cucumber, *Apostichopus japonicas*. In: advance in sea cucumber aquaculture and management, FAO Fisheries Technical Paper 463, pp. 287-288.
McLachlan J (1979). Growth media marine. In: Stein JR (ed) Handbook of phycological methods. Culture methods and growth measurements. Cambridge University Press, Cambridge, pp. 25-51.
Nanba N, Okuda T (1992). Egg release of five Fucalean species in Tsuyazaki, Japan. Nippon Suisan Gakkaishi, 58: 659-663.
Nelson SG, Glenn EP, Conn J, Moore, D, Walsh T, Akutagawa M (2001). Cultivation of *Gracilaria parvispora* (Rhodophyta) in shrimp-farm effluent ditches and floating cages in Hawaii: a two phase polyculture system. Aquaculture, 193: 239-248.
Park PJ, Heo SJ, Park EJ, Kim SK, Byun HG, Jeon BT, Jeon YJ (2005). Reactive oxygen scavenging effect of enzymatic extracts from *Sargassum thunbergii*. J. Agric. Food Chem., 53: 6666-6672.
Ryder E, Nelson SG, McKeon C, Glenn EP, Fitzsimmons K, Napolean S (2004). Effect of water motion on the cultivation of the economic seaweed *Gracilaria parvispora* (Rhodophyta) on Molokai, Hawaii. Aquaculture, 238: 207-219.
Santelices B, Varela D (1995). Regenerative capacity of *Gracilaria* fragments: effects of size, reproductive state and position along the axis. J. appl. Phycol., 7: 501-506.
Seo Y, Lee HJ, Park KE, Kim YA, Ahn JW, Yoo JS, Lee BJ (2004). Peroxynitrite-scavenging constituents from the brown alga *Sargassum thunbergii*. Biotechnol. Bioprocess Eng., 9: 212-216.
Sun X, Wang F, Zhang L, Wang X, Li F, Liu G, Liu Y (2007). Observations on morphology and structure of receptacles and pneumathode of *Sargassum thunbergii*. Mar. Fish., Res., 28: 125-131.
Tokuda H, Ohno M, Ogawa H (1987). The Resources and cultivation of seaweeds. Midori-shobou, Tokyo, pp. 354.
Wang F, Sun X, Li F (2006). Studies on sexual reproduction and seedling-rearing of *Sargassum thunbergii*. Mar. Fish. Res., 27:1-6.
Yoshida G, Uchida T, Arai S, Terawaki T (1999). Development of

adventive embryos in cauline leaves of *Sargassum macrocarpum* (Fucales, Phaeophyta). Phycol. Res., 47: 61-64.

Zhuang C, Itoh H, Mizuno T, Ito H (1995). Antitumor active fucoidan from the brown seaweed umitoranoo (*Sargassum thunbergii*). Biosci. Biotechnol. Biochem., 59: 563-567.

Zou J, Li Y, Liu Y, Zhang T, Wang Y (2005). Studies on the biological characters and technology of raft culture of *Sargassum thunbergii*. Shandong Fish., 22: 22-28.

8

Optimization of medium conditions for efficient plant regeneration from embryo of cotton (var.narishima)

M. Guru Prasad, P. Sudhakar and T. N. V. K. V. Prasad*

Institute of Frontier Technology, Regional Agricultural Research Station, Acharya N G Ranga Agricultural University, Tirupati,-517 502, A. P., India.

An efficient and rapid regeneration protocol was developed using embryo from germinating seedlings of cultivar of cotton (var.narishima) as an explant. A vertical slit given in the apex of embryo shoot regeneration was successful on murashige and skoog medium with 2 mg/L benzyl amino purine (BAP). The shoots were sub cultured in every 15 days in regeneration medium and later they were transferred to rooting medium (1 mg/L IBA).

Key words: Cotton, embryo, tissue culture, regeneration, plant growth regulators.

INTRODUCTION

Cotton is an important crop that is grown throughout the world. It is grown as a source of fiber, food and feed. Lint, the most economically important product from the cotton plant, provides a source of high quality fibre of the textile industry. Cotton seed is important source of oil and cotton seed meal is a high protein product used as livestock feed. Although significant progress has been made in the field of cotton improvement with conventional breeding methodology with the limitations to introduce new alleles. Genetic engineering offers a directed method of plant breeding that selectively targets one or a few traits for introduction into the crop plant. Over 100 years ago, Haberlandt envisioned the concept of plant tissue culture for the cultivation of plant tissues and organs. Plant tissue culture is now a well established technology. Around the mid of twentieth century, there is a notion that plants could be regenerated or multiplied from either callus or organ culture through either micropropagation or clonal propagation. But today, these routine technologies have been expanded to include somatic embryogenesis, somatic hybridization, virus elimination as well as the application of bioreactors to mass propagation. The first transgenic cotton plants were reported by Umbeck et al. (1987). Since then many researchers have been reported insect resistant (Perlak et al., 1990; Cousins et al., 1991; Thomas et al., 1995; Li et al., 1998) or herbicide resistant (Bayley et al., 1992; Rajasekharan et al., 1996; Keller et al., 1997) transgenic cotton plants. Transgenic cotton has been mainly obtained by Agrobacterium-mediated transformation (Satyavathi et al., 2002) and also been generated by regeneration from shoot apex tissues (Gould and Magallanes-Cedeno, 1998). Another method followed to regenerate transgenic plants in cotton has been the pollen tube pathway transformation. In this paper we report on generating cotton transgenic plants by regeneration from embryo of var.narshima, a popular variety in most of the cotton growing areas in India.

To tackle the problems pertaining to regeneration in cotton and certain other recalcitrant crops, alternate methods to minimize or eliminate the steps of regeneration are being standardized. These are called in planta transformation protocols. The strategy essentially involves in plant inoculation of embryo axes of germinating seeds and allowing them to grow into seedling *ex vitro*. Research with Arabidopsis has benefited from the development of high throughput transformation method that avoid plant tissue culture (Azipiroz-leehan and Feldmann, 1997).

MATERIALS AND METHODS

Source of seeds

The seeds of cotton (var.Narasimha) were procured from

---

*Corresponding author. E-mail:tnvkvprasad@gmail.com.

Abbreviations: MS, Murashige and Skoog's medium; BAP, benzyl amino purine; IBA, Indole butyric acid.

Table1. Effect of various concentrations of BAP with MS on shoot regeneration from embryo apex explant in cotton.

Treatments	BAP concentration (mg/L)	Cultures showing shoot regeneration (%)	No. of shoots per culture	Shoot vigour
MS+ T1 (control)	0BAP	ND	ND	ND
MS+ T2	1BAP	12±1.27	2.3	Poor
MS+ T3	2BAP	90±8.82	18.3	Good
MS+ T4	3BAP	70±6.67	16.2	Good
T5	MS+4%BAP	50±3.33	10.4	Moderate

Figure 1. (a) Embryo explants; (b)shoot regeneration and sub-culturing of embryo explant (After 20 days from embryo); (c) *in vitro* rooting (After 32days from embryo) (d) Regenerated plantlet growing in pots (After 45 days from embryo).

Regional Agricultural Research Station, Acharya N G Ranga Agricultural University Nandyal.

Seeds of breeding line of cotton which is Narshima (Female) were soaked overnight in distilled water and were surface sterilized first with 1% Bavastin for 10 min and thoroughly washed with distilled water. Later the seeds were transformed to laminar air flow and treated with 0.1%HgCl$_2$ for 10 to 15 min and washed thoroughly with distilled water for three to four times after treatment with each sterilant. This confirms the removal of mercuric chloride (sterilant). The seeds are soaked in distilled water overnight and water is removed. Then seeds are transformed to petri plates where seeds are pressed with forceps so that embryo is pushed out. Embryo was split longitudinally to about half of its length from the top. The explant was placed in contact with the medium. The medium used was MS containing, 3% sucrose, 0.8% agar with different concentrations (1,2,3,4 mg/L) of the plant growth regulator, Benzyl amino purine (BAP) were tested. The culture was maintained in dark for two days and transferred to light (85 µmol m^{-2} s^{-2}) at 25°C±1°C for 15 days with one sub culture was maintained for every 15 days thereafter. Observations and regeneration frequencies, number of shoots and shoot length were recorded 15 days after transferring to regeneration media. Elongated shoots were rooted on MS+IBA (indole butyric acid) 1 mg/L+3% sucrose+0.8% agar. Plantlets transferred to plastic buckets containing soil rite were irrigated with water and/or half strength Hoagland solution alternatively. For histological studies, embryo were fixed in acetic acid and alcohol (3:1) for 24 h, stained with hematoxylin and viewed under Olympus CX31 light microscope.

RESULTS, DISCUSSION AND CONCLUSION

The type and concentration of plant growth regulators strongly influenced the organogenic potential of the embryo apex explant of variety Narshima. The responding frequency of embryo seemed to depend more on concentration of benzyl amino purine. The regeneration response of embryo apex explant in cotton to the concentration of BAP is shown in Table 1. The regeneration of cotton plant from embryo at different stages are shown in Figure 1. It is evident that without

BAP there is no response of regeneration and maximum shoot regeneration response (90±8.82) has been observed with 2 mg/L BAP concentration along with MS. The further increment in BAP concentration to 3 and 4 mg/L along with MS showed decreased shoot regeneration response of 70±6.67, 50±3.33 respectively. Maximum number of shoots has been recorded (18.3) against the BAP concentration 2 mg/L compared to control (2.3).

Our work demonstrated the role of medium preparation and conditions for efficient regeneration of cotton plant from embryo. This method is cost-effective and successful in commercial perspective.

REFERENCES

Azipiroz-Lee H, Feldmann KA, (1997). T-DNA Insersition mutagenesis in Arabidopsis: Going back and forth. Trends Genet., 13: 152-156.

Bayley CN, Trolinder C, Ray MM, Morgan JE, Euisenberry, Ow DW (1992). Engineering 2,4-D resistance into cottan. Theor. Appl. Genet., 83: 645-649.

Cousins YJ, Lyon BR, Liewellyn DJ (1991). Transformation of an Australian cottan cultivar: Prospects for cottan improvement through genetic engineering. Aust. J. Plant Physiol., 18: 481-494.

Gould JH, Magallance-Cedeon M (1998). Adaptation of cottan shoot apex culture to Agrobacterium-mediated transformation. Plant Mol. Rep., 10: 35-38.

Keller GL, Spatola D, McCaba B, Martinelli W, Swain, Jhon ME (1997). Transgenic cottan resistant to herbicide bialophos. Transgenic Res., 6: 385-392.

Li YE, Chen ZX, Wu X, Jiao SJGJ, Wu JH, Fan XP, Merg JH, Zhu Z, Wang W, Zhu Y, Xu HL, Xaio GF, LI XH (1998). Obtaining transgenic cottan plants with Cowpea trypsin inhibitor gene. Acta Gossypii Sin., 10: 237-243.

Perlak FJ, Deaton RW, Armstrong RL, Fuschs RL, Sims SR, Greenplate JT (1990). Fishhoff, Insect resistant cottan plants. Biotechnology, 8: 939-943.

Rajeshekaran K, Grula JW, Hudspeth RL, Pofelis S, Anderson DM (1996). Herbicide resistant Acala and Coker cottans transformed with a native gene encoding mutant forms of acetohydroxyacid synthase. Mol. Breed., 2: 307-319.

Satyathavathi VV, Prasad V, Lakshmi G, Lakshmi GS (2002). High efficiency transformation protocol for three Indian cottan varities via Agrobacterium tumefaciens. Plant Sci., 162: 215-223.

Thomas JC, Adams DG, Keppenne VD, Washmann CC, Brown JK, Kanost MR, Bohnert HJ (1995). Protease inhibitors of Manduca sexta expressed in transgenic cottan. Plant Cell Rep., 14: 758-762.

Umbeck P, Tohnson G, Barton K, Swain W (1987). Genetically transformed cottan (Gossypium hirsutum L.). Plant Biotechnol., 5: 263-266.

9

In vitro proliferation of shoot regeneration from embryo of *Cajanus cajan* L (var.LGG-29)

M. Guru Prasad, T. N. V. K. V. Prasad* and P. Sudhakar

Institute of Frontier Technology, Regional Agricultural Research Station, Acharya N G Ranga Agricultural University, Tirupati-517 502, A. P. India.

An efficient and direct shoot bud differentiation and multiple shoot induction from embryo explants of pigeon pea (*Cajanus cajan* L.) has been achieved. The frequency of shoot bud regeneration was influenced by the type of explant, genotype and concentrations of cytokinin. Explant embryo were cultured on Murashine and Skoog (MS) medium augmented with different concentrations of Benzyl amino purine (BAP). Among the various concentrations tested, 1.0 mg/l BAP and 0.1 mg/l naphthalene acetic acid (NAA) were found to be the best for maximum shoot bud differentiation. Percentage, as well as the number of shoots per explant showing differentiation of shoot buds was higher on MS media supplement with BAP. The optimal BAP concentration for shoot regeneration was 1.0 mg/l. Elongation of multiple shoots was obtained in MS medium with the concentration 0.4 mg/l gibberillic acid (GA3). The elongated shoots were successfully rooted on MS medium containing different concentrations of auxins. Among them indole buteric acid (IBA) at 1.0 mg/l induced maximum frequency of rooting followed by NAA and indole acetic acid (IAA). Regenerated plants were successfully established in soil where 90 to 95% of them have been developed into morphologically normal and fertile plants. This method can thus be advantageously applied in the production of transgenic pigeon pea plants.

Key words: Tissue culture, redgram, embryo, multiple shoots.

INTRODUCTION

The importance of grain legume is multipurpose and their seeds are mostly used to supply vegetative proteins for humans. Red gram or pigeon pea was high among the grain legumes of India, consumed by large population of the country. Genetic improvement through molecular techniques (Lawrence et al., 2001) has been considered for a wide range of grain legumes. The availability of a genetic transformation system would facilitate the agronomic traits (Singh et al., 2003) affecting production efficiency as well as the nutritional quality of redgram. This paper describes the regeneration protocol for genetic improvement for redgram (Venkatachalam et al., 1999).

MATERIALS AND METHODS

Seeds of *Cajanus cajan* L. (var.LGG-29) are procured from Directorate of oil seed research, Hyderabad were used in this study. Seeds were sterilized on surface with 0.1% mercuric chloride solution for 10 min and then rinsed 5 times with sterile distilled water. The seeds were soaked for overnight and the embryo was excised and inoculated in dark for 2 days and then transfer to photoperiod at 25±2°C with a light intensity of 60 µEm^{2}s^1. After 5 days embryos are cut into 5 mm (Geetha et al., 1998) and they are inoculated in regeneration medium (1 mg/l BAP and 0.1 mg/l NAA). After one week shoot buds appear and later they are convert into shoots. These shoots are sub cultured in shoot elongation medium

*Corresponding author. E-mail: tnvkvprasad@gmail.com.

Table 1. Effect of BAP and NAA against embryo explant for shoot regeneration.

BAP	NAA	No. of explants kept for regeneration	No. of responding explants	No. of shoots per explants
0	0	8	NR	NR
0.5	0	8	NR	NR
1	0	8	5	NR
1.5	0	8	4	3
2	0	8	2	1
0	0.1	8	NR	NR
0.5	0.1	8	NR	NR
1	0.1	8	5	7
1.5	0.1	8	2	1
2	0.1	8	3	4

NR: Not responded.

Table 2. Effect of IBA against root induction.

Growth regulator	Rooting frequency	Days of rooting	Mean no. of roots per shoot	Mean length of roots per shoot
0	ND	ND	ND	ND
0.5	ND	ND	ND	ND
1	70	21	5.5 ± 0.03	3.7 ± 0.1
1.5	60	17	3.2 ± 0.03	2.3 ± 0.2
2	40	14	3.5 ± 0.03	1.6 ± 0.3

ND: Not detected.

(0.4 mg/l GA3). Then they are transferred into rooting medium 1 mg/l IBA. The adventitious root appeared within 2 weeks and developed further in 4 weeks. Then plants are ready for transplantation to pots.

Culture media and conditions

The embryo was placed on MS medium with 3% (w/v) sucrose supplemented with BAP (1 to 5 mg/l) for direct shoot bud regeneration. Similar conditions of the media have been reported by (Prakash et al., 1994; Kumar et al., 1983). All media were adjusted to pH 5.8 prior to the addition of 0.8% (w/v) agar and autoclaved at 121°C and 15l b for 15 min. The culture was maintained at 25±2°C in the culture room with 16 h photoperiod with 60 µEm²s² light intensity provided by cool white fluorescent tubes.

The regenerated explant were taken and sub-cultured in the regeneration media for every 15 days. After regeneration upto 2 cm long, they were placed in the shoot elongation media (SEM) supplement with MS salts and hormone GA3 (0.4 mg/l) concentration (Mohan et al., 1998). Then shoots were sub cultured for every 15 days. The sub-cultured shoots were then transferred to the rooting medium supplemented with 1 mg/l IBA (Daal et al., 2003). For every 15 days, the sub-culture repeated. After getting the roots, they were transferred to the soilrite for hardening (George et al., 1998). Later plants were transferred to the green house.

RESULTS AND DISCUSSION

The shoot regeneration was observed from embryo after

3 to 4 weeks in the regeneration medium (MS) of various concentrations (0 to 2) (Frankalin et al., 1998). From the Table 1, it is evident that with the BAP at concentrations 1 and 0.1 mg/l in combination with NAA showed maximum number of shoots per explant. Along with the shoot buds, some buds were found to be greenish in colour. Green and healthy shoots were measured to be 2 to 3 cm in length were excised and sub-cultured into the shoot elongation medium (GA3, 0.4 mg/l). For every 15 days the sub-culture must be repeated. When shoots grow to their maximum lengths, then they were sub-cultured into the rooting medium. From the Table 2, it is clear that IBA at concentration 1 mg/l showed the maximum root length in 21 days compared to other concentrations of NAA and IAA. After reaching maximum root length it has been acclimatized in sand clay and vermiculate in the ratio of 1:1:1. It has also been acclimatized in soilrite, vermiculate and peat. Similar *in vitro* regeneration and transformation of pigeon pea were reported by Thu et al. (2003).

Conclusions

Our study proved that efficient regeneration could be done with embryo as an explant in redgram (var.LRG-41). However, proper medium conditions should be employed for better shoot and root growth after regeneration. Further studies are in progress in

a   b

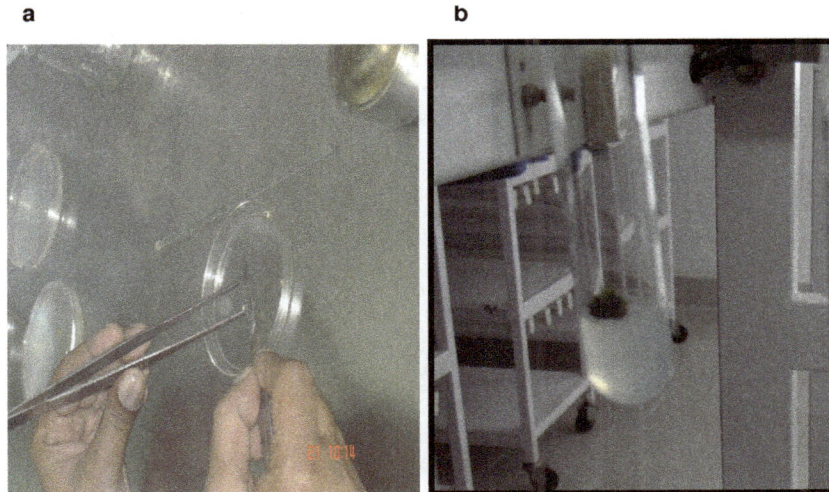

Figure 1. (a) Excised embryo of redgram (var.LGG- 29), (b) Multiple shoots regeneration from embryo.

a   b

Figure 2. (a) Shoots sub-cultured from embryo (b) Rooted shoot let from embryo.

transgenics for the exploitation of the results obtained from this experiment (Figures 1 and 2).

REFERENCES

Geetha N, Venkatachalam P, Prakash V, Lakshmi Sita G (1998). High frequency induction of multiple hoots and plant regeneration from seedling explant of pigeon pea *(Cajanus cajan* L.). Curr. Sci., 10: 10 36-1041.

Kumar AS, Reddy GM (1983). Plantlet regeneration from different callus cultures of pigeon pea (*Cajanus cajan* L.) Plant Sci. Lett., 32: 271-278.

Prakash SN, Pental D, Sarin NB (1994). Regeneration of pigeon pea (*Cajanus cajan*) from cotyledonary node via multiple shoot formation. Plant Cell Rep., 13: 623-627.

Mohan ML, Krishnamurthy KV (1998). Plant regeneration in pigeonpea (*Cajanus cajan* (L.) Millsp.) by organogenesis. Plant Cell Rep., 17: 705-710.

Daal S, Lavana, Devi P, Sharma KK (2003). An efficient protocol for shoot regeneration and genetic transformation of pigeonpea (*Cajanus cajan* (L.) Millsp.) by using leaf explants. Plant Cell Rep., 21: 1072-1079.

Eapen S, George L (1998). Thidiazuron –induced shoot regeneration in pigeon pea (*Cajanus cajan* L.).Plant Cell Tiss. Organ. Cult., 53: 217-220.

Frankalin G, Jayachandran R, Melehias G (1998). Ignacimuthu,s. Mutiple shoot induction and regeneration of pigeon pea (*Cajanus cajan* (L.) Millsp.) cv. Vamban 1 from apical and axillary meristem. Curr. Sci., 74: 936-937.

Lawrence PK, Koundal KR (2001). *Agrobacterium tumefacians-*mediated transformation of pigeonpea (*Cajanus cajan* (L.) Millsp,) and molecular analysis of regenerated plants Curr. Sci., 80: 1428-1432.

Singh ND, Sahoo L, Sonia, Jaiwal PK (2003). *In vitro* shoot organogenesis and plant regeneration from cotledonary node and leaf explants of pigeonpea (*Cajanus cajan* (L.) Millsp.), Mol. Breed., 11: 159-168.

Thu TT, Mai TTX, Dewaele E, Farsi S, Tadesse Y, Angenon G, Jacobs M (2003*). In vitro* regeneration and transformation of pigeonpea (*Cajanus cajan* (L.) Millsp.) Mol. Breed., 11: 159-168.

10

Recent studies on the pullout strength behavior of spinal fixation

Mohammed Shahril Azwan* and Inzarulfaisham Abd Rahim

School of Mechanical Engineering, USM, Engineering Campus, 14300 Nibong Tebal, Pulau Pinang, Malaysia.

Spinal fixations with pedicle screws are widely used nowadays in order to provide spine stability and correcting spinal deformity. The pullout strength of pedicle screws can be evaluated by means of pullout strength testing. This paper reviews recent experimental and finite element analysis evidence concerning the factors that affecting the pullout strength of pedicle screws in various bone materials. Cadaveric bones and synthetic foam blocks were used by researchers in the recent experiment. Types of screw, screw designs, insertion technique, bone mineral density and bone-screw interface have their own significant effects to the fixation strength and will be further discussed.

Key words: Spinal fixation, pedicle screws, pullout strength, cadaveric bones, synthetic foam blocks.

INTRODUCTION

Spinal fixation with pedicle screw has become the most commonly used methods in spinal instrumentation system. Their capability in providing fixation stability and the effectiveness in correcting spinal deformity causes the surgeons to widely use pedicle screws fixation. They have been used also for solid bone fusion in patients with deformity like scoliosis or kyphosis, fracture, spondylolisthesis, degenerative arthritis or tumor (Yilmaz et al., 2009). To appraise the stability of the fixation system, a pullout test was done to evaluate the performance of the pedicle screws. In this test, the fixation strength or the holding power of the screw to the bone thread surfaces was determined. Many studies were done by pulling out the embedded screws in either synthetic bone or cadaver bone at a certain rate of withdrawal to determine the pullout strength of particular pedicle screws.

However, loosening or failure of pedicle screws was also reported in many cases. It was due to inadequate of fixation strength of the screw especially in patients with osteoporosis. A lot of efforts have been made in practice to increase the pullout strength of pedicle screw include increasing the length or diameter of the screw, coatings the screw with hydroxyapatite, or for severe bone loss,

using materials as polymethylmethacrylate (PMMA) and calcium phosphate bone cement (Lei et al., 2006).

The purpose of the present paper is to summarize recent studies that show how the pullout performance of various pedicle screws were influenced by several factors including screw designs, cement augmentation, screws coating, pilot hole and tapping, insertional torque and bone density.

Types of pedicle screws

Numerous types of pedicle screws were available in the market as shown in Figure 1. Many studies were done in order to identify the ability and mechanical performances of each pedicle screw. The holding strength of pedicle screws in vertebral bones is influenced by many factors. The bone-screw interface is the crucial part. There was a report on the capability of conical screws to improve thread purchase by compacting the cancellous bone at the cancellous-cortical interface throughout the pedicle (Hsu et al., 2005). However, there was another study comparing conical and cylindrical screws wherein which screws provide better pullout resistances. The holding strength of conical and cylindrical screws was compared under ideal conditions and after a compromising event; cyclic loading and 180° turn back was applied. Cyclic loading was applied because clinical failure of spinal implants is produced most often by fatigue.

*Corresponding author. E-mail: mohammedshahril@gmail.com.

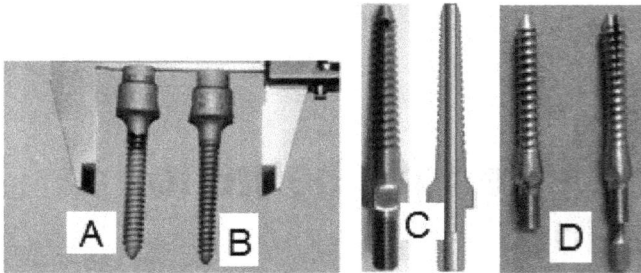

Figure 1. Several types of typical pedicle screws available [2, 5, 9]: A) cylindrical, B) conical, C) cannulated with radial hole and D) expandable.

Pedicle screws are subjected to complex cyclic forces that combine tranverse bending and axial pullout loads. The results showed that conical screws failed more easily when pullout test were performed compared with cylindrical screws for both ideal and compromising event (Lill et al., 2000). Conversely, Abshire et al. (2001) found inversely results by conducting the same experiments. They declared that conical screws provided an increase in the pullout strength compared with cylindrical screws without loss of pullout strength, stiffness or energy-to-failure when conical or cylindrical screws were backed out 180° or 360° from full insertion. They also stated that the conical screws engage more of the pedicle cortex as well as the cancellous bone at the corticocancellous margin than cylindrical screws. This contradictory fact maybe due to the conical and cylindrical screws designs that were used in particular studies.

Besides, cortical and cancellous screws were compared in osteoporotic bone material with axial and angled pullout test (Patel et al., 2008). The authors found that cancellous screw had a significantly higher pullout force than the cortical screw and they suggest that only screws placed axially or up to 10° angle may increase its holding power. In this study, failure of screw fixation was observed from the stripping of the internal screw threads within the bone material. No failure was observed at the screw treads even after fully pullout of screws. In addition, the bone material itself had a significant effect on the failure of implants. However, limitation existed in this study wherein the screws diameter used are not identical even though the length is equal.

A comparative study between cervical pedicle screws and lateral mass screws was done in order to prove that cervical pedicle screws have higher pullout strength using the same screws dimension. Screw loosening has been documented as a failure mechanism for lateral mass screws, where the decreasing size of the lateral masses results in lower pullout strength (Johnson et al., 2006). In this study, all screws were repetitively cycled at a rate of 25 mm/min for 200 cycles. As hypothesized, pedicle screws have superior pullout strength than the lateral screws. Lateral screws loosen more than the pedicle

screws over time, although both screw types have similar initial stability. Thompson et al. (1997) demonstrated that noncannulated screws had better holding power compared with cannulated screws.

Cannulated screws are designed with enlarged core diameter to accommodate guide pins. Hence, the ratio of major diameter to core diameter will be decreased and this will affects the pullout strength of cannulated screws. Despite, pullout strength for cannulated, pedicle screws could be increased with bone cement augmentation. Chen et al. (2009) presented the effects of radial holes for cannulated screws with cement injection on the bone-screw interface. The pullout strength for the cannulated screws increase as the number of radial holes increased. It is cause by large amount of exuded cement injection from the radial holes and providing better bone-screw holding strength. They also found that tapping the pilot holes significantly reduced the pullout strength of the screws. A comparison was also made between small-diameter cannulated and solid-core screws by Kissel et al. (2003). They concluded that cannulated and solid core screws of similar dimension and thread length have similar holding strengths. Therefore, these cannulated screws make an attractive alternative for surgical fixation applications.

Various methods have been used to improve pedicle screw fixation including increasing the diameter of screws. Yet, this may not always be possible because of anatomic constraints. It may increase the risk for pedicle fracture with possible neural injury. A better solution is to use an expanding pedicle screws. Expanding pedicle screw may improve bone fixation by increasing the screw tip diameter, allowing greater bone contact with no increase in pedicle insertion diameter. The screw tip diameter increased by inserting a smaller gauge screw into the threaded interior of expanding screw and opens the fins concentrically as it advances. Many studies were done to evaluate the pullout performance of expansive pedicle screws. Apparently, expansive pedicle screws had significantly improved pullout strength when compared with self-tapping screws in both cases of low and high bone mineral density (BMD) (Cook et al., 2000).

An effort was made by Cook et al. (2004) to investigate whether polymethymethacrylate (PMMA) bone cement augmentation of an expandable pedicle screw can further improve fixation strength compared to the expandable screw alone in severely osteoporotic bone. They found that there was an increase in pullout strength with the cemented expandable screw compared with a noncemented expandable screw including a greater than twofold increase in pullout strength in the most severely osteoporotic bone. The stiffness and energy absorbed to failure was also significantly increased. The cement augmentation may be especially useful in the severely osteoporotic patient. Significantly, the removal and replacement of a pedicle screw in a revision procedure substantially decreases the mechanical fixation strength

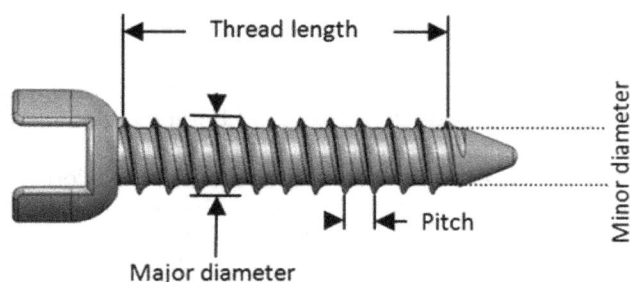

Figure 2. Parts of typical orthopaedic screw.

of the screw.

Furthermore, the turning back of the screws becomes necessary when surgeons cannot successfully insert screws into the proper position during the first attempt, which reduces the holding strength. Lei et al. (2006) have show that the turning back torque and pullout strength of expansive screws were significantly greater than those conventional screws. Similarly, their pullout strength was higher in the revision test. These findings suggest that expansive screw is ideal in problematic situations where the bone integrity is compromised by either osteoporosis or pedicle screw revision by providing biomechanical parameters similar to those expected for normal bone and in primary surgery.

PEDICLE SCREW DESIGN

Noteworthy, the designs of pedicle screws are also important for providing sufficient fixation strength of the implant. Note that, major diameter, minor diameter, pitch, thread length, thread shape and thread depth of the screws are affecting the pullout strength of the screws (Figure 2). Chapman et al. (1996) conducted a study on factors that affected the pullout strength of cancellous bone screws. In general, cancellous screws are designed to have greater thread depth and decreased thread cross-sectional thickness in comparison to cortical screws, to provide more holding power in porous material such as cancellous bone. Other than that, major diameter of the screw also gives potential effects on the holding power wherein larger diameter provides greater holding power.

However, the diameter of screw is limited by the size of pedicles especially the immature pedicles. Yilmaz et al. (2009) have done a study to evaluate the effect of dilation of immature pedicles on the pullout strength of the screws. They found that the dilation does not affect the pullout strength of the screws after 3 months in *in vivo* model. Besides, larger diameter provided a more rigid construct than PMMA augmented pedicle screw fixation in revision spinal instrumentation. The less stiff and more viscoelastic of PMMA would result in reduced stiffness of

the construct and increase the ability of distributing creep deformations during fatigue than the large diameter construct (Kiner et al., 2008).

Studies have been done by numerous researches via finite element analysis. Zhang et al. (2004) created a three-dimensional finite element model to simulate the behavior of bone and screw during screw pullout. From the simulation, bone experienced significant shear stress at the thread root during screw pullout. At maximum force, only those elements at the thread root region reached the yield point and failed. They declared that screws with larger major diameter and smaller minor diameter and pitch will lead to an increase in the pullout strength. Moreover, it is obvious that a linear correlation between the pullout strength and thread numbers can be obtained. More thread numbers can resist higher forces or in other words, longer purchase length can effectively increase pullout strength.

Chatzistergos et al. (2009) also preferred finite element analysis compared to experimental study. Several factors affecting the pullout forces of screw have been demonstrated. They also found that, the major diameter or outer radius was the most important factors and it was proven that larger outer radius yielded greater pullout forces. Without any doubt, they stated that inclination of the thread or thread angle is considered to be the less important ones. However, they did show that an increase of the thread angle resulted in an increase of the pullout force.

Bone cement augmentation

Recently, bone cement augmentation was used to enhance the strength of osteoporotic bones and to enhance screw fixation. The technique of bone cement augmentation was crucial aspect since it was reported that the technique is affecting the fixation strength. The techniques for screw fixation with cement augmentation include inserting screws when cement is soft, inserting screws when cement is curing characterized by its doughy consistency and inserting screws after drilling and tapping hardened cement. In the study of Flahiff et al. (1995), the fixation strength of polymethymethacrylate (PMMA) bone cement was evaluated among those cement augmentation techniques. They concluded that inserting cortical screws into plastic femur sawbones when the PMMA was in a doughy consistency prior to curing (approximately 8 to 10 min after mixing) produced the strongest cement-screw construct. Drilling and tapping holes after the cement had hardened resulted in the weakest cement-sawbone construct.

A biomechanical cadaveric analysis of PMMA augmented pedicle screw fixation was done by Frankel et al. (2007). They highlighted that the screw augmentation procedures were performed using fenestrated bone tap system which can prevent backflow of cement toward neural elements and allowing custom foam for

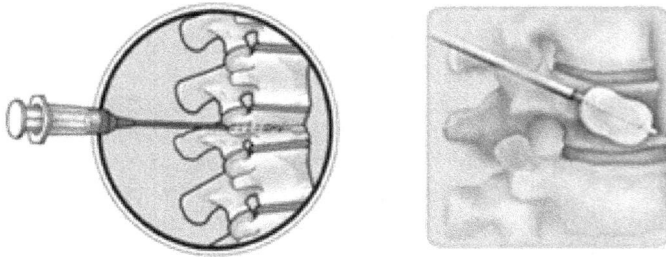

Figure 3. Cement augmentation method; vertebroplasty (left) and kyphoplasty (right).

subsequent screw placement. The results showed that PMMA did increase the pullout strength of screw fixation for both osteoporotic and nonosteoporotic bones. In fact, the pullout strength did not significantly change with increased cement volume usage. Thus, they recommended using the lower range of cement volumes in pedicle screw augmentation procedures to perhaps reduce the likelihood of cement toxicity.

Previous method of cement augmentation is the standard transpedicular approach. In addition, other methods of cement augmentation are vertebroplasty augmentation approach and balloon kyphoplasty augmentation as shown in Figure 3. Bone cement is injected through a small hole into a fractured vertebra for vertebroplasty augmentation whilst in kyphoplasty augmentation a balloon is used to create a void which is filled with PMMA. A series of comparison between them has been done by Becker et al. (2008). Vertebroplasty augmented screws showed a significant higher pullout force than others.

Kyphoplasty technique also increased the pullout force compared with nonaugmented screws. However, surgeons may favor balloon kyphoplasty for screw augmentation since vertebroplasty has considerable risks regarding cement leakage and a slightly higher perioperative morbidity. By the study of Burval et al. (2007), they evaluated the performance of kyphoplasty augmentation with an exposed to cyclic fatigue loading prior to pullout testing. 5000 cycles of caudalcephalad loading were applied perpendicular to the pedicle screws. Their results showed that pedicle screw augmentation using kyphoplasty technique increased the pullout failure 2 to 3 fold in osteoporotic vertebrae. Chang et al. (2008) agreed that PMMA has advantages like readily availability, inexpensiveness, short application time and fixation strength that are practical for clinical applications. Sequential dilatation with K-wires during screw tract preparation can be used to prevent leakage of PMMA cement. They also declared that the sufficient usage volumes of PMMA cement for lumbar and thoracic screw are 3 and 2 ml, respectively.

As hypothesized by Blattert et al. (2009) removal of screws, if needed, might cause problems. Among

screw designs, the cannulated-fenestrated screws might cause problems especially during revision, because of their winglike cement interconnection between the screw core and surrounding tissue. However, they have successful showed that revision characteristics of these screws following cement augmentation are not problematic even for osteoporosis cases. Obviously, the cement interconnection between the screw core and surrounding bone tissue is fragile enough to break off in the event of extraction torque and to release the screw.

Other types of bone cements instead of PMMA are calcium sulfate paste and particulate calcium phosphate. Rohmiller et al. (2002) suggested that PMMA should be replaced by calcium sulfate paste since they demonstrated similar pullout strength. PMMA has potential dangers if leakage into spinal canal occurs. In addition, the calcium sulfate paste does not have an exothermic reaction in its curing phase and this will eliminates the risk of thermal damage to spinal canal. The paste also improves safety around spinal cord since it is biodegradable and designed to be resorbed by body rather than becoming a permanent space-occupying lesion in the canal in the event of leakage. Particulate calcium phosphate is alternative bone cement to PMMA. The special ability of the calcium phosphate in the fixation application is to speed natural tissue healing and then be replaced by the patient's own bone tissue. It has been demonstrated that the pullout strength of screws with augmentation of calcium phosphate is high. Interestingly, the application of calcium phosphate to normal bone will reduce screw pullout strength. Calcium phosphate improved the pullout strength of failed screw in low density polyurethane blocks (osteoporotic bone) (Hashemi et al., 2004).

Pedicle screws surface coating

As described early, PMMA may be associated with complication such as leakage, exothermic damage to the bone and adjacent tissue and long term screw loosening resulting from nonbiologic bonding. The use of hydroxyapatite (HA) coated implants (Figure 4), on the other hand, may improve the stability of the bone-metal interface without the disadvantages of PMMA (Stea et al., 1995; Moroni et al., 1998). Hasegawa et al. (2005) showed that HA-coated pedicle screws had higher pullout resistance than uncoated screws and it is clearly revealed that the inter-spaces between the screw threads had filled with new bone and good bonding was present between the bone and the apatite coating of the screw. Even under loaded conditions, HA coating improves fixation of pedicle screws, with increased pullout resistance and reduced risk of loosening.

As stated by Sanden et al. (2001) higher pullout resistance by HA-coated screws at early stage was mainly caused by differences in surface roughness of the coatings, while the difference at later stage was due to

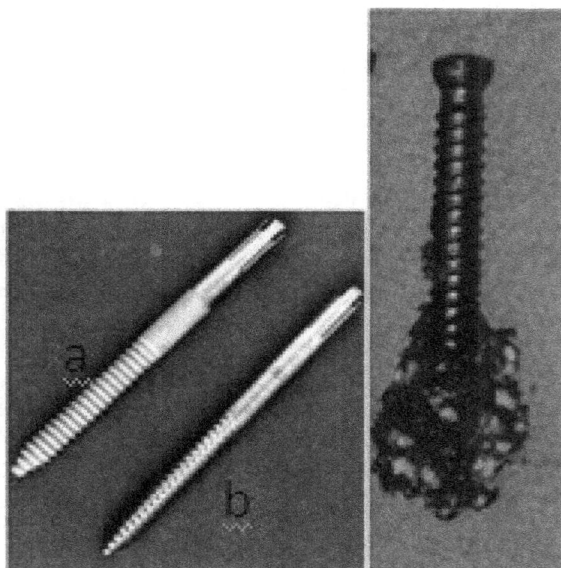

Figure 4. a) Coated and b) uncoated screws (left) and PMMA augmented screw (right) [19, 30].

bone reaction around the HA-coated screws. A study on improving extraction torque of HA-coated pedicle screws was also been done. The surface roughness of plasma-sprayed HA-coated implants is generally three to four times greater than the roughness of machined metal implants (Wennerberg et al., 1993). The fixation strength of HA-coated screws was highly correlated to its surface roughness. By using fully coated screws, extraction was extremely difficult compared to extraction of conventional stainless steel screws, which were regularly loose. By reducing the area of the screws that is coated, it may be possible to achieve an enhanced purchase while extraction will be easier (Sanden et al., 2000).

Titanium coated on the surface of titanium alloy implants have also been widely used in orthopaedics and dentistry with great success because of their good mechanical properties and biocompability rather than HA coatings. However, titanium coating is bioinert, which cannot bond chemically to bone tissue. Therefore, attempts have been made including modification of titanium coating with alkali and heat treatment to improve the surface properties of titanium coating implants (Kokubo et al., 1996). Absolutely, alkali-modified implants shows higher shear strength which can be attributed to the changes in surface topography and chemistry. Significantly, it can improve and accelerate the early in growth of bone and osseointegration to reduce clinical healing times and thus, to improve implant success rates (Xue et al., 2005).

Upasani et al. (2009) pointed the benefits of hydroxyapatite – titanium plasma sprayed (HA-TPS) composite coating that may leverage the advantages of both individual coatings to further improve screw fixation.

Generally, HA-TPS did improved peak torque during screw extraction. However, the composite coating may not provide a significantly greater benefit compared with HA alone. They clarified that improved osseointegration with HA coatings may result in a decreased incidence of screw loosening and improved outcomes of transpedicular spinal instrumentation in nonfusion procedures.

OTHER FACTORS INFLUENCING FIXATION STRENGTH

Few reports have addressed the effects of a pilot hole tapping on the bone/screw interfacial strength. However, the reported results have been inconsistent (Thompson et al., 1997). Johnson et al. (2006) examined the effect of tapping on pullout strength in synthetic polyurethane foam. Their results indicated there was no demonstrable effect on holding power when screws were inserted with or without tapping. Chapman et al. (1996) and Cook et al. (2004) indicated that, tapping in porous materials decreased screw pullout strength because the removal of materials by the tap enlarged the hole considerably, reducing the holding power of the screw threads.

Ronderos et al. (1997) analyzed the axial pullout strength of tapped vesus untapped pilot holes for bicortical screws in the anterior cervical spine. They concluded that tapping a pilot hole neither weakens nor strengthens the axial pullout strength of fully threaded cortical bone screws. Carmouche et al. (2005) found that tapping the pedicles in the lumbar spine did decrease the pullout resistance but not in thoracic spine. One explanation for this is that in severe osteoporosis, there is simply less cancellous bone available to be compressed. Thus, the process of tapping does not cause a significant net loss of cancellous bone and fixation is more dependent on cortical purchase.

A pilot hole is drilled with smaller diameter to make a path so that screws can easily be inserted in correct direction. Pilot hole also has its own affects on pullout resistance and insertion torque. Leite et al. (2008) found that higher insertion torque can be resulted as long as the pilot hole used was smaller than inner or minor diameter of screw. In other words, with the increased diameter of the pilot hole, the insertion torque was reduced. That phenomenon correlated with bone removal during pilot hole drilling which would affect the anchorage of the implant. As the pilot hole's diameter increases, a larger amount of bone is removed, and a smaller amount of bone is available to be compacted around the implant, thus, reducing the insertion torque. During the surgical procedure, insertion, removal and re-insertion of the screws often happen to determine the pathway of pilot hole and to detect possible violations of the lateral wall of the pilot hole that might damage adjacent structures of vertebrae (Kim and Lenke, 2005).

However, this procedure might interfere with the screw

holding strength. Therefore, Defino et al. (2009) investigated the effect of repetitive pilot hole use on the insertion torque and pullout strength of screws. They found that insertion torque and pullout strength did reduce between the first and following insertion. Despite, the pattern of reduction of pullout strength was not similar to the pattern of reduction of the insertion torque. However, the study of Foley et al. (1990) revealed that there was no significant difference in pullout strength as noted in pretapped or self-tapped screws inserted into the same hole one, two or three times before pullout testing.

Besides, the effect of insertional torque to pullout resistance is uncertain. Leite et al. (2008) stated that, the insertion torque of implants showed a correlation with pullout resistance only for perforation values of pilot hole smaller than the inner diameter of the screw. Inceoglu et al. (2004) found that, there was no significant correlation between pullout strength and insertional torque. They concluded that insertional torques are not good predictors of pullout strength and stiffness, particularly in nonstandard screw and thread designs.

Bone mineral density (BMD) is another factor that influenced pullout strength especially in osteoporosis cases. Many studies have shown the correlation between BMD and pullout strength. Hsu et al. (2005) agreed that both pullout strength and insertion torque in the foam with the higher density was consistently higher than that in the foam with the lower density. Battula et al. (2006) concluded that the depth of insertion of the tip of the screw for adequate fracture fixation in normal bone is 1 mm or more past the far cortex and in osteoporotic bone it is at least 2 mm past the far cortex. It shows that osteoporotic bone need more screws surface area interface with the bone in order to provide sufficient holding power. Osteoporotic bones usually related to screw loosening that mainly caused by cyclic caudocephalad toggling at the bone-screw interface when an axial compression load was transmitted through the plate or rod to the screw.

A clinical study was done by Okuyama et al. (2001) to investigate the screw loosening behavior among osteoporosis patients. After several years followed up, loosening of pedicle screws occurred and it affect the implant stability. Thus, BMD is supposed to be a very important parameter influencing the stability of pedicle screws. Suzuki et al. (2001) found that pedicle screw coupling increased the pullout strength in osteoporotic spine. However, the improvement is only subjected to BMD of more than 90 mg/ml but not for BMD of less than 90 mg/ml. Zhang et al. (2006) noted that screw pullout strengths were directly proportional to the shear strength of the foam material. Low BMD will have lower pullout strength as their shear strength is very low. Ramaswamy et al. (2009) studied the holding power of screws in osteoporotic, osteopenic and normal bone. They found that the pullout strengths of all screws were correlated to the foam density and significantly it was better in higher density foam.

CONCLUSION

The main factors affecting the pullout force of a bone screw are its design, the material properties of the bone and the insertion technique followed by the surgeon. Conflicts still exist whether to perform experimental study using cadaver or synthetic foam blocks since results may vary within the materials. A continuous study is needed to gather information and knowledge as much as possible to enhance more stable and rigid spinal fixation system. Extra concern must be put on to the osteoporosis cases since major problems of fixation stability and rigidity are referring to them.

REFERENCES

Abshire BB, McLain RF, Valdevit A, Kambic HE (2001), Characteristics of pullout failure in conical and cylindrical pedicle screws after full insertion and back-out, Spine, 1: 408-414.

Battula S, Schoenfeld A, Vrabec A, Njus GO (2006). Experimental evaluation of the holding power/stiffness of the self-tapping bone screws in normal and osteoporotic bone material. Biomech., 21: 533-537.

Becker S, Chavanne A, Kropik K, Aigner N, Ogon M, Redl H (2008). Assessment of different screw augmentation techniques and screw designs in osteoporotic spines. Eur Spine, pp. 1462-1469.

Blattert TR, Glasmacher S, Riesner HJ, Josten C (2009). Revision characteristics of cement-augmented, cannulated-fenestrated pedicle screws in the osteoporotic vertebral body: A biomechanical in vitro investigation. Neurosurg. Spine, 11: 23-27.

Burval DJ, McLain RF, Milks R, Inceoglu S (2007). Primary pedicle screw augmentation in osteoporotic lumbar vertebrae. Spine, 32: 1077-1083.

Carmouche JJ, Molinari RW, Gerlinger T, Devine J, Patience T (2005). Effects of pilot hole preparation technique on pedicle screw fixation in different regions of the osteoporotic thoracic and lumbar spine. Neurosurg. Spine, 3: 364-370.

Chang MC, Liu CL, Chen TH (2008), Polymethylmethacrylate augmentation of pedicle screw for osteoporotic spinal surgery. Spine, 33: 317-324.

Chapman JR, Harrington RM, Lee KM, Anderson PA, Tencer AF, Kowalski D (1996). Factors affecting the pullout strength of cancellous bone screws. Biomech. Eng., 118: 391-398.

Chatzistergos PE, Magnissalis EA, Kourkoulis SK (2009). A parametric study of cylindrical pedicle screw design implications on the pullout performance using an experimentally validated finite-element model. Med. Eng. Phys., 32(2):145-54.

Chen LH, Tai CL, Lai PL, Lee DM, Tsai TT, Fu TS, Niu CC, Chen WJ (2009). Pullout strength for cannulated pedicle screws with bone cement augmentation in severely osteoporotic bone: Influences of radial hole and pilot hole tapping. Clin. Biomech., 24(8):613-8.

Cook SD, Salkeld SL, Stanley T, Faciane A, Miller SD (2004). Biomechanical study of pedicle screw fixation in severely osteoporotic bone. Spine, 4: 402-408.

Cook SD, Salkeld SL, Whitecloud III TS, Barbera J (2000), Biomechanical evaluation and preliminary clinical experience with an expansive pedicle screw design. Spinal Disord., 13: 230-236.

Defino HLA, Rosa RC, Shimano AC, Volpon JB, Paula FJA, Schleicher P, Schnake K, Kandziora F (2009). The effect of repetitive pilot- hole use on the insertion torque and pullout strength of vertebral system screws. Spine, 34: 871- 876.

Flahiff CM, Gober GA, Nicholas RW (1995). Pullout strength of fixation screws from polymethylmethacrylate bone cement. Biomater., 16:

533-536.

Foley WL, Frost DE Tucker MR (1990). The effect of repetitive screw hole use on the retentive strength of pretapped and self-tapped screws. Oral Maxillofac Surgery, 48: 264-267.

Frankel BM, Agostino SD, Wang C (2007). A biomechanical cadaveric analysis of polymethylmethacrylate-augmented pedicle screw fixation. Neurosurg. Spine, 7: 47-53.

Hasegawa T, Inufusa A, Imai Y, Mikawa Y, Lim TH, An HS (2005), Hydroxyapatite-coating of pedicle screws improves resistance against pullout force in the osteoporotic canine lumbar spine model: A pilot study. Spine, 5: 239-243.

Hashemi A, Bednar D, Ziada S (2009). Pullout strength of pedicle screws augmented with particulate calcium phosphate: An experimental study. Spine, 9: 404-410.

Hsu CC, Chao CK, Wang JL, Hou SM, Tsai YT, Lin J (2005). Increase of pullout strength of spinal pedicle screws with conical core: biomechanical tests and finite element analyses. Orthop. Res., 23: 788-794.

Inceoglu S, Ferrara L, McLain RF (2004), Pedicle screw fixation strength: pullout versus insertional torque. Spine, 4: 513-518.

Johnson TL, Karaikovic EE, Lautenschlager EP, Marcu D (2006), Cervical pedicle screws vs. lateral mass screws: Uniplanar fatigue analysis and residual pullout strengths. Spine, 6: 667-672.

Kim YJ, Lenke LG (2005). Thoracic pedicle screw placement: free hand technique. Neurol. India, 53: 512-519.

Kiner DW, Wybo CD, Sterba W, Yeni YN, Bartol SW, Vaidya R (2008). Biomechanical analysis of different techniques in revision spinal instrumentation: Larger diameter screws versus cement augmentation. Spine, 33: 2618-2622.

Kissel CG, Friedersdorf SC, Foltz DS, Snoeyink T (2003). Comparison of pullout strength of small-diameter cannulated and solid-core screws. Foot Ankle Surg., pp. 334-338.

Kokubo T, Miyaji F, Kim HM, Nakamura T (1996). Spontaneous formation of bonelike apatite layer on chemically treated titanium metals. Am Ceram Soc., 79, 1127-1129.

Lei W, Wu Z (2006). Biomechanical evaluation of an expansive pedicle screw in calf vertebrae. Eur. Spine, 15: 321-326.

Leite VC, Shimano AC, Goncalves GAP, Kandziora F, Defino HLA (2008). The influence of insertion torque on pedicular screw's pullout resistance, Acta Ortop. Bras., 16: 214-216.

Lill CA, Schlegel U, Wahl D, Schneider E (2000), Comparison of the in vitro holding strengths of conical and cylindrical pedicle screws in a fully inserted setting and backed out 180°. Spinal Disord., 13: 259-266.

Moroni A, Toksvig-Larsen S, Maltarello MC (1998), Comparison of hydroxyapatite-coated, titanium-coated, and uncoated tapered external-fixation pins. Bone Joint Surg., 80A: 547-554.

Okuyama K, Abe E, Suzuki T, Tamura Y, Chiba M, Sato K (2001), Influence of bone mineral density on pedicle screw fixation: A study of pedicle screw fixation augmenting posterior lumbar interbody fusion in elderly patients. Spine, 1: 402-407.

Patel PSD, Shepherd DET, Hukins DWL (2008). Axial and angled pullout strength of bone screws in normal and osteoporotic bone material. ICBME Proceedings, 23: 1619-1622.

Ramaswamy R, Evans S, Kosahvili Y (2009). Holding power of variable pitch screws in osteoporotic, osteopenic and normal bone: Are all screws created equal?, Injury. 41(2):179-83.

Rohmiller MT, Schwalm D, Glattes RC, Elalayli TG, Spengler DM (2002). Evaluation of calcium sulfate paste for augmentation of lumbar pedicle screw pullout strength. Spine, 2: 255-260.

Ronderos JF, Jacobowitz R, Sonntag VKH, Crawford NR, Dickman CA (1997). Comparative pull-out strength of tapped and untapped pilot holes for bicortical anterior cervical screws. Spine, 22: 167-170.

Sanden B, Olerud C, Johansson C, Larsson S (2000). Improved extraction torque of hydroxyapatite-coated pedicle screws. Eur. Spine, 9: 534-537.

Sanden B, Olerud C, Larsson S (2001). Hydroxyapatite coating enhances fixation of loaded pedicle screws: A mechanical in vivo study in sheep. Eur. Spine, 10, 334 – 339.

Stea S, Visentin M, Savarino L (1995). Microhardness of bone at the interface with ceramic-coated metal implants. Biomed. Mater. Res., 29: 695-699.

Suzuki T, Abe E, Okuyama K, Sato K (2001), Improving the pullout strength of pedicle screws by screw coupling. Spinal Disord., 14: 399-403.

Thompson JD, Benjamin JB, Szivek JA (1997). Pullout strengths of cannulated and noncannulated cancellous bone screws. Clin. Orthop. Relat. Res., 341: 241-249.

Upasan VVi, Fransworth CL, Tomlinson T, Chambers RC, Tsutsui S, Slivka MA, Mahar AT, Newton PO (2009). Pedicle screw surface coatings improve fixation in nonfusion spinal construct. Spine, 34: 335-343.

Wennerberg A, Albrektsson T, Andersson B (1993). Design and surface characteristics of 13 commercially available oral implants systems. Oral Maxillofac Implants, 8: 622-633.

Xue W, Liu X, Zheng XB, Ding C (2005). In vivo evaluation of plasma-sprayed titanium coating after alkali modification. Biomater., 26, 3029-3037.

Yilmaz G, Demirkiran G, Ozkan C, Daglioglu K, Pekmezci VM, Alanay A, Yazici M (2009). The effect of dilation of immature pedicles on pullout strength of the screws. Spine, 34: 2378-2383.

Zhang QH, Tan SH, Chou SM (2006). Effects of bone materials on the screw pull-out strength in human spine, Med. Eng. Phys., 28: 795-801.

Zhang QH, Tan SH, Chou SM (2004). Investigation of fixation screw pull-out strength on human spine. Biomech., 37: 479-485.

11

Synthesis, characterization and biological activity of some novel azetidinones

A. Rajasekaran*, M. Periasamy and S. Venkatesan

Department of Pharmaceutical Chemistry, KMCH College of Pharmacy, Coimbatore, Tamilnadu, India.

A series of seven novel azetedinones 4a-g have been synthesized by cyclocondensation of various Schiff bases of phenothiazine with chloroacetyl chloride in presence of triethylamine. Various Schiff bases of phenothiazine were synthesized by condensation of 2-hydrazinyl-1-(10H-phenothiazin-10-yl)ethanone 2 with various aryl aldehydes. Compound 2 was synthesized by reacting 2-chloro-1-(10H-phenothiazin-10-yl) ethanone 1 with hydrazine hydrate. The synthesized compounds were characterized by IR, ^{1}H-NMR and mass spectra. The titled compounds were evaluated for anti-tubercular, anti-bacterial, anti-fungal and anti-inflammatory activity by Lowenstein-Jensen medium method, cup plate method, disc diffusion method and carrageenan induced paw edema method respectively. All the seven compounds 4a-g at a concentration of 100, 10 and 1 µg/L showed inhibition against the growth of Mycobacterium tuberculosis. Compounds showed good anti-bacterial activity against Staphylococcus aureus and Bacillus subtilis. Compound 4e showed equipotent antibacterial activity with streptomycin standard and showed better anti-bacterial activity against gram positive bacteria profile than gram negative bacteria. Zone of inhibition was not found for the compounds against Escherichia coli and Pseudomonas aeruginosa. Compounds 4a-g exhibited good antifungal activity against Aspergillus species and no activity against Candida albicans. None of the reported compounds showed promising anti-inflammatory activity.

Key words: Azetidinones, anti-tubercular, anti-bacterial, anti-fungal, inflammation.

INTRODUCTION

Azetidin-2-one, a four-membered cyclic lactam (ß-lactam) skeleton has been recognized as a useful building block for the synthesis of a large number of organic molecules by exploiting the strain energy associated with it. The Staudinger reaction ([2+2] ketene-imine cycloaddition reaction) is regarded as one of the most fundamental and versatile methods for the synthesis of structurally diverse 2-azetidinone derivatives, although many synthetic methods have been developed to date (Van der steen et al., 1991; Palomo et al., 1999; Jiaxi, 2009). Azetidin-2-ones can also be synthesized by enolate-imine condensations (Maurizio et al., 2000), Kinugasa (Jose, 2004), annulations (Anthony et al, 1987) and cyclization reactions (Benito and Alberto, 1998). Also it is used in the synthesis of a variety of ß-lactam antibiotics (Deshmukh et al., 2004). Efforts have been made in exploring such new aspects of ß-lactam chemistry versatile intermediates for the synthesis of aromatic ß-amino acids and their derivatives, peptides, polyamines, polyamino alcohols, amino sugars and polyamino ethers (Alcaide and Almendros, 2001) the cyclic 2-azetidinone skeleton has been extensively used as a template to build the heterocyclic structure fused to the four-membered ring. This provides an access to diverse structural type of synthetic target molecules lacking ß-lactam ring structure (Brickner et al., 1992).

Azetidinones are of great biological interest, especially as anti-tubercular (Kagthara et al., 2000), antibacterial (Singh et al., 2005, Bhanvesh and Desai 2004; Patel and Patel, 2004; Pratibha et al., 2004; Ashok et al., 2003; Devendra and Sharma, 2002; More et al., 2002; Choudhari and Mulwad, 2003; Oza et al, 2003; Padam and Saksena., 2003; Freddy and Sushil, 2004), antifungal

*Corresponding author, E-mail: rsekaran2001in@yahoo.co.in,

(Pandey et al., 2005; Mehta et al., 2006) and as anti-inflammatory agents (Srivastava et al. 1999, 2002; Gdupi et al, 1996; Shalabh et al., 2006). Patel and Mehta (2006) carried out the synthesis of azetidinone and thiazolidinone derivatives from 2-amino-6-(2-naphthalenyl)thiazolo[3,2-d]thiadiazole. Singh (2004) has also reviewed beta latcams in the new millennium, that is, monobatcams and carbapenems. Wang et al. (2009) synthesized fourteen derivatives of 2-azetidinones and reported for cholesterol absorption inhibitory action. Singh et al. (2007) have prepared some new 2-azetidones from N-(salicylidene) amine and 2-diazo-1,2-diarylethanones.

The chemical structure of phenothiazine provides a most valuable molecular template for the development of agents able to interact with a wide variety of biological processes. Phenothiazine derivatives possess potential biological activities as antinociceptive (Gildasio et al., 2004) anticonvulsant (Mia et al., 1998), anti-tumour (Andreani et al., 1991), antimalarial (Martha et al., 2002), anti-tubercular (Aaron et al., 2007), anti-emetic (Manish et al., 2003), antihistaminic (Oliver et al., 2006) and antipsychotic (Bateman, 2003) agents. The synthesis of phenothiazines fused with azetidinones are not reported so for. Hence, it was thought worthwhile to synthesize new congeners by incorporating phenothiazine and azetidinone moieties in a single molecular frame work and to evaluate their antimicrobial and anti-inflammatory activity.

MATERIALS AND METHODS

All melting points were taken by open capillary tubes and were uncorrected. Thin layer chromatography was performed on precoated Silica Gel 60 F_{254} plates from E.Merck using chloroform and benzene as mobile phase (75:25) and visualized by exposure to iodine vapors. IR spectra were recorded on a Perkin Elmer IR spectrophotometer, using KBr pellets. ^{1}HNMR spectra were recorded on Bruker DRX 300 (300 MHz) NMR spectrophotometer in DMSO-d_6 using TMS as internal standard and Mass spectra on Jeol SX 102 (FAB) Mass spectrophotometer.

Synthesis of 2-chloro-1-(10H-phenothiazin-10-yl) ethanone (1)

Phenothiazine (0.1 mol) and chloroacetyl chloride (0.15 mol) in dry benzene were refluxed for 2.5 h on a water bath. The solid product was filtered off and recrystallized from benzene. The IR spectrum of compound 1 revealed a sharp strong absorption band around 1680 cm^{-1} due to the presence of the carbonyl function in the structure. The methylene protons resonate as singlet at δ 4.3 and the aromatic protons resonate as multiplets at δ 6.70-6.92 ppm, its mass spectra revealed a molecular ion peak at m/z 275 (M$^+$) corresponding to the molecular formula $C_{14}H_{10}ClNOS$.

Synthesis of 2-hydrazinyl-1-(10H-phenothiazin-10-yl) ethanone (2)

Compound 1 (0.1 mol) and hydrazine hydrate (0.1 mol) in benzene

was refluxed for 3 h on a water bath. Solid mass obtained was filtered and recrystallized from benzene. The IR spectrum of 2 showed the absence of acid chloride stretching frequency, instead it gave a band at 1657 cm^{-1} for carbonyl group and showing two broad bands in the region of 3300-3400 and at 3100-3400 cm^{-1} for NH$_2$ and NH frequencies, respectively. ^{1}H NMR spectrum of compound 2 exhibited signals at δ 10.12 and δ 4.62 ppm for -NH and -NH$_2$ (D$_2$O exchangeable) of hydrazide respectively. The structure was further confirmed by recording its mass spectra. It gave the molecular ion peak at m/z 271 (M$^+$) corresponds to molecular formula $C_{14}H_{13}N_3OS$.

Synthesis of Schiff's bases (3a-g)

A mixture of equimolar quantities of compound 2 (0.01 mol) and appropriate aryl aldehyde (benzaldehyde, 2, 4-dimethoxybenzaldehyde, anisaldehyde, 4-nitrobenzaldehyde, 4-chlorobenzaldehyde, 4-hydroxybenzaldehyde and 3, 4-dichlorobenzaldehyde) (0.01 mol) were dissolved in ethanol (95%). The contents were refluxed for a period of 3 h on a steam bath. The solid obtained was separated out and recrystallized from ethanol. The yield and melting point were reported in Table 1. The IR spectra of compounds 3a-g showed strong absorption bands for carbonyl group (1657 cm^{-1}), aromatic C-H Stretching (3100 cm^{-1}) and aromatic C=C Stretching (1600 and 1500 cm^{-1}). Compound 3d showed absorption bands for nitro group (1315 & 1515 cm^{-1}). Compound 3f showed absorption bands for hydroxyl group (3280 - 3450 cm^{-1}). ^{1}H NMR spectrum of compounds 3a-g showed a quartet for methine protons at δ 7.8 (1H, N=CH-R), multiplets at 6.9-8.1 (Ar-H). Compounds 3b and 3c showed a singlet at δ 3.82 due to the signals of methoxyl protons. Compound 3f showed a singlet for hydroxyl protons at δ 5. Compounds 3a-g gave molecular ion peak at m/z 359, 419, 389, 404, 393, 375 and 428 (M+) respectively for their corresponding molecular formulae (Table 2).

Synthesis of azetidinones (4a-g)

Triethylamine (0.01 mol) in 1,4-dioxane, chloroacetyl chloride (0.01 mol) was added drop wise to a solution of the compounds 3a-g (0.005 mol) and at room temperature. The reaction mixture was stirred for 30 min. The mixture was then refluxed for 3 h on a water bath. The solid obtained after removal of 1,4-dioxane was recrystallized from ethanol. The yield, melting point and spectral characterization of the compounds were reported in Tables 3, 4 and Figure 1.

Anti-tubercular activity

Synthesized compounds 4a-g were screened for antitubercular activity against Mycobacterium tuberculosis H$_{37}$Rv strain using Lowenstein-Jensen medium method (Cambau et al. 2000). Ten milligram of each synthesized compound was dissolved in 10 ml of dimethyl sulfoxide to get a concentration of 1000 µg/L. Further dilutions were made with dimethyl sulphoxide to get different concentrations such as 100, 10 and 1 µg/mL. 0.8 ml of each concentration was used for the study. To this, 7.2 ml of Lowenstein-Jensen medium was added. Pyrazinamide (M/s Sigma Chemical Co.) was used as the standard drug. The dilution of pyrazinamide was made with dimethyl sulphoxide to get different concentrations of 100, 10 and 1 µg/mL. 0.8 ml of each concentration was used for the study. A sweep from M. tuberculosis H$_{37}$Rv strain culture was discharged with the help of nichrome wire loop with a 3 mm external

Table 1. Physical and analytical data of newly synthesized compounds.

Compound	R	M.P. (°C)	Yield (%)	Rf value*	Molecular formula
3a	(phenyl)	228-230	72	0.78	$C_{21}H_{17}N_3OS$
3b	(dimethoxyphenyl, OCH$_3$, H$_3$CO)	168-170	62	0.67	$C_{23}H_{21}N_3O_3S$
3c	(OCH$_3$ phenyl)	198-200	68	0.67	$C_{22}H_{19}N_3O_2S$
3d	(NO$_2$ phenyl)	278-280	70	0.50	$C_{21}H_{16}N_4O_3S$
3e	(Cl phenyl)	240-242	60	0.70	$C_{21}H_{16}ClN_3OS$
3f	(OH phenyl)	290-292	59	0.57	$C_{21}H_{17}N_3O_2S$
3g	(Cl, Cl phenyl)	205-207	64	0.73	$C_{21}H_{15}Cl_2N_3OS$

*methanol and benzene as mobile phase, spot detection-Iodine vapour.

diameter, into a sterile distilled bijou bottle containing six 3 mm glass beads and 4 ml of sterile distilled water.

The bottle was shaken with the help of a mechanical shaker for 2 min. Then using nichrome wire loop, 3 mm external diameter, a loopful of suspension was inoculated on the surface of each of Lowenstein-Jensen medium containing the test compounds. Lowenstein-Jensen medium containing pyrazinamide as well as control were inoculated with Mycobacterium tuberculosis H$_{37}$R$_V$ strain. The inoculated medium was incubated at 37°C for 4 weeks. At the end of 4 weeks readings were taken and recorded in Table 5.

Anti-bacterial activity

Antibacterial activity was evaluated by agar cup plate method (Shanmugakumar et al., 2006). Nutrient agar medium was used for the study. After sterilization the nutrient agar medium was melted, cooled and inoculated with two Gram-positive organisms S. aureus, B. subtilis and two Gram-negative organisms E. coli, P. aeruginosa and poured into sterile Petri dish to get a uniform thickness of 6 mm. Cups were made out in the other plate using sterile cork borer (6 mm dia). The standard antibacterial agent streptomycin (100 µg L^{-1}), solvent control (0.5% v/v Tween 80) and the synthesized compounds in a concentration of 100 µg L^{-1} were added with the sterile micro pipette into each cup. The plates were then incubated at 37°C for 24 h and the diameter of zone of inhibition were measured and recorded in Table 5.

Anti-fungal activity

Anti-fungal susceptibility test was done by disc diffusion method (Mahoto et al., 2002) using Sabouraud's dextrose agar medium. After sterilization the medium was inoculated with Candida albicans and Aspergillus fumigatus. The standard antifungal agent clotrimazole (100 µg L^{-1}), solvent control (0.5% v/v Tween 80) and the synthesized compounds in a concentration of 100 µg L^{-1} were then added by sterile micro pipette. The plates were the incubated

Table 2. Spectral data of newly synthesized compounds.

Compound	IR (ν, cm⁻¹)		^{1}H-NMR in DMSO (δ, ppm)	Mass spectra m/z value
3a	1690	C=O	1.8 (d, 1H, N-CH-C)	359
	3454	N-H	6.6-7.1 (m, 13H, Ar-H)	
	3080, 1600, 790	Aromatic	10.1 (s, 1H, N=CH)	
	1569	N=CH-		
3b	1690 cm⁻¹	C=O	1.8 (d, 1H, N-CH-C)	419
	3454	N-H	3.82 (s, 6H, OCH₃)	
	3080, 1600, 790	Aromatic	6.9-8.0 (m, 11H, Ar-H)	
	1360	C-N	10.1 (s, 1H, N=CH)	
	1569	N=CH-		
3c	1690 cm⁻¹	C=O	1.8 (d, 1H, N-CH-C)	387
	3454	N-H	3.82 (s, 3H, OCH₃)	
	3080, 1600, 790	Aromatic	6.9-8.1 (m, 12H, Ar-H)	
	1360	C-N	10.1 (s, 1H, N=CH)	
	1569	N=CH-		
3d	1315, 1515	NO₂	1.8 (d, 1H, N-CH-C)	404
	1690 cm⁻¹	C=O	6.9-8.1 (m, 12H, Ar-H)	
	3454	N-H	10.1 (s, 1H, N=CH)	
	3080, 1600, 790	Aromatic		
	1569	N=CH-		
	1360	C-N		
3e	1690 cm⁻¹	C=O	1.8 (d, 1H, N-CH-C)	393
	3454	N-H	6.9-7.4 (m, 8H, Ar-H)	
	3080, 1600, 790	Aromatic	7.7 – 7.8 (m, 2H +2H, Ar-H) p-	
	1569	N=CH-	chloro phenyl ring	
	1360	C-N	10.1 (s, 1H, N=CH)	
3f	3280 - 3450	OH	1.8 (d, 1H, N-CH-C)	375
	1690 cm⁻¹	C=O	5 (s, 1H, OH)	
	3454	N-H	6.9-8.5 (m, 12H, Ar-H)	
	3080, 1600, 790	Aromatic	10.1 (s, 1H, N=CH)	
	1569	N=CH-		
	1360	C-N		
3g	1690 cm⁻¹	C=O	1.8 (d, 1H, N-CH-C)	428
	3454	N-H	6.9-8.4 (m, 11H, Ar-H)	
	3080, 1600, 790	Aromatic	10.1 (s, 1H, N=CH)	
	1569	N=CH-1360 C-N		

Table 3. Physical and analytical data of newly synthesized compounds.

Compound	R	M.P. (°C)	Yield (%)	Rf value*	Molecular formula
4a	(phenyl)	211-215	71	0.66	$C_{23}H_{18}ClN_3O_2S$
4b	(OCH₃, H₃CO substituted phenyl)	160-163	64	0.59	$C_{25}H_{22}ClN_3O_4S$
4c	(OCH₃ substituted phenyl)	180-187	67	0.63	$C_{24}H_{20}ClN_3O_3S$
4d	(NO₂ substituted phenyl)	195-197	65	0.56	$C_{23}H_{17}ClN_4O_4S$
4e	(Cl substituted phenyl)	117-120	60	0.73	$C_{23}H_{17}Cl_2N_3O_2S$
4f	(OH substituted phenyl)	209 -213	64	0.76	$C_{23}H_{18}ClN_3O_3S$
4g	(Cl, Cl substituted phenyl)	173-175	55	0.72	$C_{23}H_{16}Cl_3N_3O_2S$

*methanol and benzene as mobile phase, spot detection-Iodine vapour.

at 37°C for 24 h and the diameter of zone of inhibition were measured and recorded in Table 5.

Acute toxicity study

This involves the estimation of the median lethal dose (LD_{50}), which is the dose that will kill 50% of the animal population within 24 h post treatment of the test substance. The method of Miller and Tainter (1944) was adopted. The animals used for the studies were in accordance with principles of laboratory animal care and were approved by Institutional animal ethical committee. Swiss Albino mice were starved of feed but allowed access to water 24 h prior to the study and were then grouped (five mice per group). They were treated intraperitoneally with different doses of the test compounds (200, 400, 600, 800 and 1000 mg kg⁻¹). The animals were then observed for 24 h for any behavioral effects such as nervousness, excitement, dullness, in-coordination or even death. The LD_{50} was estimated from the geometric mean of the dose that caused 100% mortality and the dose which caused no lethality at all.

Anti-inflammatory activity

The anti-inflammation activity was evaluated by carrageenan induced paw edema method (Winter et al., 1962). Albino rats of Wistar strain weighing 100-200 g of either sex were divided into nine groups each of six animals. The animals were maintained under normal environmental conditions. They were fed ad libitum with standard feed and water. Tween 80 suspension (0.5% v/v) of the test compounds were administered intraperitoneally in a dose of 100 mg kg⁻¹. The control group was given only 0.5% v/v Tween 80 (0.5 ml) suspension. One group was administered with phenylbutazone as standard, intraperitoneally in a dose of 100 mg kg⁻¹. After 30 min of the administration of test compounds and phenylbutazone, paw edema was induced in albino rats by injecting 0.1 ml of carrageenan (0.9% v/v in normal saline) suspension, into subplantar region of the left hind paw of each rat. After 3 h of carrageenan injection, the increase in paw volume was measured by a plethysmometer. The anti-inflammatory activity was measured in terms of percentage inhibition of edema and is analyzed statistically by students "t" test and recorded in Table 6.

RESULTS AND DISCUSSION

The present protocol describes a simple and efficient method for the synthesis of azetidinones by different Schiff bases of phenothiazines. It has been demonstrated

Table 4. Spectral data of newly synthesized compounds.

Compound	IR (v, cm⁻¹)		¹H-NMR in DMSO (δ, ppm)	Mass spectra m/z value
4a	1657	C=O azetidinone	1.8 (d, 1H, N-CH-C)	436
	3420	N-H	2.5 (d, 1H, C-CH-Cl)	
	3100, 1500,770	Aromatic	6.9-7.8 (m, 13H, Ar-H)	
	727	C-Cl	10.45 (s, 1H, CONH)	
4b	1657	C=O azetidinone	1.8 (d, 1H, N-CH-C)	496
	3420	N-H	2.5 (d, 1H, C-CH-Cl)	
	1360	C-N	3.82 (s, 6H, OCH₃)	
	3100, 1500,770	Aromatic	6.9-8.0 (m, 11H, Ar-H)	
	727	C-Cl	10.45 (s, 1H, CONH)	
4c	1657	C=O azetidinone	1.8 (d, 1H, N-CH-C)	466
	3420	N-H	2.5 (d, 1H, C-CH-Cl)	
	1360	C-N	3.82 (s, 3H, OCH₃)	
	3100, 1500,770	Aromatic	6.9-8.1 (m, 12H, Ar-H)	
	727	C-Cl	10.45 (s, 1H, CONH)	
4d	1315,1515	NO₂	1.8 (d, 1H, N-CH-C)	481
	1657	C=O azetidinone	2.5 (d, 1H, C-CH-Cl)	
	3420	N-H	6.9-8.1 (m, 12H, Ar-H)	
	1360	C-N	10.45 (s, 1H, CONH)	
	3100, 1500,770	Aromatic		
	727	C-Cl		
4e	1657	C=O azetidinone	1.8 (d, 1H, N-CH-C)	470
	3420	N-H	2.5 (d, 1H, C-CH-Cl)	
	1360	C-N	6.9-7.4 (m, 8H, Ar-H)	
	3100, 1500,770	Aromatic	7.7 – 7.8 (m, 2H +2H, Ar-H) p-chloro phenyl ring	
	727	C-Cl	10.45 (s, 1H, CONH)	
4f	3280 - 3450	OH	1.8 (d, 1H, N-CH-C)	452
	1657	C=O azetidinone	2.5 (d, 1H, C-CH-Cl)	
	3420	N-H	5 (s, 1H, OH)	
	1360	C-N	6.9-8.5 (m, 12H, Ar-H)	
	3100, 1500,770	Aromatic	10.45 (s, 1H, CONH)	
	727	C-Cl		
4g	1657	C=O azetidinone	1.8 (d, 1H, N-CH-C)	505
	3420	N-H	2.5 (d, 1H, C-CH-Cl)	
	1360	C-N	6.9-8.4 (m, 11H, Ar-H)	
	3100, 1500,770	Aromatic	10.45 (s, 1H, CONH)	
	727	C-Cl		

that cyclocondensation of Schiff bases with chloroacetyl chloride in triethylamine revealed with fairly high yields in a relatively short reaction time and easy work-up procedures. These conditions enable this method to be applicable for the synthesis of 2-azetidinone based heterocyclic. The purity of the synthesized compounds were confirmed by performing TLC, which gave single spot (Tables 1 and 2). IR absorption band at 1569 cm⁻¹

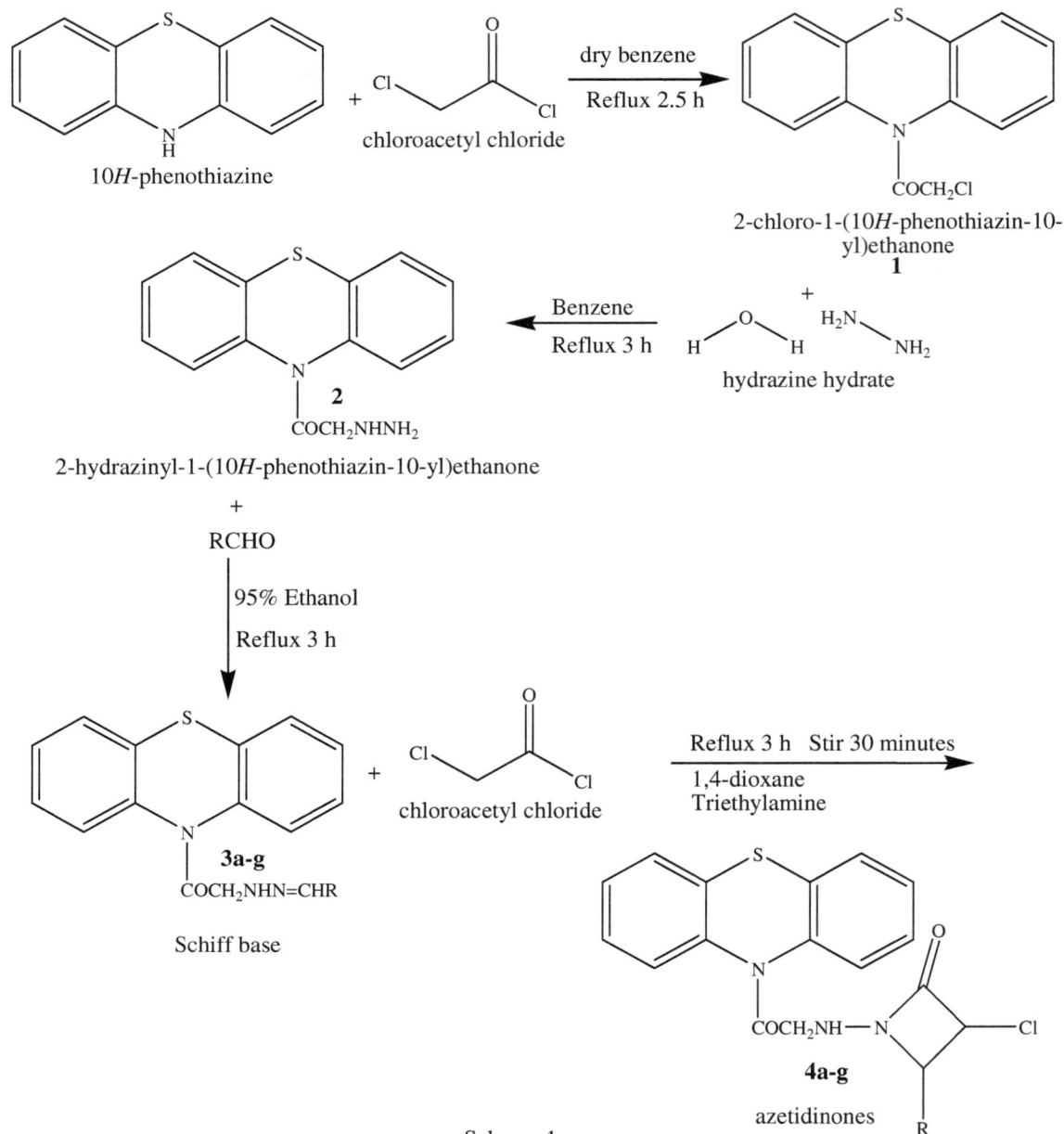

Scheme 1

Figure 1. Synthesis of azetidinones (4a-g).

for stretching vibration of -CH=N- and ^{1}H NMR signals for the presence of one imine proton (CH=N-) at 10.0357 ppm (1H, s), confirms the condensation of reactants to form Schiff-base. Similarly IR, ^{1}HNMR and mass spectraldata obtained were in correlation with synthesized azetidinones.

The synthesized compounds 4a-g was evaluated for their *in vitro* anti-tuberculosis activity against M. tuberculosis strain H$_{37}$Rv at 1, 10 and 100 µg mL^{-1} concentration. Pyrazinamide was used as the standard drug for comparison. The antitubercular data revealed that the all synthesized azetidinones proved to be active against the test organism M. tuberculosis, H$_{37}$R$_V$ strain, at 100, 10 and 1 µg mL^{-1} levels.

The compounds tested for anti-bacterial activity proved to be effective particularly against the Gram-positive organisms like *S. aureus* and *B. subtilis*. None of the compounds produced inhibition zone against Gram-

Table 5. Antimicrobial activity data of the titled compounds.

| Compound | Zone of inhibition in mm | | | | | | Concentration in µg mL^{-1} | | |
| | Bacteria | | | | Fungi | | Mycobacterium tuberculosis | | |
	S.a	B.s	E.c	P.a	C.a	A.f	100	10	1
Control			No zone inhibition				++	++	++
Streptomycin	19	18	19	20	--	--	--	--	--
Clotrimazole	---	--	-	--	20	21	--	--	--
Pyrazinamide	--	--	--	--	--	--	-ve	-ve	-ve
4a	09	08	00	00	00	12	-ve	-ve	-ve
4b	10	08	00	00	00	12	-ve	-ve	-ve
4c	08	10	00	0	00	11	-ve	-ve	-ve
4d	08	09	00	00	00	09	-ve	-ve	-ve
4e	15	18	00	00	00	10	-ve	-ve	-ve
4f	09	07	00	00	00	11	-ve	-ve	-ve
4g	09	08	00	00	00	12	-ve	-ve	-ve

S.a-*Staphylococcus aureus*, B.s-*Bacillussubtilis*, E.c-*Escherichia coli*, P.a-*Pseudomonas aeruginosa*, C.a- *Candida albicans*, A.f-*Aspergillus fumigatus*, -ve indicates complete inhibition of H$_{37}$ RV, ++ indicates intensive growth of *M. tuberculosis*.

Table 6. Evaluation of anti-inflammatory activity of the synthesized compounds by carrageenan induced paw edema method.

Compound (100 mg/kg^{-1})	Paw volume (mean ± SEM)	Percentage Inhibition after 3 h
Control	0.44 ± 0.190	---
Phenylbutazone	0.17 ± 0.020	61.07*
4a	0.42 ± 0.020	13.26*
4b	0.49 ± 0.030	14.92*
4c	0.38 ± 0.038	12.98*
4d	0.47 ± 0.030	14.32*
4e	0.44 ± 0.020	13.88*
4f	0.36 ± 0.030	12.44*
4g	0.52 ± 0.220	15.12*

*$p < 0.05$ represent significant difference when compared with control groups.

negative organisms like *E. coli* and *P. aeruginosa*. Compound 4e having 4-chloro phenyl substitution showed potent anti-bacterial activity, which is equipotent activity with the reference standard streptomycin against *B. subtilis*. It also exhibited a good anti-bacterial activity against *S. aureus*. The anti-fungal activity was tested against the fungal species *A. fumigatus* and *C. albicans* at 100 µg mL^{-1} concentration. All the compounds 4a-g showed almost equal anti-fungal activity against *A. fumigatus*, but none of the compounds produced inhibition zone against *C. albicans*.

All the doses employed for acute toxicity studies found to be non toxic, since there is no mortality observed up to the dose of 1000 mg kg^{-1}. The compounds 4a-g afforded 12-15% protection against carrageenan induced paw edema, where as the standard drug phenylbutazone under similar conditions showed 61% inhibition after 3 h

of carrageenan injection. All the tested compounds have demonstrated poor anti-inflammation properties.

REFERENCES

Aaron BB, Jay HK, Eric MF, Erica LA, Cristofer MP, Heather MW, Michael JR, Miguel, Sang SC and Yuehong W (2007). Synthesis and anti-tubercular activity of quaternized promazine and promethazine derivatives. Bio. Med. Chem. Lett. 17: 1346-1348.

Alcaide B, Almendros P (2001). 4-Oxoazetidine-2-carbaldehydes as useful building blocks in stereocontrolled synthesis. Chem. Soc. Rev. 30: 226-240.

Andreani A, Rambaldi M, Locatelli A, Aresca P, Bossa R, Galatulas I (1991). Potential antitumor agents XVIII (1). Synthesis and cytotoxic activity of phenothiazine derivatives. Eur. J. Med. Chem. 26: 113-116.

Anthony GMB, Gregory GG, Michal S, Sven JT (1987). Beta.-Lactam annulation using (phenylthio)nitromethane. J. Org. Chem. 52: 4693-4702.

Ashok K, Pratibha S, Ravi S, Pankaj M (2003). Synthesis and

anti-microbial screening of N-substituted-3-chloro-4-dithiocarbamate azetidin-2-ones. Indian J. Chem. 42: 416-420.

Bateman DN (2003). Antipsychotics. Medicine 31: 34-35.

Benito A, Alberto RV (1998). New intramolecular cyclization and rearrangement processes based on the radical aryl-aryl coupling of arylsubstituted 2-azetidinones. Tetrahedron Letters 36:6589-6592

Bhanvesh, DN, Desai KR (2004). Synthesis of some heterocyclic Schiff base and azetidinone compounds and their anti-bacterial activity. Asian J. Chem. 16: 1749-1452.

Brickner SJ, Gaikema JJ, Zurenko GE, Greenfield IJ, Manninen PR, Ulanowicz DA (1992). N-acyl 3-alkylidenyl- and 3-alkyl azetidin-2-ones: a new class of monocyclic beta-lactam antibacterial agents. 1. Structure-activity relationships of 3-isopropylidene and 3-isopropyl analogs. J. Antibiot. (Tokyo) 45: 213-226.

Cambau E, Truffot-Pernot C, Boulahbal F, Wichlacz C, Grosset J, Jarlier V (2000). Mycobaterial growth indicator tube versus the proportion method on Lowenstein-Jensen medium for antibiotic susceptibility testing Mycobaterium tuberculosis. Clin. Micro. Inf. Dis. 12: 938-942.

Choudhari BP, Mulwad VV (2003). Synthesis and antimicrobial screening of N-[6'-coumarinyl amino-3-chloro-4-aryl azetidine-2-ones. Ind. J. Heterocycl. Chem. 2: 197-200.

Deshmukh AR, Bhawal BM, Krishnaswamy D, Govande VV, Shinkre BA, Jayanthi A (2004). Azetidin-2-ones, synthon for biologically important compounds.Curr. Med. Chem. 11: 1889-1920.

Devendra K, Sharma RC (2002). Synthesis and antimicrobial profile of 5- imidazolinones, sulphonamides, azemethine, 2-azetidinones and formazans derived from 2-amine, 3-cyano-5(5'-chloro-3-methyl-1-phenyl pyrazol-4'-yl vinyl) 7,7'-dimethyl-6,7-dihydrobenzo(b) thiophenes. J. Ind. Chem. Soc. 79: 284-285.

Freddy H, Sushil Kumar M (2004). Synthesis of some azetidin-2-ones and thiazolidin-4-ones as potential antimicrobial agents. Ind. J. Heterocyclic Chem. 13: 197-200.

Gdupi RH, Jeeson M, Bhat AR (1996). Synthesis of 2(2'-carboxy-5'-nitro phenyl)-3,4-substitute azetidin-2-ones as anti-inflammatory and anti-microbial agents. Ind. J. Heterocycl. Chem. 6: 199-196.

Gildasio AS, Luciana MMC, Fernanda CFB, Ana LPM, Eliezer JB, Carlos AMF (2004). New class of potent anti-nociceptive and anti-platelet 10H-phenothiazine-1-acylhydrazone derivatives. Bio. Med. Chem. 12: 3149-3158.

Jiaxi Xu (2009). Stereoselectivity in the synthesis of 2-azetidinones from ketenes and amines via Staudinger reaction. Arkivoc. 4: 21-44.

Jose MC (2004). ß-Lactam Synthesis by the Kinugasa Reaction. Ang. Chem. Int. Ed. 43: 2198-2200.

Kagthara P, Teja S, Rajeev D, Parekh HH (2000). Synthesis of some 2-azetidinones as potential antitubercular agents, Indian J. Heterocycl. Chem. 10: 9-12.

Manish S, Norman BD, Mellar PD, Marie L, Ruth L (2003). Olanzapine as an antiemetic in refractory nausea and vomiting in advanced cancer. J. Pain. Symptom Manage. 25: 578-582.

Martha K, Nectarios K, Leann T, Leslie WD (2002). Novel phenothiazine antimalarials: synthesis, antimalarial activity, and inhibition of the formation of β-haematin. Biochem Pharmacol. 63: 833-842.

Maurizio B, Mauro C, Franco C (2000). The S-Thioester Enolate/Imine Condensation: A Shortcut to ß-Lactams. Eur. J. Org. Chem. 4: 563-572.

Mehta PD, Sengar NPS, Subrahmanyam EVS, Satyanarayana D (2006). Synthesis and biological activity studies of some thiazolidinones and azetidinones. Indian J. Pharm. Sci. 68: 103-106.

Mia LL, Ralph RR, Jesse MN, Raymond B, James PS, Angela MN, Carlynn AS, Scott AR (1998). Synthesis, characterization and anticonvulsant activity of enaminones. Part 5: Investigations on 3-carboalkoxy-2-methyl-2,3-dihydro-1H-phenothiazin-4[10H]-one derivatives. Bioorg. Med. Chem. 6: 2289-2299.

Miller LC, Tainter ML (1944). Estimation of the ED and its error by means of logarithmic-probit graph paper. Proc. Soc. Exp. Biol. Med. 58: 261-266.

More SV, Dongarkhadekar DV, Charan RN, Jadhav RWN, Bhusare P (2002). Synthesis and antibacterial activity of new Schiff base 4-thiazolidones and 2-azetidinones. J. Indian Chem. Soc. 79: 768-769.

Oliver HW, Alexander S, Frank LH, Wolfram H, Stephan G (2006). The human Ca^{2+}-activated K^+ channel, IK, can be blocked by the tricyclic antihistamine promethazine. Neuropharmacol. 50: 458-467.

Oza HB, Datta NJ, Joshi DG, Parekh HH (2003). Synthesis of some new-2-azetidinones as potential antimicrobial agents. 3: 275-276.

Padam K, Saksena RK (2003). Synthesis and antimicrobial activity of some new 2-phenyl-3-p(2'-methyl-3'-aryl-4'-oxo thiazolin-2'yl)phenyl quinazolin-4-ones and 2-phenyl-3-p(1'- ayrl-3-phthalimido-4'-methyl azetidin-2'-one-2'yl)phenyl quinazolin-4-ones, Indian J. Chem. 12: 315-318.

Palomo C, Aizpurua JM, Ganboa I, Oiarbide M (1999). Asymmetric Synthesis of beta-Lactams by Staudinger Ketene-Imine Cycloaddition Reaction. Eur. J. Org. Chem. 12: 3223-3235.

Pandey VK, Gupta VD, Upadhyay M, Singh VK, Tandon M (2005). Synthesis, characterization and biological activity of 1,3,4-substituted 2-azetidinones, Indian J. Chem. 44: 158-162.

Patel, HS, Patel VK (2004). Synthesis and anti-microbial activity of some new N(4,4'- disubstituted amino sulphonyl phenyl)-3-chloro-4-phenyl azetidine-2-ones, Oriental J. Chem. 17: 425-28.

Patel KH, Mehta AG (2006). Synthesis and antifungal activity of azetidinone and thiazolidinone derivatives of 2-amino-6-(2-naphthaleneyl)thiazolo[3,2-d]thiadiazole. Eur. J. Chem. 3: 267-273.

Pratibha S, Ashok K, Shikha S (2004). Bisheterocyclic Synthesis and anti-microbial studies of some biologically significant 2(N-3-chloro-4-substituted azetidinone-2)amino-4-hydroxy purines, Indian J. Chem. 43: 385-388.

Shalabh S, Tripti S, Rajan M, Saxena KK, Virendra Kishore S, Ashok K (2006). A study of anti-inflammatory activity of some novel α-amino naphthalene and β-amino naphthalene derivatives, Archiv der Pharmazie 339: 145-152.

Shanmugakumar SD, Amerjothy S, Balakrishna K (2006). Pharmacognostical, antibacterial and antifungal potentials of the leaf extracts of Pithecellobium dulce Benth. Phcog Mag. 2: 163-167.

Singh GS, Mbuka E, Pheko T (2007). Synthesis and antimicrobial activity of new 2-azetidinones from N-(Salicylidene) amines and 2-diazo-1,2-diarylethanones. Arkivoc pp. 80-90.

Singh GS, Mmolotsi BJ (2005). Synthesis of 2-azetidinones from 2-diazo-1, 2-diarylethanones and N-(2-thienylidene) imines as possible antimicrobial agents, Farmaco 60: 727-730.

Singh GS (2004). ß-Lactams in the New Millennium. Part-I: Monobactams and Carbapenams. Mini Rev. Med. Chem. 4: 69-72.

Srivastava SK, Srivastava S, Srivastava SD (1999). Synthesis of new carbazoyl-thiazol-2-oxo-azetidines antimicrobial, anticonvulsant and anti-inflammatory agents, Indian. J. Chem. 38: 183-187.

Srivastava SK, Srivastava S, Srivastava SD (2002). Synthesis of 1, 2, 4-triazolo-thiadiazoles and its 2-oxoaze-tidines as antimicrobial, anticonvulsant and anti-inflammatory agents, Indian. J. Chem. 41: 2357-2363.

Wang Y, Zhang H, Huang W, Kong J, Zhou J, Zhang B (2009). 2-Azetidinone derivatives: Design, synthesis and evaluation of cholesterol absorption inhibitors. Eur. J. Med. Chem. 44: 1638-1643.

Winter CA, Risley EA, Nuss GW (1962). Carrageenan-induced edema in hind paws of the rats as an assay for anti-inflammatory drugs, Exp. Biol. Med. 111: 544-547.

Van der Steen FH, Van Koten G (1991). Synthesis of 3-amino-2-azetidinones: A literature survey. Tetrahedron 47: 7503-7524.

12

Micropropagation of *Maerua oblongifolia*: A rare ornamental from semi arid regions of Rajasthan, India

Mahender Singh Rathore* and Narpat Singh Shekhawat

Biotechnology Unit, Department of Botany, Jai Narain Vyas University, Jodhpur, Rajasthan-342033 India.

A method for *in vitro* regeneration of *Maerua oblongifolia* (Capparaceae) from nodal shoot explants is outlined. Percent shoot response with multiplication rate (21.1± 2.33) shoots per explant (30 mm length) was achieved when cultured on semisolid Murashige and Skoog (MS) medium containing 3% sucrose and supplemented with 2.0 mgl^{-1} of (Benzylaminopurine) BAP + additives (25.0 mgl^{-1} adenine sulphate + 25.0 mgl^{-1} citric acid + 50.0 mgl^{-1} ascorbic acid). Further amplification of shoots was achieved when concentration of BAP was lowered (0.25 mgl^{-1}) and Kinetin (0.25 mgl^{-1}) along with 0.1 mgl^{-1} IAA was incorporated in the MS medium. A maximum of 58.1 ± 3.88 shoots of length 4-5 cm were obtained. The *in vitro* regenerated shoots rooted *in vitro* on half-strength MS medium containing 3.0 mgl^{-1} of IBA. About 85% of shoot rooted (4.04 ± 0.96 roots per shoot) on this medium. Other auxins such as NOA also promoted rooting but, the response in terms of percentage of rooting (75%) and shoot number (2.9 ± 1.59 roots per shoot) was low as compared to IBA. *In vitro* regenerated shoots of length 4 to 5 cm having 1-2 nodes were excise individually and pulse treated with 200.0 mgl^{-1} of IBA for 3.0 min for *ex vitro* rooting. After an initial acclimatization period of 2-3 months in a green house, about 80% plants were successfully hardened and were then transferred to earthen pots in nursery. Protocol developed is highly reproducible and economical as commercial agar and sugar cubes has been used. Multiplication rate is very high *in vitro* reported so far for this plant species. This standard protocol of mass propagation of *M. oblongifolia* eliminates the dependence on natural stands for seed production and will also serve for conservation of this threatened species.

Key words: Acclimatization, *ex vitro* rooting, *in vitro*, micropropagation, *Maerua oblongifolia,* soilrite.

INTRODUCTION

Maerua oblongifolia (Forsk) A. Rich in Guill and Perr. (Capparaceae) is locally known as *Orapa* (Bhandari, 1990). It is rare liana of the Thar Desert, and is a large, scabrous, unarmed, woody climbing shrub with pale-brown smooth bark. This plant bears greenish-white flowers, in corymbs on short lateral or terminal shoots.

*Corresponding author. E-mail: mahendersr@gmail.com.

Abbreviations: BAP, Benzylaminopurine; **IAA,** indole-3- acetic acid; **IBA,** indole-3-butyric acid; **Kn,** kinetin; **MS,** Murashige and Skoog (1962) medium; **NOA,** naphthoxyacetic acid; **PGR,** plant growth regulator; **RH,** relative humidity; **SFP,** spectral flux photon.

Fruits are pale-brown, 8 to 12 cm long, constricted between the seeds, forming an elongated, twisted and knotted berry; each knot is one-seeded. This plant exhibits wide variation in fruit size and morphology (Rathore et al., 2005). *M. oblongifolia* is a threatened plant in the area. Plant is woody climber and climbs over *Prosopis cineraria, Maytenus emarginata, Tecomella undulata, Salvadora* spp. and resembles *Cocculus* species. Plants produce aromatic flowers during summers. The ripe fruits are sugar rich, sweet with high calorific value and are rarely seen as these are eaten by squirrels and birds. *M. oblongifolia* due to its attractive aromatic flowers can be developed as garden and ornamental plant. *M. oblongifolia* is highly drought and high temperature resistant therefore it can be a potential target for gene prospecting. Plant provides shelter and

Nodal Shoot Explant

↓

Surface Sterilization

↓

Bud breaking

↓

Amplification of shoots
(By repeated transfer)

↓

Nodal segments of *in vitro* produced shoots

↓

Subculturing
MS medium + additives

↓

Basal clump further divided and
amplified on same medium

↓

Individual shoot pulsed with root inducing PGRs

↓

Transferred to glass bottles containing soilrite

↓

Shoots rooted *ex vitro* in green-house

↓

Plantlets acclimatized in green-house

Hardened plants

↓ Field transfer

Plants growing profusely and flowered

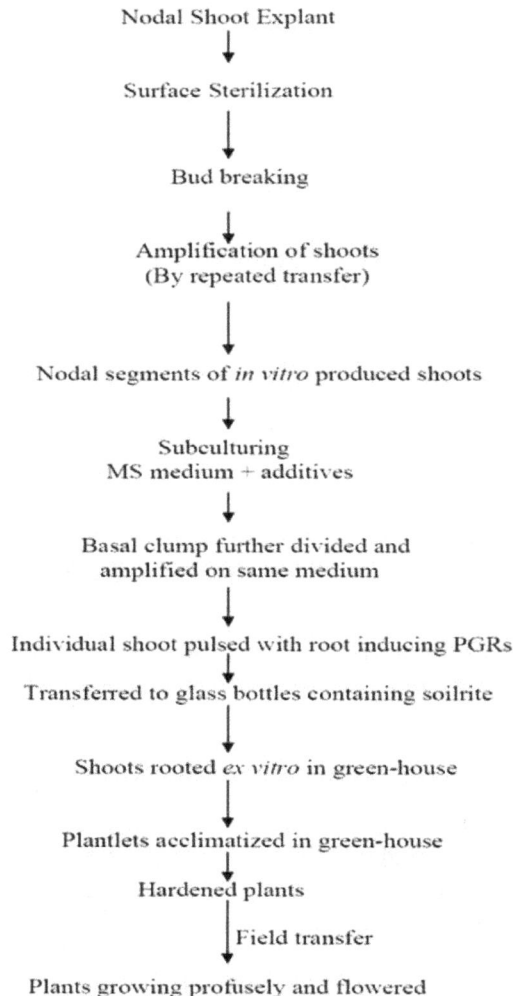

Figure 1. Schematic diagram of different steps performed for micropropagation.

food to birds and animals (Bhandari, 1990). This plant always grow in association with some specific tree species, as the population of these trees is decreasing day by day due to their over exploitation for fodder, food, timber and for medicinal purposes the plant of *M. oblongifolia* is also facing threat of extinction. Moreover, the plant propagates in nature through seeds. The viability of these seeds is less and most of the seeds are eaten up by the rodents. So, there is an urgent need for development of non conventional methods for mass propagation of *M. oblongifolia*. Plant tissue culture as means of non-conventional method of propagation is being applied for conservation and propagation of plant germplasm (Deora and Shekhawat, 1995; Rathore et al., 1993; 2007). We report protocol for mass scale clonal propagation of *M. oblongifolia*. The protocol developed is highly reproducible and the multiplication rate is very high as compared to previously reported studies (Rathore et al., 2005).

MATERIAL AND METHODS

Surface sterilization and inoculation of explant

Explants of *Maerua oblongifolia* were harvested from mature plant growing at the campus of Central Arid Zone Research Institute (CAZRI), Jodhpur, India. Explants were harvested in all quarters of the year 2005. Nodal shoot segments were used for culture initiation. Nodal shoot segments with 1-2 nodes (4 to 5 cm in length) were surface sterilized with aqueous solution of 0.1% w/v $HgCl_2$ (Mercuric chloride) solution for 4 to 6 min. The surface-sterilized explants were washed thoroughly 6 to 8 times with sterile water. These were then kept in chilled aqueous solution of ascorbic acid (0.1% w/v) and citric acid (0.05% w/v) for 5.0 min to prevent leaching of phenolics.

The surface sterilized explants were inoculated vertically on MS semisolid medium (Murashige and Skoog, 1962) containing a range of concentrations (0.5, 1.0, 2.0, 3.0, 4.0 and 5.0 mgl^{-1}) of BAP and Kn separately for the activation of axillary meristem. Initiation medium was supplemented with 25.0 mgl^{-1} adenine sulphate + 25.0 mgl^{-1} citric acid + 50.0 mgl^{-1} ascorbic acid as additives (Rathore et al., 2009). Cultures were incubated in culture rooms with constant temperature (28±2°C), 60% Relative Humidity and 25 to 30 μmol m^{-2} s^{-1} SFP for 12 h photoperiod day night cycle.

Multiplication of shoots

The cultures were multiplied by (i) Repeated transfer of mother explants along with regenerated shoots on fresh medium, and (ii) Subculture of *in vitro* produced shoots, were cut into nodal shoot segments with 1-2 nodes and cultured on fresh media. Different concentrations (0.10, 0.25, 0.50, 1.0 and 2.0 mgl^{-1}) of BAP and Kinetin were used separately. BAP (1.0 mgl^{-1}) in combination with varying concentrations of Kn (0.10, 0.25, 0.50, 1.0 mgl^{-1}) and 0.1 mgl^{-1} IAA was also incorporated in the MS medium (both liquid and semisolid) along with additives in order to optimize culture conditions for shoot amplification.

In vitro rooting of *in vitro* produced shoots

The cloned shoots rejuvenated *in vitro* were excised individually and transferred on to root induction media. The media evaluated for this purpose was half-strength MS semisolid medium containing NAA, NOA or IBA in concentration range of 0.5 to 4.0 mgl^{-1} along with 0.1% activated charcoal. These individual inoculated shoots were then placed under diffused light.

Ex vitro rooting and hardening

Healthy and strong *in vitro* produced shoots were harvested and washed with autoclaved water to remove adhered nutrient medium. Individual shoot was then pulse treated with various concentrations (50.0, 100, 200, 300, 400 and 500.0 mgl^{-1}) of either IBA or NOA for different time durations (1 to 10 min). These were inoculated on soilrite moistened with MS macro salts dissolved in distilled water. The bottles containing pulse treated shoots inoculated on soilrite were then capped with polycarbonate and placed near pad section of green house. After 7 to 8 weeks the plantlets were transferred to black polybags and were placed near fan section of green house for gradual acclimatization and hardening. Hardened plants were then transferred to earthen pots containing mixture of garden soil, organic manure, sand, and vermicompost in 3:1:1:1 ratio and were placed in nursery. The outline of whole process of *invitro* production of cloned plants is represented diagrammatically in Figure 1.

Figure 2A. *In vitro* buds break from nodal segment of M. oblongifolia on MS with 2.0 mgl-1 BAP + additives.

Data analysis

All experiments were set up in completely randomized block design (RBD) for single factor experiments (Compton and Mize, 1999) and repeated three times. Each treatment had minimum ten replicates. The observations on number of shoots, height of shoots and percentage of rooted shoots were scored after a regular time interval of 15 days. The data were subjected to the standard deviation of the mean and single factor ANOVA (Gomez and Gomez, 1984).

RESULTS AND DISCUSSION

In vitro culture initiation

The explants collected during the months of July-August (rainy seasons) responded the best *in vitro* as compared to explants harvested in any other months of the year. It was also recorded that mature nodal explants selected from the adult plants showed poor response as compared to nodal stem segments prepared from fresh shoots sprouts. It was recorded that (i) these mature/old stem segments carried recalcitrant contamination in cultures, and (ii) the response of explants in terms of bud breaking was very low even if these were harvested during rainy seasons. Among all these explants tested, juvenile shoot segment of length 4 to 5 cm with 2-3 nodes were found to be most suitable for culture initiation. These shoots when surface sterilized with 0.1% $HgCl_2$ for 5.0 min did not showed any kind of contamination in cultures. A 100% bud break occurred after 10 days of inoculation on MS semisolid medium. On MS medium supplemented with 2.0 mgl^{-1} BAP + additives, a maximum of 21.1±2.33 shoots of length 3.85±0.80 cm were obtained (Figure 2A and Table 1). This rate of shoot regeneration is very high as compared to previous report (Rathore et al., 2005). Fewer shoots (4.2±1.03) of length 2.81±0.82 cm were differentiated on MS medium with higher concentration of Kinetin along with additives. Thus, rejuvenation of instant meristem was achieved in *M. oblongifolia* by selection of explant type, and season for explant harvest and by treatment of cytokinins. Such treatments were recommended for micropropagation of woody trees (Aitken-Christie and Connett, 1992; Rathore et al., 1993; Shekhawat et al., 1993). Higher light conditions (25 to 30 μmol m^{-2} s^{-1}) favored early bud breaking which has not been reported by any of the previous workers for this plant.

In vitro multiplication of cultures

Shoots initiated from activation of axillary meristem were further amplified by two ways (i) repeated transfer of mother explant (Boulay, 1987; Franclet and Boulay, 1989; Deora and Shekhawat, 1995) or (ii) subculturing of *in vitro* rejuvenated shoots. Repeated transfer was carried out on MS semisolid medium supplemented with 0.25 mgl^{-1} of BAP and 0.1 mgl^{-1} IAA + additives. On this medium 17-21 shoots of length 2.0 to 3.0cm were obtained after 15-20 days (Figure 2B). This medium was supportive for shoot development up to fourth culture cycle. For subculturing of *in vitro* rejuvenated shoots liquid MS medium was also tried for further amplification of shoots but, that was not found suitable for multiplication and maintenance of cultures. Multiplication of cultures was achieved by subculturing shoots on MS semisolid medium supplemented with 1.0 mgl^{-1} of BAP + 0.25 of Kinetin + 0.1 mgl^{-1} IAA and additives proved to appropriate medium for mass multiplication of shoots. About 58.1 ± 3.38 shoots of length 4.19 ± 0.29 cm were obtained after 20-22 days of inoculation (Figure 2C and Table 2). Increase in IAA resulted in callus formation at the base of shoots. Among the cytokinins BAP was found superior to Kinetin (Bonga and Von-Aderkas, 1992) when taken along with additives and placed in 12 h photoperiod. This treatment was not reported by other workers (Rathore et al., 2005). It was also observed during the investigation that culture medium containing more than the average cytokinin, number and length of shoots was

Table 1. Effects of cytokinins on multiple shoots induction from nodal explant of *Maerua oblongifolia* on MS medium + additives.

	PGR concentration (mg/l)	Response (%)	Shoot number ± SD	Shoot length ± SD (cm)
Control		0	0.1±0.31	0.15±0.47
BAP	0.5	65	11.8±0.63	1.31±0.16
	1.0	85	13.8±0.78	2.12±0.62
	2.0	100	21.1±2.33	3.85±0.80
	3.0	100	17.7±1.56	2.49±0.58
	4.0	100	14.6±0.69	2.01±0.31
	5.0	95	13.9±0.99	1.78±0.41
Kinetin	0.5	45	0.5±0.70	0.61±0.80
	1.0	65	1.9±0.99	1.37±0.31
	2.0	85	3.2±1.13	1.98±0.48
	3.0	90	4.2±1.03	2.81±0.82
	4.0	95	2.1±0.56	1.82±0.42
	5.0	95	1.5±0.70	1.47±0.35
Computed F				
BAP	Replication		1.198265^{ns}	1.745911_{ns}
	Treatment		92.6342**	51.51383**
	CD		1.079681	0.4438736
Kinetin	Replication		1.451614_{ns}	2.02361^{ns}
	Treatment		32.2407**	26.90421**
	CD		0.7153642	0.4673683

^{ns}- Non-significant; *- significant ($p \leq 0.05$); **- highly significant ($p \leq 0.01$).

Figure 2B. Repeated transfer of explant on MS supplemented with 0.25 mgl-1 BAP and 0.1 mgl-1 IAA + additives.

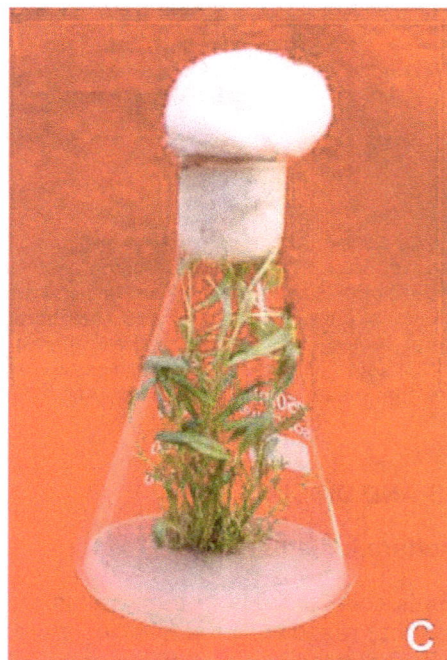

Figure 2C. Subculturing of *in vitro* produced shoots on MS supplemented with 1.0 mgl-1 of BAP + 0.25 of Kinetin + 0.1 mgl-1 IAA + additives.

Table 2. Amplification of shoots on MS medium supplemented with different concentrations of cytokinins + 0.1mgl $^{-1}$ IAA and additives.

	PGR concentration (mgl $^{-1}$)	Shoot number ± SD	Shoot length ± SD (cm)
Control	0.00	4.9±1.19	1.36±0.22
BAP	0.10	14.3±1.15	2.41±0.50
	0.25	21.1±2.76	3.61±0.66
	0.50	17.2±1.98	3.24±0.51
	1.00	13.8±1.75	2.97±0.28
	2.00	12.5±1.58	2.78±0.31
Kinetin	0.10	06.1±1.28	1.83±0.35
	0.25	06.8±1.13	2.06±0.30
	0.50	10.2±1.69	2.91±0.54
	1.00	13.2±1.54	3.23±0.35
	2.00	10.1±0.87	2.89±0.36
BAP 1.0 + Kinetin	0.10	48.2±3.70	3.57±0.29
	0.20	51.2±3.61	3.96±0.35
	0.25	58.1±3.38	4.19±0.29
	0.50	54.4±4.83	3.85±0.54
	1.00	51.5±4.19	3.64±0.17
Computed F			
BAP	Replication	0.7687483[ns]	1.582865[ns]
	Treatment	84.31517**	34.03079**
	CD	1.681016	0.3841529
Kinetin			
	Replication	0.8292583[ns]	0.7227636[ns]
	Treatment	56.34505**	37.20142**
	CD	1.193614	0.3406382
BAP 1.0 + Kinetin	Replication	0.5734175[ns]	0.5811815[ns]
	Treatment	96.61724**	142.802**
	CD	3.438248	0.2490872

[ns]- Non-significant; *- significant ($p \leq 0.05$); **- highly significant ($p \leq 0.01$).

reduced (Rathore et al., 1993; Shekhawat et al., 1993).

Rooting and hardening

In vitro rooting of individual shoots was achieved with ½ strength semisolid MS medium containing 3.0 mgl^{-1} of IBA + 0.1% activated charcoal. A 80% response was recorded in terms of *in vitro* rooting. About 2-3 roots of length 2 to 3 cm were obtained after 20-25 days of inoculation. The *in vitro* root induction was low on medium supplemented with NOA (1-2 roots per shoot). Activated charcoal is said to promote *in vitro* rooting as it provides darkness and adsorbs PGRs (Thomas, 2008). Diffused light (10 -20 μmolm^{-2}s^{-1} SFP) also favored *in vitro* root induction. Delayed rooting was observed under high light intensity (30 to 50 μmol m^{-2} s^{-1} SFP). *In vitro*

produced shoots also rooted under *ex vitro* conditions. Shoots of length 4 to 5 cm having 2-3 nodes were harvested and when pulse treated with 200.0 mgl^{-1} of IBA for 3.0 min, 85% of the shoots rooted after 20-25 days of this treatment. About 4.9 ± 0.87 roots of length 4.03 ± 0.49 cm were obtained. Delayed and poor rooting was recorded if shoots were pulse treated with NOA. IBA was found more suitable for *ex vitro* rooting of shoots (Rathore et al., 2010) (Table 3). Induction of rooting is affected by several intrinsic and extrinsic factors (Schiefelbein and Benfey, 1991; Shimizu-Sato et al., 2009; Wilson and Van Staden, 1990). The concentration of IBA and way of its treatment influenced the root induction (Van der Krieken et al., 1993). Higher concentrations (400 to 500 mgl^{-1}) of each PGR (IBA and NOA) produce a decrease in root number and reduce the root length. These rooted plantlets were then subjected to

Table 3. Effect of treatments of various root inducing auxins on *ex vitro* root induction from *in vitro* produced shoots of *Maerua oblongifolia*.

	PGR concentration (mgl^{-1})	Response (%)	Root number ± SD	Root length ± SD (cm)
Control		0	0.00±0.00	0.00±0.00
IBA				
	50	40	2.5±0.84	2.33±0.78
	100	55	2.9±0.73	2.93±0.43
	200	85	4.9±0.87	4.03±0.49
	300	80	4.4±0.96	3.96±0.57
	400	80	3.6±0.69	3.16±0.55
	500	75	3.2±0.63	2.74±0.71
NOA				
	50	20	1.4±0.51	0.86±0.35
	100	35	1.9±0.73	1.58±0.34
	200	55	2.5±0.84	2.47±0.33
	300	75	2.9±1.59	3.23±0.68
	400	80	1.9±0.87	2.61±0.36
	500	75	1.4±0.51	2.13±0.33
Computed F				
IBA				
	Replication		2.169284[ns]	0.9480646[ns]
	Treatment		53.65695**	58.07076**
	CD		0.6144642	0.5034854
NOA				
	Replication		0.5714286[ns]	1.260765[ns]
	Treatment		11.20408**	83.54815**
	CD		0.7888107	0.3441053

[ns] Non-significant; * significant ($p \le 0.05$); ** highly significant ($p \le 0.01$).

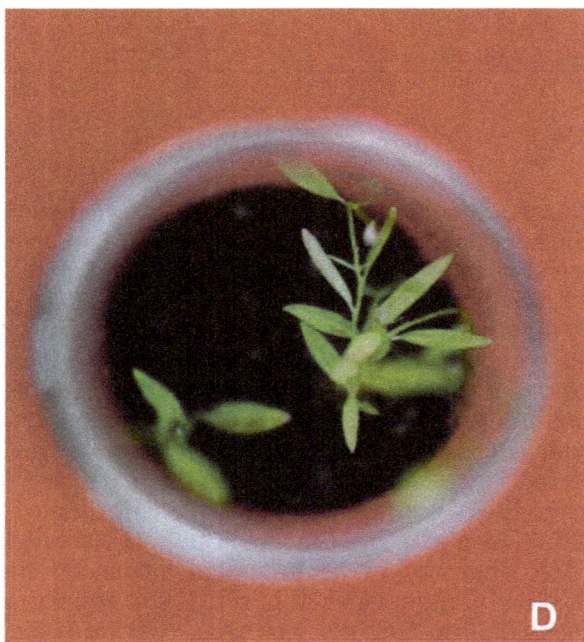

Figure 2D. Plantlets of *M. oblongifolia* under hardening phase in a green house.

different regions of temperature and humidity in a green house (Figure 2D). About 85% plantlets were successfully hardened (Figure 2E).

Significant findings

In micropropagation of *M. oblongifolia* some important attributes were recorded. The culture medium, type of cytokinins used in it, type of explants and harvesting period were found to be critical factors for *in vitro* regeneration of *M. oblongifolia*. The explants harvested during July-August months were found to be most suitable for culture establishment. Our findings hold the suggestion that in woody plants, axillary meristems from shoots of a severely pruned source plant are more amenable to physiological reinvigoration and respond better to tissue culture similar to other workers on tree species (Rathore et al., 2007). Analyses of data reveal that concentration of BAP in the culture initiation medium significantly increased the shoot numbers. On still higher (< 5.0 mgl^{-1}) concentration of this synthetic cytokinin, the difference in shoots length was found to be non-significant. Thus for both subculture and repeated transfer cytokinins

Figure 2E. Hardened and acclimatized plants in polybags.

requirement was significantly low. It was observed that, once the axillary meristem is activated the cytokinin requirement is reduced for proliferation (Rathore et al., 2009). Rate of shoot multiplication achieved in the present study is very high as compared to earlier reports (Rathore et al., 2005). This facilitates high volume shoot production per vessel and thus reducing the cost of production as commercial agar and sugars have been used. The shoots of *M. oblongifolia* rooted the best *in vitro* at diffused light conditions. *Ex vitro* rooting of shoots showed higher response than *in vitro* rooting in terms of root number length and survival rates. *Ex vitro* rooting has been reported for the first time in this plant. *Ex vitro* rooting is more beneficial as it reduces one step of hardening and acclimatization. The high rate of multiplication, successful *ex vitro* rooting, higher survival of cloned plantlets makes this protocol suitable for cloning and multiplication of selected germplasm of *M. oblongifolia*.

ACKNOWLEDGEMENTS

Research programmes in the Department of Botany are sponsored by the University Grants Commission (UGC), New Delhi under SAP-DSA Phase-II (1997-2001) and Phase-III (2002 to 2007). Micropropagation and Hardening facilities are established by the financial support from the Department of Biotechnology (DBT), Govt. of India. The Department of Science and Technology (DST), Govt. of India provided funds for infrastructure development in the Department of Botany under FIST programme.

REFERENCES

Aitken-Christie J, Connett M (1992). Micropropagation of forest trees, Transplant Production Systems. Kluwer Academic Publishers, The Netherlands, pp. 163-194.
Bhandari MM (1990). Flora of Indian Desert. MPS Repros., Jodhpur, India, pp. 42.
Bonga JM, Von-Aderkas P (1992). In vitro culture of Trees. Kluwer Academic Publishers., Dordrecht, The Netherlands.
Boulay M (1987). In vitro propagation of tree species. In: Plant Tissue and Cell Culture (Green CE, Somers DA, Hackett WP and Biesboer DD, Ed.) Alan R. Liss. Inc., New York, pp. 367-382.
Compton ME, Mize CW (1999). Statistical considerations for in vitro research: I-Birth of an idea to collecting data. In vitro Cell Dev. Biol-Plant. 35:115-121.
Deora NS, Shekhawat NS (1995). Micropropagation of Capparis decidua (Forsk.) Edgew-a tree of arid horticulture. Plant Cell Rep., 15: 278-281.
Franclet A, Boulay M (1989). Rejuvenation and clonal silviculture for Eucalyptus and forest species harvested through short rotation. Ed. Pereira JS and Lederberg, In: Biomass Production by Fast-Growing Trees. Kluwer Academic Publishers, The Netherlands, 267-274.
Gomez KA, Gomez AA (1984). Statistical procedure for agricultural research. John Wiley and Sons., New York.
Murashige T, Skoog F (1962). A revised medium for rapid growth and bioassays with tobacco tissue cultures. Physiol. Plant. 15: 473-497.
Rathore JS, Rathore MS, Shekhawat NS (2005). Micropropagation of Maerua oblongifolia-a liana of arid areas. Phytomorphology, 55: 241-247.
Rathore JS, Rathore MS, Singh M, Singh RP, Shekhawat NS (2007). Micropropagation of mature tree of Citrus limon. Indian J. Biotechnol., 6: 239-244.
Rathore MS, Rathore Mangal S, Shekhawat NS (2010). Ex vivo implications of phytohormones on various in vitro responses in Leptadenia reticulata (Retz.) Wight.& Arn.-An endangered plant. Environ. Exp. Bot., doi: 10.1016/ j.envexpbot.2010.05.009.
Rathore MS, Shekhawat NS (2009). Micropropagation of Pueraria tuberosa (Roxb. Ex Willd.) and determination of puerarin content in different tissues. Plant Cell Tissue Org. Cult., 99(3): 327-334.
Rathore TS, Deora NS, Shekhawat NS, Singh RP (1993). Rapid micropropagation of a tree of arid forestry Anogeissus acuminata. Biol. Plant., 35: 381-386.
Schiefelbein JW, Benfey PN (1991). The development of plant roots: New Approaches to underground problems. Plant Cell., 3: 1147-1154.
Shekhawat NS, Rathore TS, Singh RP, Deora NS, Rao SR (1993). Factors affecting in vitro clonal propagation of Prosopis cineraria. Plant Growth Reg., 12: 273-280.
Shimizu-Sato S, Tanaka M, Mori H (2009). Auxin-cytokinin interactions in the control of shoot branching. Plant Mol. Biol. 69: 429-435.
Thomas TD (2008). The role of activated charcoal in plant tissue culture. Biotechnol Adv., 26: 618-631.
Van der Krieken WM, Breteler H, Visser MHM, Mavridou D (1993). The role of the conversion of IBA into IAA, on root regeneration in apple: Introduction of a test system. Plant Cell Rep., 12: 203-206.
Wilson PJ, Van Staden J (1990). Rhizocaline, Rooting Co-factors, and the Concept of Promoters and Inhibitors of Adventitious Rooting-A Review. Ann. Bot., 66: 479-490.

13

Fibrin sealant as scaffold can be a suitable substitute to autograft in short peripheral nerve defect in rats

Z. Jamalpoor[1,5], M. Ebrahimi[2,6], N. Amirizadeh[3], K. Mansoori[4], A. Asgari[5] and M. R. Nourani[1,2]*

[1]Trauma Research Center, Baqiyatallah University of Medical Sciences, Tehran, Iran.
[2]Chemical Injury Research Center, Baqiyatallah University of Medical Sciences, Tehran, Iran.
[3]Iranian Blood Transfusion Organization, Tehran, Iran.
[4]Department of Physical Medicine and Rehabilitation, Shafayahyaian Rehabilitation Hospital, Iran University of Medical Sciences, Tehran, Iran.
[5]Department of Physiology, Baqiyatallah University of Medical Sciences, Tehran, Iran.
[6]Department of Organ Anatomy, Yamaguchi University Graduate School of Medicine, Ube, Japan.

There is considerable evidence that peripheral nerves have the potential to regenerate in an appropriate microenvironment. In this study, the process of nerve regeneration through a fibrin scaffold was examined. 45 male Wistar rats were randomly divided into one control, Autograft (Auto) and two experimental groups, Epineurum (Epi) and Fibrin scaffold (Fib). Right sciatic nerve was exposed of which 5 mm was cut. The nerve defect was then bridged with a nerve autograft, empty epineurium and fibrin scaffold in the corresponding groups. All animals were examined one, three and five weeks after the operation to evaluate nerve regeneration and functional recovery employing light microscopy and walking track analysis, respectively. The gastrocnemius muscle contractility was also examined at 35th day post surgery in all groups using electromyography (EMG). Histological, functional evaluation and EMG evidences show that the nerve regeneration in both groups Auto and Fib were statistically equivalent and superior to that of Epi group ($P<0.01$). The present findings indicate that a fibrin scaffold enhances nerve regeneration as effective as a nerve autograft.

Key words: Sciatic nerve, fibrin scaffold, electromyography, autograft.

INTRODUCTION

The peripheral nervous system, in contrast to the central nervous system, has the ability to spontaneously regenerate injured axons (Jubran and Widenfalk, 2003). However, functional recovery is achieved upon complete axonal regeneration, which includes remyelination and reinnervation of the appropriate muscle and sensory targets (Millesi, 1984). After nerve crush, regeneration is usually successful, because the continuity of the endoneurial tubes is preserved. In contrast, there is limited axonal growth across gaps that result from complete nerve transection (Haastert et al., 2006). Nerve autografts are currently considered as the gold standard technique for the repair of peripheral nerve injuries; however, due to the disadvantages related to autograft (e.g., formation of neuroma, limitations in availability of donor site and donor site morbidity), interest to investigate other means of repairing peripheral nerve lesions has increased (Dellon and Mackinnon, 1988). Biomaterials such as veins, arteries, silicone, or polyglycolic acid allow relatively good nerve regeneration, especially in short nerve defects (Auba et al., 2006). Narakas (1988) revived the use of fibrin in nerve repair. Since then, its use has steadily gained popularity amongst the peripheral nerve surgeons. In the repair of

*Corresponding author. E-mail: r.nourani@yahoo.com.

rat median nerve, fibrin sealants produced less inflammatory response and fibrosis, better axonal regeneration, and better fiber alignment than the nerve repairs performed with microsuture alone. In addition, the fibrin sealant techniques were quicker and easier to use (Ornelas et al., 2006). Bozorg et al. (2005) have reported promising results with fibrin sealant in the repair of facial nerve in human beings.

Subsequent studies clarify the application of fibrin sealant as a glove but in this study the main aim was to examine the efficacy of fibrin sealant as scaffold in repair of cut rat sciatic nerve, as well as a comparison with the use of empty epineurium and nerve autograft.

MATERIALS AND METHODS

Animal grouping

The study was performed in 45 male Wistar rats, each weighing 200 to 250 g. The animals were randomly divided into three equal groups and underwent right sciatic nerve cut followed by repair using one of the following methods: Nerve autograft (group Auto), empty epineurium (Epi) and fibrin scaffold (Fib). In all animals, a five millimeter nerve gap was bridged at the time of surgery. The animals were then sacrificed one, three and five weeks after surgery for histological assessment. Therefore, groups of five animals underwent repair for 5 mm gap length at each time point. For surgical procedures and electrophysiological studies, the rats were anesthetized with a mixture of ketamine (60 mg/kg) and xylazine (10 mg/kg) given by intraperitoneal injection, and repeated as needed.

Preparation of fibrin scaffold

Fibrin albeit a filling material elsewhere, it serves as a scaffold in the present study due to its three dimensional (3D) structure.

Plasma was from human source. It is not necessary to remove plasma from cryoprecipitate and it should be dissolve into 10 ml of remaining supernatant plasma. Cryoprecipitate is a well defined blood product and is rich in fibrinogen. After addition of thrombin into the concentrated fibrinogen, fibrin would be formed. Concentration of fibrinogen could be adjusted to an optimum consistency. The fibrin that is formed from effect of thrombin on fibrinogen can be used as a biologic semi-solid scaffold and a few studies showed that it can support growth of cells. Prepared thrombin and fibrinogen were aliquoted and stocked. Whenever we needed them, we thawed and mixed them into a glass tube with 2 mm diameter therefore the consistency was the same.

The production of fibrin scaffold had four preparatory steps: Plasmaphoresis, cryo, fibrinogen and thrombin preparation.

Plasmaphoresis

About 230 ml of human plasma, obtained from donated blood from Iranian Blood Transfusion Organization was divided into three bags using plasmaphoresis. An anti-coagulant was added according to hematocrit level, from 1:12 to 1:8 prior to plasmaphoresis. The first bag received 50 ml of the end product and the second bag, received the rest. The bags were then frozen at -80°C and kept at -20°C for further analysis.

Cryo preparation

The second bag containing 180 ml of plasma was defrosted in a vertical position at 4°C for at least 12 h. The plasma was then centrifuged at 3500 rpm for 10 min. The supernatant was carefully and completely transferred to the third fresh bag using an extractor; the remaining 10 to 15 ml was left in the second bag to dissolve the yellow cryo precipitate.

Fibrinogen preparation

Enough protamin sulfates was added to the Falcon tube containing an equal amount of cryo. Then, the tube was centrifuged at 3000 rpm for 10 min. The supernatant was precisely and completely removed and enough 0.2 M sodium citrate was used to dissolve the precipitated fibrinogen.

Thrombin preparation

The first bag bearing fresh frozen plasma was defrosted at 37°C. Thrombin Processing Device (TPD) was turned on and kept running at standby position for 5 min at 45°C before thrombin was injected. Eleven milliliters of plasma was introduced into the TPD, we also made sure it was running at 90°C prior to injection. The device was turned upside down seven times to mix the plasma with kaolin and the powder of TPD. The TPD was then laid quiet in a horizontal position at laboratory temperature for 30 min. Subsequently, it was mixed for another five minutes and left in horizontal position again. Next, the solution was mixed and the device put in a vertical position. Finally, the thrombin-containing plasma was extracted.

To prepare fibrin scaffold during surgery, an equal amount of thrombin and fibrinogen was mixed, 1 ml each, to form a gel. Five minutes later, under aseptic condition, pieces of about 5 mm length × 2 mm thick of fibrin were then quickly obtained and added as scaffold to the sciatic gap already induced (Dresdale, 1985; Rock, 2007; Buschta, 2004; Weibrich, 2002; Zimmermann, 2003).

Surgical procedures

Once, the animals were acquainted with the animal house, after a 2-week stay, their right lateral thigh was shaved and the skin prepared with povidone iodone solution. The rat was placed in the prone position on a warming mat and under sterile condition a dorsolateral skin incision was made from the posterior thigh to the knee of the right hindleg and sciatic nerve exposed between the sciatic notch and the popliteal bifurcation by a gluteal muscle splitting incision. Under the operating microscope, after longitudinal incision in the epineurium, 5 mm of sciatic nerve below the sciatic notch (10 mm) was removed while epineurium was kept intact. In group Epi after cutting sciatic nerve, epineurium was sutured with 10-0 nylon suture. In group Fib, 5 mm sciatic gap was filled by fibrin scaffold (5 mm length and 2 mm diameter) and then epineurium was sutured with 10-0 nylon suture. In the group Auto, 5 mm of sciatic nerve distal to sciatic notch (10 mm) transected and then rotated for 180° before suturing it in place.

The skin wounds were closed with 4-0 nylon suture. After surgery, the rats were returned to individual cages for recovery and were permitted to mobilize freely with unlimited access to food and

Table 1. Electromyographic results of nerve repairs in different groups (Mean ± SD).

Group	EMAP (mV)	Latency (ms)
Auto	12.3 ± 3.2	1.44 ± 0.08
Epi	5.4 ± 0.89 [ab]	2.4 ± 0.55 [ab]
Fib	11.02 ± 3.6	1.5 ± 0.13

[a] denotes statistical significance P<0.01 compared to Auto group;
[b] denotes statistical significance P<0.01 compared to Fib group.

water.

Electromyographic assessment

To evaluate nerve regeneration, electromyographical measurements were performed 5 weeks after surgery in all three groups. The motor distal latency (DL) and the evoked muscle action potential (EMAP) of gastrocnemius muscle were measured using an electromyographic recorder (Biomed 3250). The sciatic nerve proximal to the site of cut was stimulated with an electric monophasic stimulus using needle electrodes. To reduce any possible interference, a ground electrode was placed inside the muscle adjacent to the nerve. The gastrocnemius response was recorded by cap electrodes placed on the gastrocnemius muscle. The distance between the site of stimulation and the muscle was 2 cm and this distance was kept constant in repeated studies.

Histomorphological study

One, three and five weeks postoperatively and after EMG assessment, the animals were sacrificed with an overdose of ketamine (200 mg/kg). Immediately after the operation site, the distal segment of sciatic nerve was harvested and fixed in a 2% glutaraldehyde solution at 4°C for 12 h. Then, the nerve was dehydrated in increased concentrations of ethanol, passed through propylene oxide and embedded in epon resin and cross-sections (0.5 μm), distal to the neurorrhaphy site, using a microtome (Leica, Germany). All specimens were stained with 0.1% (w/v) toluidine blue in preparation for light microscopy. Morphometry was performed with an image analysis program (Image_ J, Ver 1.42, http://rsbweb.nih.gov/ij/download.html) (Abramoff et al., 2004). Video images were obtained with a digital camera (Nikon, Ds-Fil-L2, Japan) attached to a light microscope (Nikon, 50i, Japan). By using a modified version of Etho method, a manual count of myelinated fibers (MFs) was undertaken for five randomly selected square area (total = 0.02 mm^2) (Eto et al., 2003). These counts were then averaged to produce a mean estimate of myelinated fibers per one squared millimeter field.

Walking-track analysis

Walking-track analysis was performed 7, 21 and 35 days after the operation. After dipping the hindlimbs in ink, each rat was made walk down a 130×25 corridor freely. The foot prints were then analyzed to measure (1) distance from the heel to the third toe, called as the print length (PL); (2) distance from the first to the fifth toe, the toe spread (TS); and (3) distance from the second to the fourth toe, the intermediate toe spread (ITS) (De Medinaceli and Wyatt, 1982). All these measurements were taken from the experimental and normal sides. The measurements were used to calculate the factors as follows: Print length factor (1) (PLF) = (EPL-NPL)/NPL; (2) toe spread factor (TSF) = (ETS-NTS)/NTS; (3) intermediate toe spread factor (ITF) = (EIT-NIT)/NIT. These factors were then incorporated into Bain sciatic functional index-formula (Bain et al., 1989). SFI = -38.3 × PLF + 109.5 × TSF + 13.3 × ITF - 8.8. A SFI value of 0 is considered normal; a SFI of -100 was indicated to be total impairment, as it would be for a complete transaction of the sciatic nerve. If there is functional recovery due to regeneration techniques, the SFI should increase from -100 towards 0. So, SFI can be used as a parameter for measuring post-injury functional recovery.

Statistical analysis

In this study, all numerical data are presented as mean± standard deviation (Mean ± SD). Results were tested for statistical differences among the groups by using two-way ANOVA, followed by Tukey test as a post hoc test.

RESULTS

Electromyographic finding

By 35 days post repair, the EMAPs from the gastrocnemus muscle were reported in the groups of nerve repair with interposition autograft (Auto), empty epineurium (Epi) and fibrin scaffold (Fib). The difference between Fib and Auto groups with Epi group was significant (p<0.01), but no statistical difference was noted between Auto and Fib group. The results of the electromyographic studies of each group are shown in Table 1.

Histomorphometry finding

Regeneration of myelinated axons was observed in the distal segment of all groups. Examples of neural architecture are shown in Figure 2, distal to repair site for each group. The number of myelinated fibers (MFs) in all groups did not significantly differ one week after surgery. At 3 and 5 weeks post repair, histomorphometric analysis demonstrated a statistically significant differences (p>0.5) in number of MFs in nerve treated with the Fib and Auto groups compared to Epi group (P<0.01). There was no difference between the data of Auto and Fib group at these times (p>0.05). The average number of myelinated fibers in different treatment groups is shown in Figure 1.

Walking- track analysis

By the end of experiments, SFI values had increased in

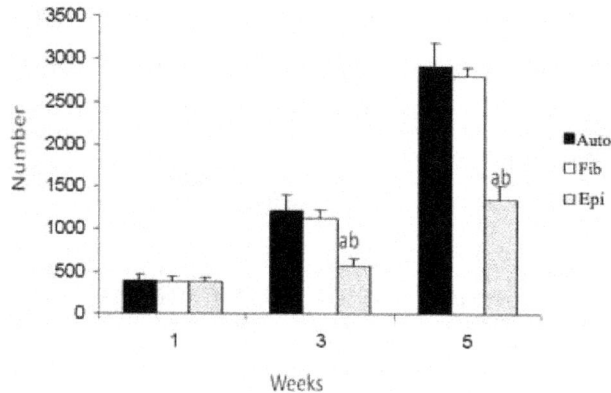

Figure 1. Number of MFs in Epineurum (Epi) and Fibrin scaffold (Fib) groups compared with Autograft (Auto) group. [a] denotes statistic al significance, P<0.01 compared to Auto group, and b denotes statistical significance P<0.01 compared to Fib group.

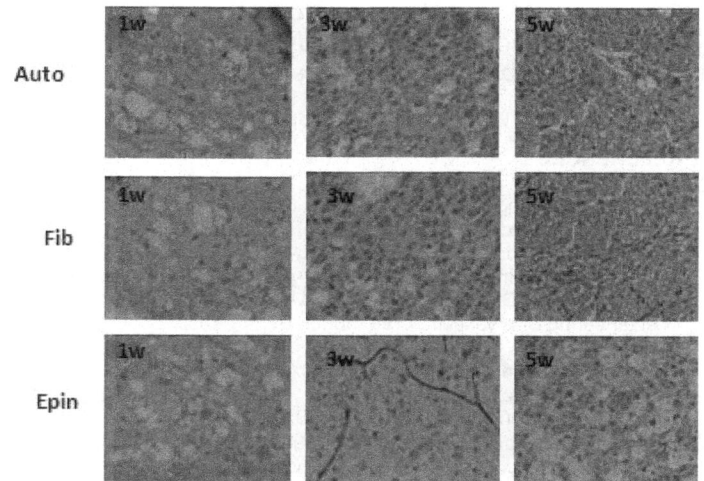

Figure 2. Photomicrographs of semithin sections at 1 week (1w), 3 (3w) and 5 weeks (5 w) post surgery in Auto., Fib. and Epi groups. Scale (bottom right) indicates of 50 µm (Toluidine blue staining, × 100).

Table 2. Results of walking track analysis in the three groups (Mean±SD).

	Weeks after surgery		
Group	1	3	5
Auto	-98.5±1.13	-85.2±3.46	-72.24±3.49
Epi	-98.1±1.1	-95.4±1.2 [ab]	-82.4±1.9 [ab]
Fib	-98.4±1.02	-88.6±5.4	-75.01±6.5

[a]denotes statistical significance, P<0.01 compared to Auto group. [b] denotes statistical significance P<0.01 compared to Fib group.

all groups, although not reaching the normal level. There was no significant difference in functional recovery among the three groups one week postsurgery; however, SFI value in Auto and Fib groups, as compared to Epi group significantly increased in weeks 3 and 5 after the operation (P<0.01). At the same weeks, there was no significant difference between SFI value in Auto and Fib group. The results of walking track analysis are presented in Table 2.

DISCUSSION

In this study, the effectiveness of the repair of 5 mm gap in the sciatic nerve with autograft, fibrin scaffold and empty epinurium was investigated. The results of this study showed that fibrin scaffold can repair gap peripheral nerve, as well as autograft. This finding is supported by the studies published by other researchers (Pittier, 2005; Galla, 2004; Martins, 2006). Previously fibrin was utilized as tissue glue to attach dissected nerve

fiber end to end instead of suture (Ornelas, 2006; Jubran and Widenfalk, 2003). Therefore, it is not surprising why many reported researchers regarding fibrin scaffold in repair of gap peripheral nerve could not be found. Fibrin supports angiogenesis and tissue repair. Also naturally, it contains sites for cellular binding and has been shown to have excellent cell seeding effects and good tissue development (Amrani, 2001; Ye et al., 2000). Fibrin scaffold is a three-dimensional structure that can mimic the extra cellular molecules (ECM) to promote and guide actively the newly formed cells and tissues (Aper et al., 2004). Simon and colleagues demonstrated that EMAP four weeks after repair of a 4 mm gap in sciatic nerve, by direct microsurgical suture, significantly greater than repair with nerve autograft and collagen based nerve guide conduit (Simon et al., 2004). It seems that some properties of collagen, such as variability in cross-linking density and fiber size, unpredictable enzymatic degradation, possible side effects and mineralization (Lee et al., 2005) delay nerve regeneration, however no toxic degradation or inflammatory reactions were detected in the fibrin scaffold (Ye et al., 2000; Aper et al., 2004). These benefits can enhance nerve regeneration.

Bioactive signaling molecules, like growth factor, play a significant role in the cellular growth, proliferation and differentiation in the ECM *in vivo*. Fibrin scaffold allows incorporation of growth factor, bioactive peptides and proteins and thus can also function as a kind of delivery system for added biologically active substances such as vascular endothelial growth factor and basic fibroblast growth factor (Jeon et al., 2005; Bhang et al., 2007).

In conclusion, the data in this study demonstrate that

fibrin scaffold can be as proper as autograft both functionally and morphometrically and thus fibrin scaffold can be a suitable alternative for autogtaft.

REFERENCES

Abramoff MD, Magelhaes PJ, Ram SJ (2004). Image processing with Image. J. Biophoton. Int., 11: 36-42.

Amrani DL, Diorio Y, Delmotte (2001). Wound healing: Role of commercial fibrin sealants. Ann. NY. Acad. Sci., 936: 566–579.

Aper T, Teebken G, Steinhoff A (2004). Use of a fibrin preparation in the engineering of a vascular graft model. Eur. J. Vasc. Endovasc. Surg., 28: 296–302.

Auba C, Hontanilla B, Arcocha J (2006). Peripheral nerve regeneration through allograft compared with autografts in FK506-treated monkeys. J. Neurosurg., 105: 602-609.

Bain J, Mackinnon S, Hunter DA (1989). Functional evaluation of complete sciatic, peroneal and posterior tibial nerve lesions rat. Plast. Reconstr. Surg., 83: 129-132.

Bhang SH, Jeon O, Choi CY, Kwon YH, Kim BS (2007). Controlled release of nerve growth factor from fibrin gel. J. Biomed. Mater. Res. A., 80: 998–1002.

Bozorg GA, Mosnier I, Julien N, El Garem H, Bouccara D, Sterkers O (2005). Long term functional outcome in facial nerve graft by fibrin glue in the temporal bone and cerebellopontine angle. Eur. Arch. Otorhinolaryngol., 262: 404-7.

Buschta,C, Dettle M, Funovics PT (2004). Fibrin sealant produced by Cryoseal FS System:product chemistry, material properties and possible preparation in the autologous preoperative setting. Vox sanguinis, 86: 257-262

De Medinaceli L, Freed, WJ, Wyatt RJ (1982). An index of functional condition of rat sciatic nerve based on measurements made from walking tracks. Exp. Neural., 77: 634-643.

Dellon AL, Mackinnon SE (1988). An alternative to the classical nerve graft for the management of the short nerve gap. Plast. Reconstr. Surg., 82(5): 849–856.

Dresdale A (1985). Preparation of fibrin glue from single-donor fresh-frozen plasma. Surgery, 97: 750-754.

Eto M, Yoshikawa H, Fujimura H (2003). The role of CD36 in peripheral nerve remyelination after crush injury. Eur. J. Neurosci., 17: 2659-2666.

Galla TJ, Vedecnik SV, Halbgewachs J (2004). Fibrin/Schwann cell matrix in poly-epsilon-caprolactone conduits enhances guided nerve regeneration (an abstract). Int. J. Artif. Organs, 27(2): 127-136.

Haastert K, Christina M, Cordula M (2006). Autologous adult human Schwann cells genetically modified to provide alternative cellular transplants in peripheral nerve regeneration. J. Neurosurg., 104: 778-786.

Jeon O, Ryu SH, Chung JH, Kim BS (2005). Control of basic fibroblast growth factor release from fibrin gel with heparin and concentrations of fibrinogen and thrombin. J. Control Release, 105: 249–259.

Jubran M, Widenfalk J (2003). Repair of peripheral nerve transections with fibrin sealant containing neurotrophic factors. Exp. Neural., 181: 204-212.

Lee CH, Singla A, Lee Y (2005). Biomedical applications of collagen. Int. J. Pharm., 221: 1–22.

Martins RS, Siqueira MG, Ciro FS, Jose Pindaro PP (2006). Correlation between parameters of electrophysiological, histomorphometric and sciatic functional index evaluations after rat sciatic nerve repair. Arq. Neuropsiquiatr., 64(3-B): 750-756.

Millesi H (1984). Nerve grafting. Clin Plast Surg. 11: 105-113.

Narakas A (1988). The use of fibrin glue in repair of peripheral nerves. Orthop. Clin. North Am., 19: 187-198.

Ornelas L, Padilla L, Di Silvio M, Schalch P, Esperantes, Innfante RL, Bustamante JC, Avalos P, Varela D, Lopez M (2006). Fibrin glue: An alternative technique for nerve coaptation - PartII: Nerve regeneration and histomorphometric assessment. J. Reconstr. Microsurg., 22: 123-128.

Pittier RF, Sauthier JA, Hubbell H (2005). Hall, Neurite extension and in vitro myelination within three-dimensional modified fibrin matrices. J. Neurobiol., 63: 1–14.

Rock G, Neurath D, Semple E (2007). Preparation and characterization of human thrombin for use in thrombin glue. Transfusion Med., 17: 187-191.

Simon J, Christian K, Jeremy S (2004). A collagen-based nerve guide conduit for peripheral nerve repair: An electrophysiological study of nerve regeneration in rodents and non human primates. J. Comp. Neurol., 306(4): 685-696 .

Weibrich G, Kleis WK, Hafner G (2002). Growth factor levels in platelet-rich plasma and correlation with donor age, sex, and platelet count. J. Craniomaxillifac Surg., 30: 97-102.

Ye Q, Zund P, Benedikt S, Jockenhoevel SP, Hoerstrup S, Sakyama JA (2000). Fibrin gel as a three dimensional matrix in cardiovascular tissue engineering. Eur. J. Cardiothorac. Surg., 17: 587–591.

Zimmermann R, Arnold D, Strasser E (2003). Sample preparation technique and white cell content influence the detectable levels of growth factors in platelet concentrates. Vox Sang., 85: 2.

14

Ultrasonic analyses algorithm on an *ex vivo* produced oral mucosal equivalent

Frank Winterroth[1]*, Junho Lee[2], Shiuhyang Kuo[3], J. Brian Fowlkes[1,4], Stephen E. Feinberg[1,3], Scott J. Hollister[1] and Kyle W. Hollman[1,5]

[1]Department of Biomedical Engineering, University of Michigan, Ann Arbor, MI, USA.
[2]School of Dentistry, University of Michigan, Ann Arbor, MI, USA.
[3]Department of Oral and Maxillofacial Surgery, University of Michigan, Ann Arbor, MI, USA.
[4]Department of Radiology, University of Michigan, Ann Arbor, MI, USA.
[5]Sound Sight Research, Livonia, MI, USA.

This study examines the use of high-resolution ultrasound with an analyses algorithm to accurately monitor development of an *ex vivo* produced oral mucosal equivalent (EVPOME). We used ultrasonic profilometry to examine EVPOME development as seeded cells on its surface proliferate. As these engineered structures develop, seeded cells stratify from their differentiation and produce a keratinized protective upper layer. Some of these transformations could alter backscatter of ultrasonic signals and produce scattering of the signal similar to an unseeded scaffold. Developing non-invasive *in vitro* ultrasonic monitoring allows adjusting tissue cultivation in-process, accounting for biological variations in the development of the EVPOME.

Key words: Acoustic microscopy, oral mucosa, non-invasive assessment, tissue engineering, ultrasound.

INTRODUCTION

The need for soft tissue replacements of oral mucosa in cases of disease, injury, or defect is enormous. Hence, developing a practical and cost - effective engineered tissue device which is compatible with the patients is essential toward proper treatment of soft tissue conditions. We have found high-frequency acoustic microscopy to be an advantageous method to study the physical and anatomical characteristics of tissues, both engineered and natural.

This study is a complement to our previous examination: using scanning acoustic microscopy (SAM) to study changes in radiofrequency (RF) data between tissues which have undergone periods of elevated thermal incubation (stressed) to those which were incubated continuously at physiological temperature (unstressed) (Winterroth et al., 2011). Here, we examine these same

tissues using SAM, coupled with filtering and an analyses algorithm to effectively study their surface characteristics. The advantages of using SAM over conventional optical microscopy include being able to image the cells and tissues without doing any preparations which could potentially kill or alter the tissues; this provides a more accurate representation of the tissues' natural properties (Cohn et al., 1997a). It will also provide evidence as to the degree of differentiation which the cells undergo without chemically affecting their properties (Kolios et al., 2002; Saijo et al., 2004).

Previously, we used SAM to examine and compare surface characteristics between the commercially available cellular cadaveric dermis (AlloDerm®) and an in-house developed *ex vivo* produced oral mucosal equivalent (EVPOME) (Winterroth et al., 2009). We further copared changes in the RF data in EVPOME specimens undergoing differentiation, apoptosis, and keratinization by studying the reflectivity off their surfaces and analyzing the degree of surface irregularities (Winterroth

*Corresponding author. E-mail: fwinterr@umich.edu.

et al., 2009). We then compared these results to EVPOME specimens which underwent standard histological preparation and examination under optical microscopy: a strong linear correlation was found between the optical and SAM images when quantifying these surface characteristics. By correlating changes in the RF data to the EVPOME (and natural tissue cells in general) undergoing differentiation, apoptosis, and keratinization, we can better understand the physiological processes of these cells as they evolve.

For this study, we examined and compared the surface acoustic profile characteristics (ultrasonic profilometry) between two EVPOME specimens: one underwent a thermal elevation for 24 h (stressed) while the other was an unstressed control. Ultrasonic profilometry uses an analysis algorithm to remove any major debris from the scanned specimens – which could potentially alter the image of the scanned surface; it is a promising method to analyze tissue types and their respective characteristics on the basis of acoustic transmission and scatter properties (Matsuyama et al., 1989; Zuber et al., 1999).

MATERIALS AND METHODS

The protocol for harvesting human oral mucosal tissue was approved by a University of Michigan Internal Review Board. Details of the EVPOME development are similar to those described elsewhere (Winterroth et al., 2009; Izumi et al., 2004). Briefly, oral mucosa keratinocytes were enzymatically dissociated from the tissue sample, and a primary cell culture was established and propagated in a chemically-defined, serum- and xenogeneic products-free culture medium, with calcium concentration of 1.2 mM. The AlloDerm® was soaked in 5 µg/cm^2 human type IV collagen at 4°C overnight prior to seeding cells to assist the adherence of cells, then approximately 1.5 x 10^5 cells/cm^2 of oral keratinocytes were seeded onto the type IV collagen pre-soaked AlloDerm®. The composites of the keratinocytes and the AlloDerm® were then cultured, in the submerged condition at 37°C for 4 days to form a continuous epithelial monolayer. After 4 days, the equivalents were raised to an air-liquid interface to encourage cellular stratification and cultured for another 7 days, resulting in fully-differentiated, well-stratified epithelial layers on the AlloDerm®. At Day 11 post-seeding, EVPOME samples were collected for SAM imaging. On Day 9 day post-seeding, one set of EVPOME specimens were incubated at 43°C for 24 h, then switched back to 37°C for another 24 h (thermally stressed). Another set of EVPOMES were kept at 37°C up through Day 11 post-seeding (non-stressed). The culture times for both conditions – stressed and non-stressed - were equal. The experiments were conducted as single-blind studies: we received two specimens, labeled as "1" and "2" for each study set. We were not told which specimens were stressed or unstressed until after presenting and announcing the results from the SAM scans.

Details of the scanning and ultrasound system are similar to those described elsewhere (Cohn et al., 1997 b). The transducer's parameters are: 15 µm scanning step size in both the transverse and horizontal directions, a lateral resolution of 37 µm, an axial resolution of 24 µm, and a depth of field of 223 µm. Axial resolution is the resolution in the direction of propagation and is determined by the length of the ultrasound pulse propagating in the tissue; lateral resolution is the resolution orthogonal to the propagation direction of the ultrasound wave. Sampling along the Z-axis was performed at a rate of 300 mega samples/second. B-scan images were obtained by stepping the transducer element laterally across the

desired region. At each position, the transducer fired and an RF A-line was recorded. After repeated firings at one position, the transducer moved to the next, where the image was constructed from A-lines acquired at all lateral positions. Because of the transducer's short depth of field, a composite B-scan image (the 2D cross-sectional display based on the time required for return of the echo to the transducer) was generated from multiple scans at different heights.

Surface profilometry was determined by first finding the instance of threshold value, fitting and subtracting the planar surface, then calculating root-mean-squared (RMS) height. RMS was computed in time domain:

$$RMS = \sqrt{\frac{1}{n}\sum_n x^2(t)}$$

where n is number of $x(t)$ samples. An analysis algorithm was applied to all the scans performed for each of the specimens used in this study and to 6 scans of AlloDerm® specimens. After detecting the front surface of each specimen and subtracting any tilt, a filter was applied to determine the broad underlying surface of the specimen; this filter was then subtracted from the tilt removed surface. A second filter was then applied to the resulting surface to remove any possible noise.

RESULTS AND DISCUSSION

The 2D B-scans from the SAM and their comparative histology images show a clear partitioning of the keratin layer from the apical surface of the specimens under stressed conditions. Comparing the B-scans taken of EVPOME at 11 days post-seeding, there is a distinct reflection off of the surface (Figure 1a) – a result of the seeded keratinocytes differentiation - which is not found in the stressed EVPOME specimen (Figure 1b). The histology images of these same specimens- AlloDerm® and EVPOME at 11-days post seeding (Figures 1c and 1d, respectively) - verify these findings as the keratinized layer in the latter image shows a smooth surface, with less surface irregularities compared to the former image. SAM C-Scans performed on the unstressed and stressed EVPOMEs after removing any tilt and applying the aforementioned filters show greater surface homogeneity in the unstressed control; evident of the keratinized layer (Figure 1e). In contrast, the stressed EVPOME show greater variations on the surface (Figure 1f).

Removing any tilt (which potentially occurs when mounting and positioning the specimens) lowers the measured RMS heights for all specimens but it maintains the mean difference between the stressed and non-stressed specimens in SAM C-scans of their surface (Figures 1e and 1f); the same is true when adding the weak filter. However, adding the weak filter seems to slightly increase standard deviation; the filter parameters may require further adjustments. B-scans taken along the transverse axis of the stressed specimen where the surface variations are the highest show significant changes in the surface at the point where there is a significant increase in the RMS height (Figure 2), accounting

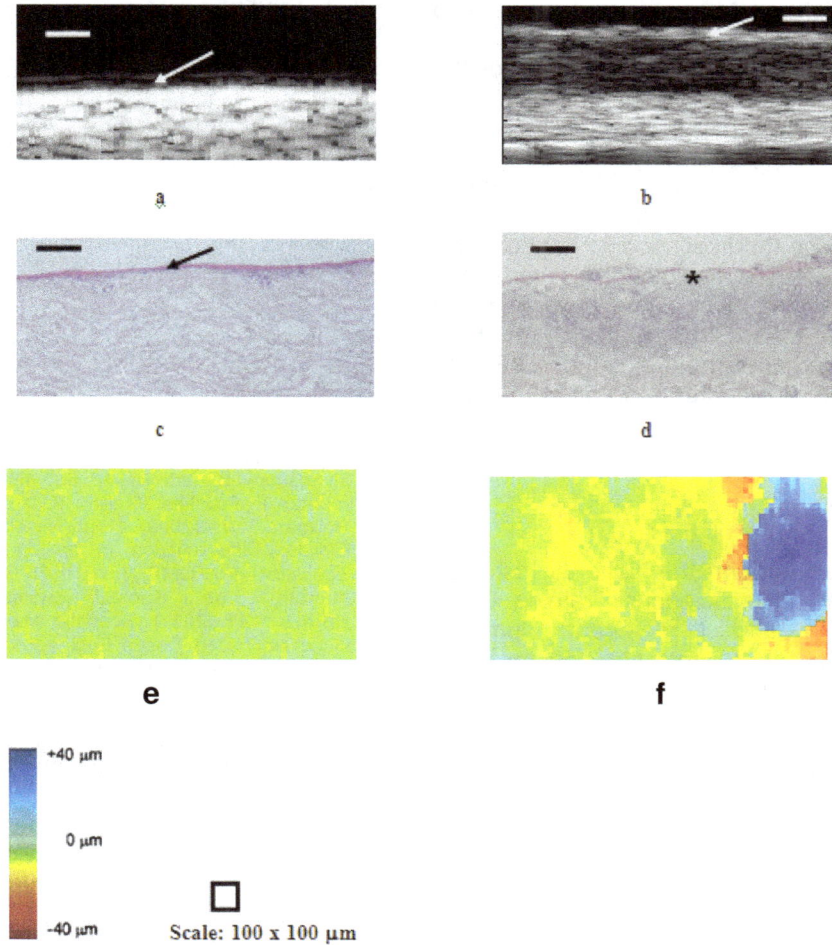

Figure 1. SAM 2D B-scans of EVPOMEs under unstressed (a) and stressed (b) conditions. Note the differences in the reflectivity on the surfaces (arrows). The histology images of the unstressed (c) and stressed (d) devices verify these findings: the keratinized layer (arrow) and the debris in the partitioned area (asterisk). The scale bars represent 100 µm. SAM C-scans of an unstressed (e) and stressed (f) EVPOME specimen. The images result from taking the image and removing any tilt, followed by fitting the planar surface and subtracting the result.

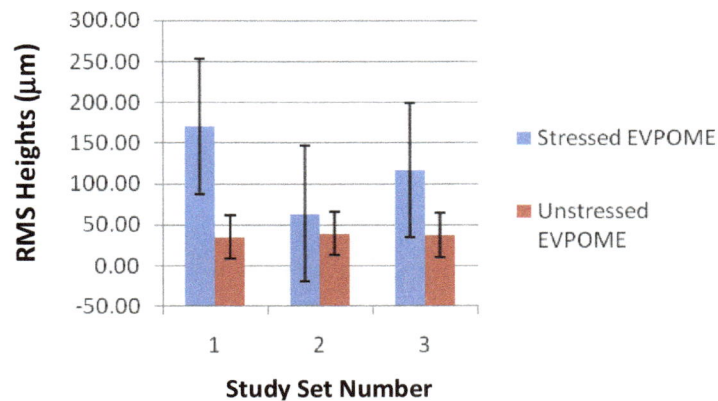

Figure 2. RMS variations in height for the stressed and unstressed EVPOME specimens for each set of studies conducted. The standard deviation of the scaffold as a percentage of the mean is 70.71.

for the higher variation in the RMS value for the stressed specimen.

This is verified by calculating the mean values for the height variations between the stressed and non-stressed specimens and determining their respective standard errors in all the study sets performed. The error bars represent the percent standard deviation of the mean. Because there are more RMS variations in the stressed specimens, there would be the likelihood for higher error (Figure 2). There is also the likelihood of the tissue samples tilting during their mounting process which would contribute to the increased RMS profiles, for both stressed and non-stressed specimens. Further EVPOME analyses will involve use of the filters and tilt corrections to validate these profilomtery findings.

ACKNOWLEDGEMENTS

This work was supported through the National Institutes of Health (NIH) Regenerative Sciences Training Grant Number 5T90DK070071 and NIH Grant Numbers R21EY018727, R01 DE13417, and NIH center core (P30) grant, EY007003. National Institutes of Health, Bethesda, MD. 20892. We gratefully acknowledge the NIH Resource Center for Medical Ultrasonic Transducer Technology at the University of Southern California (Los Angeles, CA. 90089) for designing and building the high frequency transducer used in this study.

REFERENCES

Cohn NA, Emelianov SY, Lubinski MA, O'Donnell M (1997a). An Elasticity Microscope. Part I: Methods. IEEE Trans. Ultrason. Ferr. Freq. Ctrl., 44(6): 1304-1319.

Cohn NA, Emelianov SY, O'Donnell M (1997b). An Elasticity Microscope. Part II: Experimental Results. IEEE Trans. Ultrason. Ferr. Freq. Ctrl., 44(6): 1320-1331.

Izumi K, Song J, Feinberg SE (2004). Development of a tissue-engineered human oral mucosa: from the bench to the bed side. Cells Tiss. Org., 176:134–152.

Kolios MC, Czarnota GJ, Lee M, Hunt JW, Sherar MD (2002). Ultrasonic Spectral Parameter Characterization of Apoptosis. Ultrasound in Med. Biol., 28(5): 589–597.

Matsuyama T, St Goar FG, Tye TL, Oppenheim G, Schnittger I, Popp RL (1989). Ultrasonic Tissue Characterization of Human Hypertrophied Hearts In Vivo with Cardiac Cycle-Dependent Variation in Integrated Backscatter. Circulation, 80: 925-934.

Saijo Y, Miyakawa T, Sasaki H, Tanaka M, Nitta S (2004). Acoustic Properties of Aortic Aneurysm Obtained with Scanning Acoustic Microscopy. Ultrasonics, 42: 695–698.

Winterroth F, Fowlkes JB, Kuo S, Izumi K, Feinberg SE, Hollister SJ, Hollman KW (2009). High-Resolution Ultrasonic Monitoring of Cellular Differentiation in an Ex Vivo Produced Oral Mucosal Equivalent (EVPOME). Proc. IEEE Bioultrasonics Conference.

Winterroth F, Lee J, Kuo SJ, Fowlkes JB, Feinberg SE, Hollister SJ, Hollman KW (2011). Acoustic Microscopy Analyses to Determine Good vs. Failed Tissue Engineered Oral Mucosa Under Normal or Thermally Stressed Culture Conditions. Ann. Biomed. Eng., 39(1): 44-52.

Zuber M, Gerber K, Erne P (1999). Myocardial Tissue Characterization in Heart Failure by Real-Time Integrated Backscatter. Eur. J. Ultrasound, 9: 135–143.

15

Challenges of bioengineering and endodontics

Anil Dhingra, Viresh Chopra and Shalya Raj*

Department of Conservative Dentistry and Endodontics, Subharti Dental College, Meerut (U.P.), Meerut, Uttar Pradesh India.

Stem cell research and scaffolding are now the buzz words in basic science pulp researches. The endodontic community needs to enhance its clinical understanding of the vital pulp and dentin, and embrace new treatment modalities. Several regenerative techniques have been described, while therapies involving stem cells, growth factors and scaffolds have been proposed. Each technique has its own advantages and disadvantages, although some techniques are still in a hypothetical stage or in an early stage of development. It has been accepted that the regenerative therapies could revolutionize the future endodontics with the synergitic confluence of advances in signaling pathways underlying morphogenesis and lineage stem/ progenitor cells by morphogens such as bone morphogenic proteins and synthetic scaffolds.

Key words: Stem cells, scaffolds, gene therapy, endodontics.

INTRODUCTION

Endodontology is the science that deals with endodontium. An endodontist is supposed to be an expert on diseases of the endodontium, and should be able to treat the diseases thereof. Presently, the most common therapy for the exposed and/or diseased pulp is total amputation. This crude therapy is unfortunate because today there are restorative materials that rarely require post retention for large restorations, that is, the root canal could still harbor a vital pulp if treated properly (Murray et al., 2007).

In endodontics, regenerative endodontic procedure refers to the regeneration of dentin pulp complex and even the whole tooth structure. Murray et al. (2007) defined it as a "biologically based procedure designed to replace damaged structures including dentin and root structures, as well as cells of the pulp dentin complex". Objectives of the regenerative endodontics can be outlined as regeneration of dentin or pulp-like tissue; regeneration of the damaged coronal dentin and regeneration of the reabsorbed root, cervical or apical dentin. To successfully achieve these, certain hurdles were outlined by Baum and Mooney in 2000. They threw

light on some other important issues like, till now, all the researches have been done on animals like swine, mice, dogs and rabbits. Hence, the dangers of transplant rejection by the human cells are still hovering on success. Ethical concerns are another debatable topic among the researchers. Inspite of the obstacles faced, tissue engineering offers exciting opportunities for innovative collaborative research efforts, integrating the fields of medicine, developmental biology and physical sciences (Baum and O'Colonnell, 1995).

STEM CELLS

Duailibi et al. (2006) define stem cells as "Quiescent cell populations present in low numbers in normal tissue, which exhibit the distinct characteristic of asymmetric cell division, resulting in the formation of two distinct daughter cells - a new progenitor/ stem cell and another daughter cell capable of forming a differentiated tissue".

The uniqueness of the tooth and its dentin pulp complex has a natural regenerative potential leading to the formation of tertiary dentin. Odontoblasts may survive mild injury, such as attrition or early caries, and secrete a reactionary dentin matrix. In cases of trauma with greater intensity, the pre existing odontoblasts may die leading to the recruitment and differentiation of new odontoblasts

*Corresponding author. E-mail: shalyabhatnagar@gmail.com

and synthesis of an atubular dentin (Duailibi et al., 2006).

Dental stem niches and other stem cell sources for the development of teeth *in vitro* or *ex vivo* have been discovered. As tooth formation results from epithelial-mesenchymal interactions, two different populations of stem cells have to be considered: epithelial stem cells, which will give rise to ameloblast, and mesenchymal stem cells that will form the odontoblasts, cementoblasts, osteoblasts and fibroblasts of the periodontal ligament. Thus, tooth engineering using stem cells is based on their isolation, association and culture (Murray et al., 2007).

ISOLATION OF VARIOUS DENTAL STEM CELLS

Mesenchymal stem cells (MSC)

MSC can be isolated from different sources. It was first described in the bone marrow to have been extensively characterized *in vitro* by the expression of markers such as: STRO-1, CD146 or CD44.

Stem cells from human exfoliated deciduous teeth (SHED)

The isolation of post-natal stem cells from an easily accessible source is indispensable for tissue engineering and clinical applications. Recent findings demonstrated the isolation of mesenchymal progenitors from the pulp of human deciduous incisors. These cells were named SHED (Miura et al., 2003).

Adult dental pulp stem cells (DPSC)

After a dental injury, dental pulp is involved in a process called reparative dentinogenesis, where cells elaborate and deposit a new dentin matrix for the repair of the injured site. It has been shown that adult dental pulp contains precursors capable of forming odontoblasts under appropriate signals; among these signals are the calcium hydroxide or calcium phosphate materials, which constitute pulp-capping materials used by dentists for common dental treatments (Gronthoss et al., 2000).

Stem cells from the apical part of the papilla (SCAP)

Stem cells from the apical part of the human dental papilla (SCAP) have been isolated and their potential to differentiate odontoblasts was compared to that of the periodontal ligament stem cells (PDLSC). SCAP exhibit a higher proliferative rate and appears more effective than PDLSC for tooth formation. Importantly, SCAP are easily accessible since they can be isolated from the third molars of humans (Hargreaves et al., 2008).

Stem cells from the dental follicle (DFSC)

DFSC have been isolated from the third molars' follicle of humans and used to express the stem cell markers: Notch1, STRO-1 and nesting. These cells can be maintained in culture for at least 15 passages. STRO-1 positive DFSC that can be differentiated in cementoblasts *in vitro* have the ability to form cementum *in vivo*. However, immortalized dental follicle cells are able to re-create a new periodontal ligament (PDL) after *in vivo* implantation (Huang et al., 2008).

Periodontal ligament stem cells (PDLSC)

The PDL is a specialized tissue located between the cementum and the alveolar bone and plays the role of maintaining and supporting the teeth. Its continuous regeneration is thought to involve mesenchymal progenitors arising from the dental follicle. PDL contains STRO-1 positive cells that maintain certain plasticity since they can adopt adipogenic, osteogenic and chondrogenic phenotypes *in vitro* (Srisuwant et al., 2006).

Bone marrow derived mesenchymal stem cells (BMSC)

BMSC have been tested for their ability to recreate periodontal tissue. These cells are able to form *in vivo* cementum, PDL and alveolar bone after implantation into the defective periodontal tissues. Thus, bone marrow provides an alternative source of MSC for the treatment of periodontal diseases such as recession, vertical bone loss, etc (Abe, 2008).

In search of epithelium-originated dental stem cells

Although significant progress has been made with MSC, there is no information available for dental EpSC in humans. The major problem is that dental epithelial cells, such as ameloblast and ameloblast precursors are eliminated soon after tooth eruption. Therefore, epithelial cells that could be stimulated *in vivo* to form enamel are not present in human adult teeth (Gonclaves et al., 2007).

Epithelial stem cells from developing molars

Several studies describe the use of EpSC isolated from newborn or juvenile animals, usually from third molar teeth. In these studies, epithelia were removed and the cells were dissociated enzymatically. Precursors were then amplified and associated with MSC (originated from the same tooth) *in vitro* in contact with biomaterials such as collagen sponges or synthetic polymers (Gonclaves et al., 2007).

Epithelial stem cells from the labial cervical loop of rodent incisor

The rodent incisor is a unique model for studying dental EpSC since, in contrast to human incisors or other vertebrates, this tooth grows throughout their life time. An EpSC niche, which is located in the apical part of the rodent incisor epithelium (cervical loop area), is responsible for a continuous enamel matrix production; thus, in this highly proliferative area, undifferentiated epithelial cells migrate toward the anterior part of the incisor and give rise to ameloblast (Huang et al., 2008).

GROWTH FACTORS

Growth factors have been described by Murray et al. (2007) as proteins that bind to receptors on the cell and induce cellular proliferation and/or differentiation. Many growth factors are quite versatile, simulating cellular division in numerous cell types, while others are more cell specific (Iohara, 2002; Murray, 2007). However, what regulates the abrupt transition of stem/progenitor cells from quiescent to active state in terms of proliferation, migration, differentiation and matrix secretion is still unclear.

Platelet derived growth factor (PDGF)

It consists of 2 disulphide bonded poly-peptide chains that are encoded by 2 different genes: P.D.G.F- A and P.D.G.F.-B (Iohara et al., 2006).

The primary effect of PDGF is that of a mitogen-initiating cell division. Thus, it has been characterized as a competence factor, that is, a growth factor that makes a cell competent for cell division. A progression factor, such as I.G.F.-1, is then necessary to induce mitosis. PDGF also causes replication of endothelial cells, causing budding of new capillaries (angiogenesis) (Iohara, 2006).

Insulin-like growth factors (IGF-I, II)

They are peptide growth factors with biochemical and functional similarities to insulin. Bone cells produce and respond to IGF's, and bone is a storage house for these factors in their inactive form. They are mitogens, but in fibroblastic systems, they appear as progression factors. In bone cell systems, they stimulate both proliferation of pre-osteoblasts, as well as the differentiation of osteoblasts, including Type I collagen synthesis. Thus, IGF increases both the number of cells synthesizing bone, as well as the amount of extra-cellular matrix deposited by each cell.

Combinations of PDGF and IGF have been tested in periodontal systems. The combination could potentiate the growth of multiple tissue types by combining a competence factor with a progression factor (Murray et al., 2007).

Transforming growth factor-β- (TGF- β)

It is a multifactorial growth factor, structurally related to bone morphogenic proteins, but functionally quite different. It has been shown to be chemotactic for bone cells, and may increase or decrease their proliferation depending on the differentiation state of the cells, culture conditions and concentration of TGF-β applied. If injected in a close proximity to the bone in vivo, it produces new cartilage and / or bone; however, it does not induce new bone formation when implanted away from a bony site. In spite of its effects on the augmentation of bone, no positive data have been reported on in vivo healing in a periodontal setting (Trantor et al., 2005).

Fibroblast growth factors (FGF)

They are family of at least 9 related gene products of which 2 major members are acidic FGF (a-FGF or FGF-1) and basic FGF (b-FGF or FGF-2). It can stimulate endothelial cells and periodontal ligament cell migration and proliferation, as well as bone cell replication (Nakashima et al., 2006).

Bone morphogenetic proteins (BMPs)

Bone morphogenic proteins are secreted by the oral epithelium layer (oral ectoderm) during early stages of odontogenesis (Dental lamina) (Nakashima et al., 2006).

SCAFFOLDS

The importance of scaffold material and design for tissue engineering has long been recognized. Several studies have proved the combinations of stem cells and scaffolds in the successful regeneration of periodontium, bone, pulp dentin complex and even the whole teeth in the field of dentistry and several organs and tissue outside our field, for the better health care of patients (Abe, 2008).

Co-polymers of polylactic acid and polyglycolic acid

The apparent disadvantage of these poly (α-hydroxy) acids is their degradation by hydrolysis resulting in decomposition products that are mostly metabolised to carbon-dioxide and water through the Kreb's cycle (Kim and Mooney, 1998).

Synthetic calcium phosphate ceramics

These were implemented as matrix materials for facilitating regeneration *in vivo*. The two most widely used forms of these bioceramics are:

1) Tricalcium Phosphate (TCP): This is a porous form of calcium phosphate, the most commonly used form being ß-TCP. The potential problem with the use of this material was that it regularly underwent physio-chemical dissolution too often after implantation.
2) Synthetic Hydroxyapatite: The problems associated with TCP led to the development of this second form of bioceramic. The rationale for developing this material relied, in part, on the fact that because the mineral naturally occurring in bone was hydroxyapatite, synthetic implants of the same material would be biocompatible with osseous tissue (Kim and Mooney, 1998).

Deorganified bone or anorganic bone

This is the hydroxyapatite skeleton that retains the microporous and macroporous structure of cortical and chancellors bone, remaining after chemical or low heat extraction of the organic component. Usually, bovine bone mineral is used for this purpose. Studies have demonstrated that natural bone mineral particles implanted in defects show a greater degree of incorporation into host osseous tissue and have a composite modulus of elasticity closer to that of natural bone (Murray et al 2007).

Collagen

Collagen is defined as a protein with 3 polypeptide chains, known as α-chains, each containing at least one stretch of the repeated amino acid sequence 'Gly-Xaa-Yaa', where 'Xaa' and 'Yaa' can be any amino acid, but often proline and hydroxyproline.

Collagen constitutes almost one-third of all protein in the body, and accounts for almost 60% of gingival connective tissue and 90% of the total protein in the bone (Murray et al 2007).

Chitosan (Poly-N-Acetyl Glycosaminoglycans)

Chitosan is a carbohydrate biopolymer obtained from chitin (extracted from anthropods). Chitin is second only to cellulose as the most abundant natural biopolymer. Chitosan is made by treating chitin with hot strong alkali which results in deacetylation of chitin (Kim and Mooney, 1998).

GENE THERAPY

The goal of gene-enhanced tissue engineering is to regenerate lost tissue by the local delivery of cells that have been genetically-enhanced to deliver physiologic levels of specific growth factors. The basis for this approach lies in the presence of a population of progenitor cells that can be induced, under the influence of these growth factors, to differentiate the specific cells required for tissue regeneration, with guidance from local clues in the wounded environment.

GENE DELIVERY

The application of gene therapy to treat exposed pulp by delivering DNA, RNA or antisense sequences alters gene expression within a target cell population in pulp tissue. The gene therapy manipulates cellular processes and responses. The transfected genes stimulate immune response, modify cellular information or developmental program, or produce a therapeutic protein with specific functions (Nakashima et al., 2006).

Vectors for gene transfer

Gene transfer should achieve a stable expression of transgene in a target cell in an appropriate form without side effects, such as interaction with host genome, toxicity, carcinogenic transformation and insertional mutagenesis (Nakashima et al., 2006).

Viral and non-viral gene therapy

Both viral vectors and non-viral vectors have been employed for gene transfer. Viral vectors are derived from viruses with either RNA or DNA genomes, such as Retrovirus, Lentivirus, Adenovirus, Adeno-associated virus and Herpes simplex virus.

Non-viral methods represent a simple and safer alternative to viral vectors. Simple quantitative production, low host immunogenicity and further recent advances in sustained gene expression and efficient and long-term gene expression are now making non-viral gene therapy more of a reality for human clinical medicine (Nakashima et al., 2006).

However, recent advances include:

1. Intravenous Infection at high hydrodynamic pressure: Plasmid DNA can be delivered to tissues *in vivo* by intravenous infection at high hydrodynamic pressure. It is possible in a practical sense to use a blood-pressure cuff in the limbs to achieve high pressures to deliver plasmid DNA. The delivery of DNA by coated metal microparticles by particle bombardment into cutaneous tissues has been useful. However, attendant issues include heat generation and transfer at the site of penetration of the microparticles (Nakashima et al., 2006).
2. Electroporation: Application of regulated electric pulses

in delivering genes to cells is electroporation. Electroporation is routinely used to deliver DNA to bacteria, yeast and mammalian cells in culture. Electroporation uses electric fields to create transient pores to facilitate entry of plasmid DNA. Electroporation *in vivo* was successfully used in muscle, skin, brain and liver. One of the limitations of electroporation is the tissue damage. Although electroporation is an efficient technique, it is an invasive method (Nakashima et al., 2006).

3. Ultrasound: The application of ultrasound leads to acoustic cavitation and produces cell membrane permeabilization, thereby promoting the delivery of plasmid DNA. Ultrasound contrast agents can improve cavitation. However, microbubbles and optison which are coated by albumin and contain octafluoro-propane gas were found to be superior for cavitation using ultrasound. The use of a combination of ultrasound and electro-poration was found to be better than either of these methods alone. The recent advances in the uses of ultrasound to drug and gene delivery has multiple therapeutic applications including regenerative medicine (Nakashima et al., 2006).

CONCLUSION

Overall, the future application of regenerative and tissue-engineering techniques to dentistry is one of the immense potentials capable of meeting a variety of patient needs. High-quality basic dental research is paramount to ensuring that the development of novel clinical treatments is supported by robust mechanistic data and that such approaches are effective. These efforts reveal how successful innovations in the field of dentistry can be guided by advances in basic research, highlighting the need for close partnerships between basic research and clinical scientists. This hypothesis might be a challenge to modern bioengineering and endodontics.

REFERENCES

Abe S, Yamaguchi S, Watanabe A, Hamanda K, Amagasa T (2008). Hard tissue regeneration capacity of apical pulp derived cells (APDCs) from human tooth with immature apex. Biochem. Biophysical Res., Jul, 1; 371: 90-3.

Baum BJ, O'Colonnell BC (1995). The Impact of Gene Therapy on Dentistry. JADA FEB, 126: 179-89.

Duailibi SE, Dualibi MT, Vacanti JP, Yelick PC (2006). Prospects for tooth regeneration. Periodontology, 2000; 41: 177-87.

Gonclaves SB, Dong Z, Bramante CM, Holland GGR, Smith AJ, Nor JE (2007). Tooth slice-Based models for the study of Human dental pulp angiogenesis. JOE, 33(7): 811-4.

Gronthoss, Mankani M, Brahim J, Robey GP, Shi S (2000). Post Natal Human Dental Pulp Stem Cells (DPSCs) *in vitro* and *in vivo* Proc. Nat. Acad Sci. USA Dec, 5; 97(25): 13625-30.

Hargreaves KM, Giesler T, Henry M, Wang Y (2008). Regeneration potential of the young permanent tooth: What does the future hold? Pediatric Dent., 30(3): 253-60.

Huang GTJ, Sonoyama W, Liu Y, Liu H, Wang S, Shi S (2008). The hidden treasure in apical papilla: The potential role in Pulp/Dentin regeneration and bioroot engineering. JOE Jun, 34(6): 645-51.

Iohara K, Zheng Li, Ito M, Tomokiyo A, Matsushita K, Nakashima M (2006). Side population cells isolated from porcine dental pulp tissue with self renewal and multipotency for dentinogenesis, chondrogenesis, adipogenesis, and neurogenesis Stem cells, 24: 2493-503.

Kim BS, Mooney DJ (1998). Development of biocompatible synthetic extracellular matrices for tissue engineering. TibTech May, 16: 224-30.

Miura M, Gronthos S, Zhao M, Lu B, Fisher LW, Robey PG (2003). SHED: Stem cells from human exfoliated deciduous teeth. Proc. Nat Acad. Sci. USA May, 13; 100(10): 5807-12.

Murray PE, Gracia-Godoy F, Hargreaves KM (2007). Regenerative endodontics: A review of current status and call for action. JOE Apr, 33(4): 377-90.

Nakashima M, Iohara K, Zheng L (2006).Gene therapy for dentin regeneration with bone morphogenic proteins. Curr. Gene Therapy, 6: 551-60.

Srisuwant, Tilkorn DJ, Wilson JL, Morrison WA, Messser HM, Thompson EW (2006). Molecular Aspects of Tissue Engineering in the Dental Field. Peridontology, 2000; 41: 88-108.

Trantor IR, Messer HH, Bimer R (2005).The Effect Of Neuropeptides (calcitonin gene related peptide and substance P) on cultured human pulp cells. J. Dent. Res. APR, 74(4): 1066-71.

16

Micropropagation of wild fennel (*Foeniculum vulgare var. vulgare*) via organogenesis and somatic embryogenesis

Rehab H. Abd-Allah, Ehsan M. Abo Zeid, Mahmoud M. Zakaria and Samih I. Eldahmy

Department of Pharmacognosy, Faculty of Pharmacy, University of Zagazig, Zagazig 44519, Egypt.

Wild fennel (*Foeniculum vulgare var. Vulgare*) is a perennial aromatic herb. It is native to the Mediterranean region and currently it iscultivated as an annual or perennial herb worldwide. Dried ripe fruits are commonly used plants part for obtaining essential oil. Fruits oil of wild fennel contains many volatile oils, such as α-pinene, phellandrene, p-cymene, fenchone, estragole, anethole and anisaldehyde. Important biological activities of the volatile oil of fennel fruits are hepatoprotective, hypotensive, anticancer, antioxidant, antibacterial, antifungal, antiviral, hypoglycemic, spasmolytic, analgesic, antipyretic, anti-inflammatory and C.N.S activities. Wild fennel (*Foeniculum vulgare var. vulgare*) is categorized as rare and endangered in the Egyptian flora due to urban sprawl, especially along northern coastal area and it has turn into a retreat in Egypt. Therefore attempts were made to find *in vitro* germination of its seeds as well as to explore the ability of organogenesis and somatic embryogenesis in the produced callus. In order to continue the micropropagation process, transplantation of developed plantlets to the soil was also investigated. This study also looks into the production of different medicinally valuable volatile oils in the formed calluses and micro-propagated plants.

Key words: Wild fennel, *Foeniculum vulgare*, seed germination, callus production, somatic embryogenesis.

INTRODUCTION

*Foeniculum vulgare var. vulgare*family Apiaceae,is common or bitter fennel, which is known as "shamar" in Egypt and also this is the official fennel according to the Egyptian pharmacopeia (Egyptian Pharmacopoeia, 1984). It is native to Mediterranean region but now it is cultivated as an annual or a perennial herb in Argentina, Hungary, Bulgaria, Germany, France, Italy, Greece, china and India (Leung and Foster, 1996; Chevallier, 1996). Important compounds identified in all samples of fennel volatile oils were *trans*-anethole, estragole, fenchone, limonene, *alpha*-pinene and *gamma*-terpinene (Aprotosoaie et al., 2008). The plant has shown various pharmacological properties involving antibacterial (Gulfraz et al., 2008), antispasmodic (Alexandrovich et al., 2003), analgesic (Guang-shou et al., 2011), anti-inflammatory (Ozbek, 2005), antipyretic (Tanira et al., 1996), anxiolytic (Kishore et al., 2012), antioxidant (Singh et al., 2008), (Moon et al., 1985), diuretic (Tanira et al.,1996), antihypertensive (Haze et al., 2002), mucolytic (Mills and Bone,2000) and hepatoprotective (Mansour et al., 2011) activities.

In recent years *in vitro* techniques have received increasing importance in the conservation of threatened plants and this trend is likely to be continued as more species are expected to face the risk of extinction (Kapai et al., 2010). *In vitro* propagation offers an easy, rapid and space-efficient way for mass scale multiplication of plant species.

Available reports on micropropagation of *Foeniculum vulgare var. vulgare* do not provide exclusive information about chemical composition of induced calluses or micropropagated plants. This study is aimed to investigate the *in vitro* germination behavior of the seeds of wild fennel (*Foeniculum vulgare var. vulgare*). The ability of explants excised from growing seedlings to form a stable callus was determined representing first step for micropropagation. The ability of produced plant calluses to form somatic embryos to continue the micropropagation process and transplantation to the soil were also investigated. The study also looks into the production of different medicinally valuable volatile oils components in the formed calluses and micropropagated plantlets besides fruits of cultivated and wild fennel as well as the aerial parts of the wild fennel.

MATERIALS AND METHODS

Collection of plants materials

Whole plant and seeds of *Foeniculum vulgare var. vulgare* (wild), were collected on April 2013 from SidiBarrani in the north coast of Egypt. Identity of collected material was verified by Assistant Prof. Dr. Eman Shams, Assistant Professor of plant taxonomy, Faculty of science, Cairo University. Voucher specimens were deposited to the Herbarium of the Department of Pharmacognosy, Faculty of Pharmacy, Zagazig University, Egypt. Matured seeds were collected from the wild plant on April.

Sterilization

Seeds were sterilized through submerging them 70% ethyl alcohol for different periods 1, 2, 3 and 4 min and then shaking with 5% commercial hypochlorite solution (Clorox®) for different peroids 10, 15, 20, 25 and 30 min. Under the hood, sodium hypochlorite was poured away from seeds then seeds were rinsed thrice with sterile distilled water before applying to media for germination.

Germination of seeds

Some sterilized seeds were transferred to jars (5 seeds/jar) containing solid hormonal free (HF) media of the composition (4.4 g/l M.S.(Murashige and skoog) media, 30 g/l sucrose and 8 gm/l agar) or transferred to the same media composition in addition to 50 mg/l gibberillic acid solution as growth enhancer. The lid of each jar was wrapped with para film and incubated in a growth room at 25°C and a photoperiod of 16 h fluorescent light and 8 h dark. Germination of seeds was investigated using seed germination percentage using the following formula:

$$\text{Seed germination percentage} = \frac{\text{Number of germinated seeds}}{\text{Total number of seeds cultured}} \times 100$$

Production of callus

Seedlings of 30-35 days old obtained from seeds germinated on HF solid media, were cut aseptically into pieces of 4-6 mm length of leaf, stem and root which were used as a source of explants. Explants were cultured on sterile solid M.S. medium supplemented with different growth regulators including naphthalene acetic acid (NAA), 6-benzylaminopurine (BAP), 2,4- dichlorophenoxyacetic acid (2,4-D), kinetin (K) and thidiazuron (TDZ) in different combinations and concentrations such as (2,4-D 0.5 mg/l +K 0.1 mg/l), (2,4-D 0.5 mg/l +K 0.5 mg/l), (2,4-D 1 mg/l +K 0.5 mg/l), (2,4-D 1 mg/l + K 1 mg/l), (2,4-D 2 mg/l +K 1 mg/l), (2,4-D 2 mg/l +BAP 0.25 mg/l), (NAA 1 mg/l +K 1 mg/l), (NAA 1 mg/l +BAP 0.1 mg/l) and (TDZ 0.5 mg/l +2,4-D 1 mg/l +BAP 0.1 mg/l) for initiation of callus. Each treatment consisted of five explants per jar with four replicates. All cultures were incubated in a growth room at temperature 25°C ±1 and photo period of 16 h light and 8 h dark. Callus induction and maintenance was investigated using three parameters: callus induction time, callus induction percentage and callus growth rate. Callus induction time is the number of days passed until the callus is formed. Callus induction percentage is calculated using the following formulae:

$$\text{Callus induction percentage} = \frac{\text{Number of callus produced}}{\text{Total number of explants cultured}} \times 100$$

Callus growth rate is represented by the total fresh weight (mg) in different time intervals (days) which was calculated by adjusting the weight of all calluses obtained by all previously mentioned phytohormonal combinations to 1000 mg then increase in callus fresh weight was monitored and calculated at different time intervals (10, 20, 30 and 40 days).

In vitro regeneration

Micropropagation involves direct and indirect techniques which allow the *in vitro* clonal propagation of parts or even cells of the required plant to a whole plant. Direct techniques aim to produce clones of the plant directly from meristematic tissue and buds through direct organogenesis and somatic embryogenesis. Indirect techniques involve the production of callus through which a whole plant could be produced using the above two methods.

Direct micropropagation

Direct micropropagation is rare; it involve direct shooting or rooting from buds or other meristematic tissue. This should be under the influence of different phytohormones. *Foeniculum vulgare var. vulgare* produced seedlings were cut in to different explants of root, stem and leaf, and subjected to different types of phytohormones and their combinations as the previously mentioned hormones and cultured on M.S. solid media at 25±1°C and 16 h photo period either with sucrose containing media or sucrose free media and examined periodically macroscopically and microscopically to detect any signs of organogenesis or somatic embryogenesis.

Indirect micropropagation

After callus was successfully produced, representing stage I of indirect micropropagation, stage II began and involved trials to make shoots, roots or somatic embryos from callus cells. Different conditions were applied on different callus cells produced in order to induce organogenesis or embryogenesis. Phytohormones were used solitary and in combinations with different concentrations and hormonal free media were also used with high light intensity to induce photosynthesis in the callus cells and thus encourage the cells to differentiate into shooting cells. Media with auxins only were used to induce rooting, media with cytokinins only were used to induce shooting and also some callus were transferred to hormonal free media. In each trial, cells from callus were tested under microscope to detect the production of somatic embryos.

In vivo culturing of the micropropagated plantlets

Plantlets were placed in perforated pots containing autoclaved asbestos pertile soil and all pots were covered with transparent plastic bags to maintain high humidity and allow easy illumination. Pots were placed in a growth room at temperature 25°C ±1 and photo period of 16 h light and 8 h dark. The transparent plastic bags were perforated after two days. Each pot was irrigated every two days with a little amount of water. The transparent plastic bags were removed away after seven days.

Trial for hardening the root system

Some plantlets were put in test tubes containing hormonal liquid media of the auxin (NAA 1 mg/L) and the bottom of the tube was covered with cellophane as roots favor darkness, and tubes were incubated for 30 days in the growth room at 25±1°C and 16 h photoperiod.

Analysis of volatile oils produced by callus, micropropagated plants and fruits of cultivated, wild fennel as well as the aerial parts of the wild fennel

All extracts ofcallus and micropropagated plants were prepared by the following procedures: Callus or micropropagated plants were crushed in a mortar with double distilled n-hexane coupled with ultra-sonic waves at 40°C for 15 min, maintained in well closed jars in a shaker at low speed overnight, then filterated and concentrated to 1 ml and passed to GC-MS analysis.

Fruits of cultivated, wild fennel as well as the aerial parts of the wild fennel were hydro-distilled in the Clevenger apparatus. volatile oils of each were collected after 3 h and water traces was removed using anhydrous sodium sulphate. The oils were kept in refrigerator then passed to GC-MS analysis. GC-MS analysis was conducted in Agilent 6890 gas chromatograph equipped with an Agilent mass spectrometry detector, with a direct capillary interface and fused silica capillary column PAS -5 ms (30 m x 0.32 nun x 0.25 um film thickness). Samples were injected under the following conditions: helium was used as carrier gas at approximately 1 ml/min, pulsed split less mode. The solvent delay was 3 min and the injection size was 1.0 ul. The mass spectrophotometry detector was operated in electron impact ionization mode ioning energy of 70 e.v. scanning from m/z 50 to 500. The ion source temperature was 230 °C and the quadrupole temperature was 150°C .The electron multiplier voltage (EM voltage) was maintained 1250 v above auto tune. The instrument was manually tuned using perfluorotributyl amine (PFTBA). The GC temperature program was started at 40 or 60°C as mentioned in the tables then elevated to 280°C at rate of 8°C / min. and 10 min. hold at 280°C the detector and injector temperature were set at 280 and 250°C, respectively. Wiley and Nist 05 mass spectral data base was used in the identification of the separated peaks.

RESULTS AND DISCUSSION

The wild fennel, *Foeniculum vulgare var. vulgare* (Fam.

Apiaceae) has become a rare plant in Egyptian flora now due to many human activities such as overgrazing and urban sprawl. This situation inspired the authors to carry out this study which focused on different techniques of micropropagation to succeed in making *in vitro* plant regeneration and this was also accompanied with comparing the volatile oil content of fruits of cultivated and wild fennel as well as the aerial parts of the wild fennel to detect differences between them.

Seed germination percentage

*Foeniculum vulgare*seeds sterilized by different time slots of 70% ethyl alcohol and commercial hypochlorite (5%) were cultured on free hormonal media or with 50 mg/l gibberillic acid solution either on solid or semisolid media as well as in Petri dishes containing sterile Whatman grade number 1 filter papers moisted with either sterile distilled water or sterile liquid hormonal free media or with 50 mg/l gibberillic acid solution and incubated in growth room at 25°C and a photoperiod of 16 h light and 8 h dark. Seed germination percentage is shown in Table 1. The results shows that immersing in ethyl alcohol (70%) for 1 and 2 min along with immersing in sodium hypochlorite for 10, 15 and 20 min were done but some seeds showed bacterial or fungal contaminations with these procedures but immersing in ethyl alcohol (70%) for 3 and 4 min along with immersing in sodium hypochlorite for 25 and 30 min showed no contamination but different germination percentages. Only, immersing in ethyl alcohol (70%) for 3 min along with immersing in sodium hypochlorite for 25 min showed complete germination with no contamination. Also, it was figured that increasing time of sterilization more than that inhibits the surface of some seeds decreasing the rate of germination. The same observations were obtained upon using 50 mg/L gibberillic acid solution without any changes in the time of seed germination or germination percentage.

Callus induction

The produced seedlings were used as explants for production of callus. Induction of callus was studied using different parameters such as the percentage of callus induction (Callus capacity), days for callus induction and callus growth rate measured through callus weight and callus diameter. Different phytohormones and phyto-hormone-combinations were used in this study and the successful attempts in callus production are listed in Table 2 and Figure 1. Generally, Stem, root and cotyledonary leaves can induce formation of callus butit was obviously that stem explants were the best to initiate callus over other root and leaf explants. Also hormonal combinations (2,4-D 1 mg/l +k 1 mg/l), (NAA 1 mg/l +BAP 0.1 mg/l) and (TDZ 0.5 mg/l +2,4-D 1 mg/l +BAP 0.1

Table 1. Results of germination trials with seed germination percentage through 30 days observations.

70% Ethanol	Clorox® 5%				
	10 min	15 min	20 min	25 min	30 min
1 min	Contamination	contamination	contamination	Slight contamination	Slight contamination after 8 days
2 min	Contamination	contamination	Slight contamination	Slight contamination after 7 days	Slight contamination after 13 days
3 min	Contamination	contamination	Slight contamination after 14 days	Sterile and 100% germination	Sterile and 60% germination
4 min	contamination	Slight contamination after 5 days	Sterile and 70% germination	Sterile and 50% germination	Sterile and 0% germination

Table 2. Callus capacity of different explants of seedlings on different hormonal combinations.

Media	Root (%)	Stem	Leaf	Mean
(2,4-D 0.5+k 0.1)	100	90	100	96.66667
(2,4-D 0.5+K 0.5)	100	100	100	100
(2,4-D 1+K 0.5)	50	100	50	66.66667
(2,4-D 1+K 1)	100	100	100	100
(2,4-D 2+K 1)	100	100	90	96.6667
(2,4-D 2+BAP 0.25)	100	100	90	96.66667
(NAA 1+K 1)	80	80	80	80
(NAA 1+BAP 0.1)	100	100	100	100
(TDZ 0.5+2,4-D 1 + BAP 0.1)	100	100	100	100
Mean	92.2222	**93.3333**	90.0000	

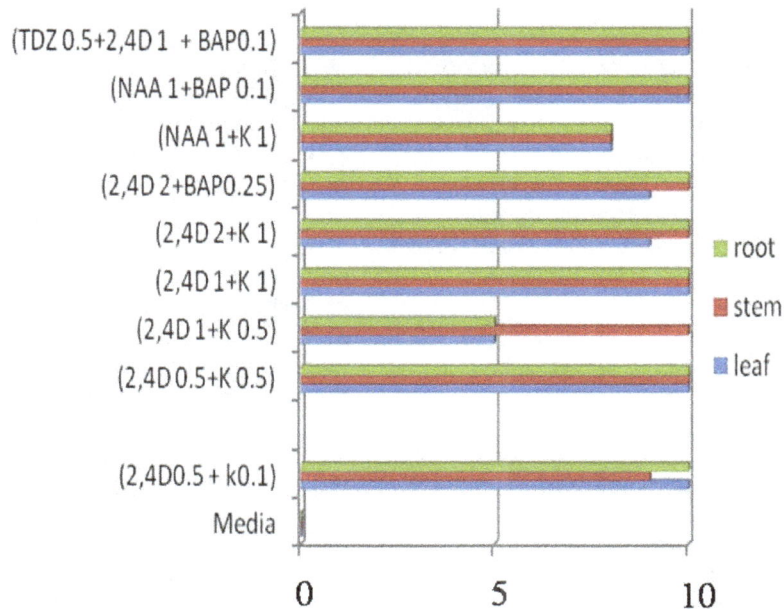

Figure 1. Callus capacity of different explants of seedling on different hormonal combinations.

Days for Callus initiation

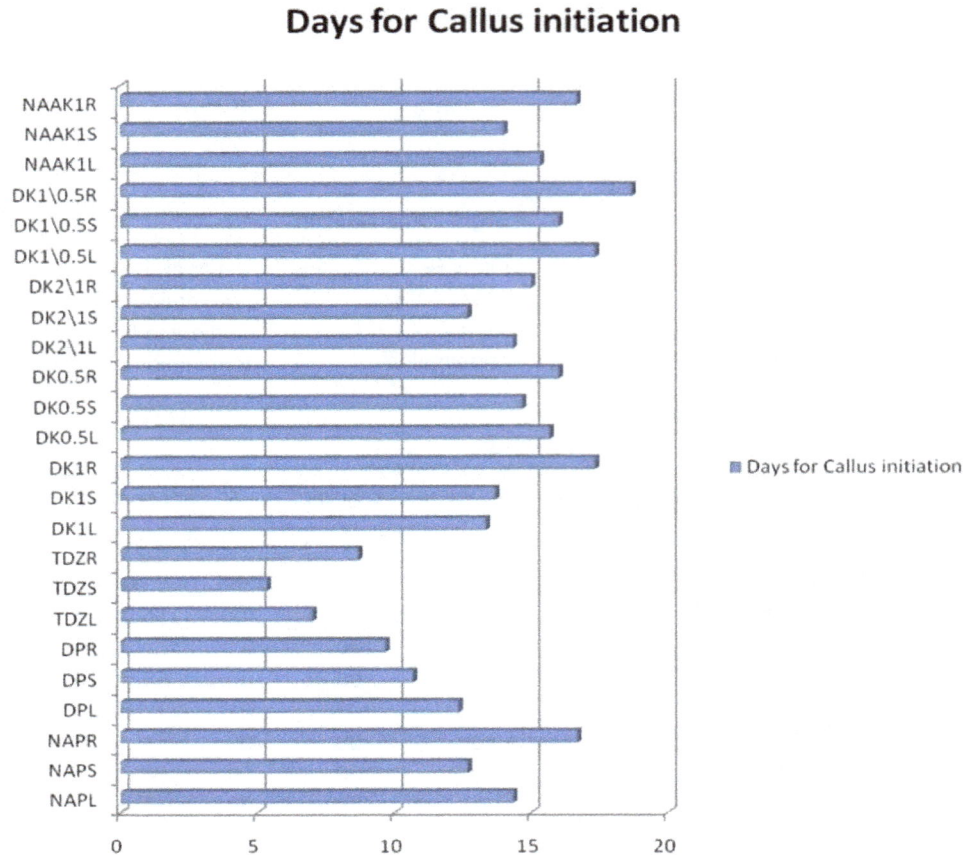

Figure 2. Callus induction days of *Foeniculum vulgare* var. *vulgare* (Fam.Apiaceae) different explants of seedlings on different hormonal combinations (2,4-D 0.5+K 0.1), (2,4-D 0.5+K 0.5), (2,4-D 1+K 0.5), (2,4-D 1+K 1), (2,4-D 2+K 1), (2,4-D 2+BAP 0.25), (NAA 1+K 1), (NAA 1+BAP 0.1) and (TDZ 0.5+2,4-D 1+BAP 0.1).

mg/l) were the best to initiate callus over other hormonal combinations. Callus induction time is illustrated in Figure 2 and Table 3. All previously mentioned hormonal combinations were able to induce callus but the hormonal combination TDZ 0.5 mg/l +2,4-D 1 mg/l mg/l +BAP 0.1 mg/l was the fastest to induce callus formation, followed by the hormonal combination 2,4-D 2 mg/l +BAP 0.25 mg/l.

Callus growth rate

Callus growth rate was monitored and calculated at different time intervals (10, 20, 30 and 40 days) on different hormonal combinations as illustrated in Table 3 and Figure 3.

From the illustrated growth curve ,the best hormonal combinations used were TDZ 0.5 mg/l +2,4-D 1 mg/l + BAP 0.1 mg/l and 2,4-D 2 mg/l +BAP 0.25 mg/l that showed significantly marked growth rate more than other hormonal combinations.

The interesting point is that the hormonal combination (TDZ 0.5 mg/l +2,4-D 1 mg/l + BAP 0.1 mg/l) has never been used or mentioned before in any literature for fennel or any umbelliferous plant and it showed highly significant results in callus formation, callus growth and also micro-propagation as we will illustrate later.

All calluses produced from the previous culture were subcultured on the same media with the same hormonal combinations initiated them and incubated in for 30 days at the same conditions. Not all hormonal combinations were able to maintain the growth of callus. Only the hormonal combinations (2,4-D 0.5 mg/l +K 0.1 mg/l), (2,4-D 0.5 mg/l +K 0.5 mg/l), (2,4-D 1 mg/l +K 0.5 mg/l), (2,4-D 1 mg/l +K 1 mg/l), (2,4-D 2 mg/l +K 1 mg/l), (2,4-D 2 mg/l +BAP 0.25 mg/l) and (TDZ 0.5 mg/l +2,4-D 1 mg/l +BAP 0.1 mg/l) were able to maintain callus growth but (NAA 1 mg/l +BAP 0.1 mg/l) and (NAA1 mg/l +K 1 mg/l) failed to maintain the growth of callus and the best growth of callus of root, leaf and stem was also noticed in the first subculture from (2,4D 2 mg/l +BAP 0.25 mg/l) and (TDZ 0.5 mg/l + 2,4-D 1 mg/l + BAP 0.1 mg/l) hormonal

Table 3. Callus induction time and callus growth rate of *Foeniculum vulgare var. vulgare* (Fam. Apiaceae). Explants were taken from plant seedling growing under sterile conditions. The experiments were repeated 3 times using 12 explants. The results show the mean and standard error of mean was not put for simplification and did not exceed 10 % of mean.

Code	Treatment					Days for callus initiation	Callus weight mg/days			
	NAA	BAP	2,4D	Kin	TDZ		mg/10 days	mg/20 days	mg/30 days	mg/40 days
NAPL	1	0.1				14.33	50.20	451.00	1103.00	1734.00
NAPS	1	0.1				12.67	62.00	367.00	1423.00	1874.00
NAPR	1	0.1				16.67	36.00	319.00	1056.00	1698.00
DPL		0.25	2			12.33	99.00	399.00	1745.00	2322.00
DPS		0.25	2			10.67	108.00	415.00	1987.00	2455.00
DPR		0.25	2			9.67	104.00	387.00	1604.00	2302.00
TDZL		0.1	1		0.5	7.00	112.00	1440.00	1450.00	2640.00
TDZS		0.1	1		0.5	5.33	130.00	514.00	2223.00	2780.00
TDZR		0.1	1		0.5	8.67	125.00	453.00	2114.00	2498.00
DK1L			1	1		13.33	59.00	206.00	949.00	1499.00
DK1S			1	1		13.67	82.00	274.00	1145.00	1546.00
DK1R			1	1		17.33	73.00	244.00	922.00	1402.00
DK0.5L			0.5	0.5		15.67	44.00	123.00	377.00	533.00
DK0.5S			0.5	0.5		14.67	79.00	156.00	412.00	587.00
DK0.5R			0.5	0.5		16.00	65.00	142.00	344.00	421.00
DK2\1L			2	1		14.33				
DK2\1S			2	1		12.67				
DK2\1R			2	1		15.00				
DK1\0.5L			1	0.5		17.33				
DK1\0.5S			1	0.5		16.00				
DK1\0.5R			1	0.5		18.67				
NAAK1L	1			1		15.33				
NAAK1S	1			1		14.00				
NAAK1R	1			1		16.67				

Explant code: L = leaf, S = stem, R = root.

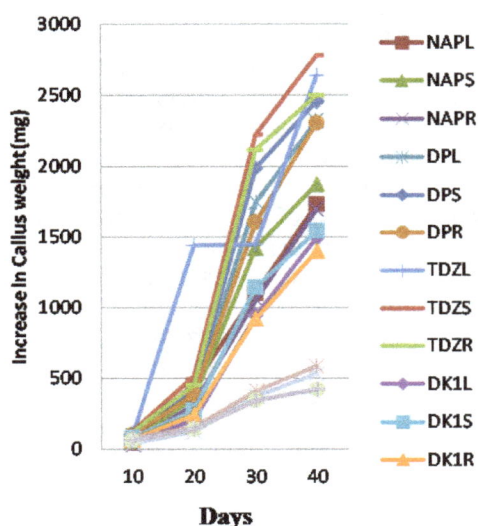

Figure 3. Callus growth rate of *Foeniculum vulgare var. vulgare* (Fam.Apiaceae) measured in the terms of increase in callus weight (mg) per days.

combinations. Subsequent subcultures were done for each 30 days old callus obtained from the first subculture from the hormonal combinations (2,4-D 2 mg/l +BAP 0.25 mg/l) and (TDZ 0.5 mg/l +2,4-D 1 mg/l + BAP 0.1 mg/l) as they showed the best growth and for the hormonal combinations (2,4-D 0.5 mg/l + K 0.5 mg/l) and (2,4-D 1 mg/l + K 1 mg/l) as they showed the best growth for the most common used hormonal combination 2,4-D and kinetin.

The growth of callus on these hormonal combinations was maintained, increased and also the best growth of callus of root, leaf and stem was noticed in the second subculture from (2,4-D 2 mg/l +BAP 0.25 mg/l) and (TDZ 0.5 mg/l +2,4-D 1 mg/l + BAP 0.1 mg/l) hormonal combinations.

In vitro regeneration

Trials to make clones of the plant either directly from the explants or indirectly after callus formation were done ultimately to achieve this point using different hormones either solely or in combinations or even hormonal free media to get clones of the whole plant.

Direct micropropagation

Foeniculum vulgare var. *vulgare* produced seedlings were cut in to different explants of root, stem and leaf, and subjected to different types of phytohormones and their combinations as the previously mentioned hormones and cultured on M.S. solid media (sucrose containing or sucrose free) at 25±1°C and 16 h photo period, however there were no signs of propagation (neither organogenesis nor somatic embryogenesis).

Indirect micropropagation

From the differently used hormones and hormonal combinations, none of them showed somatic embryogenesis but only NAA 1 mg/l +BAP 0.1 mg/l and NAA 1 mg/l +K 1 mg/l were able to produce organogenesis from their calluses and NAA 1 mg/l +BAP 0.1 mg/l hormonal combination was the only one able to continue micropropagation. Culture on the hormonal combination NAA 1 mg/l + BAP 0.1 mg/l was repeated thrice with the same conditions to ensure the occurrence of indirect organogenesis and micro-propagation each time from leaf, root and stem calluses.

All calluses of 30, 60, 90, 120 and 150 days old produced from the previously mentioned hormonal combinations from root, leaf and stem explants, were subsequently transferred to HF, K 0.5 mg/l and (K 0.5 mg/l + BAP 0.5 mg/l) solid and semi-solid MS media either sucrose containing media or sucrose free media beside their monitoring on their hormonal combinations inducing callus and incubated at 25°c ± 1 and photoperiod 16 h light and 8 h dark. In each trial, cells from callus were examined under microscope to detect the production of somatic embryos.

There were no signs of somatic embryogenesis on HF, K 0.5 mg/l and (K 0.5 mg/l +BAP 0.5 mg/l) solid and semi-solid MS media sucrose free MS media. Somatic embryogenesis has been occurred only in 90, 120 and 150 days old callus from the hormonal combination (TDZ 0.5 mg/l +2,4-D 1 mg/l +BAP 0.1 mg/l) of stem and leaf explants as well as callus from the hormonal combination (2,4-D 1 mg/l +K 1 mg/l) of stem explants, only on HF solid and semi-solid sucrose containing MS media that was detected by the formation of globular, heart, torpedo and cotyledonary forms of embryos as shown in Figure 4.

Callus that were able to give these forms of somatic embryos, started to give nodules, then dwarf shoots and roots till forming complete micropropagated plants of roots, shoots and leaflets. These plants were transferred to fresh media at every 30 days time intervals to maintain their vitality. Only plants of formed by stem explants callus

callus on the hormonal combination (TDZ 0.5 mg/l +2,4-D 1 mg/l +BAP 0.1 mg/l), were able to continue the indirect micropropagation process.

In vivo culturing of the micropropagated plantlets

After placing the micropropagated plantlets in the perforated pots and incubation, plantlets remain vital for only 10 days maximum without showing any sign of growth and start to wilt after that. This was due to weakness of the root system.

After hardening the root system by placing plantlets with their roots in contact with liquid auxin (NAA 1 mg/l) for 30 days and roots started to get thicker and formed brownish nodules as showed in Figure 5, plantlets showed a marked growth and increased in length from 9 cm to 13 cm in 15 days and maintained their vitality for 20 days but started to wilt after that.

GC-MS analysis

All prepared extracts of callus or micropropagated plants either by organogenesis or somatic embryogenesis were analysed using GC-MS technique as previously illustrated to detect the valuable produced volatile oils in each of them. The results are illustrated in Tables (4, 5, 6 and 7).

Analysis of calluses extracts showed that only calluses extracts of hormonal combinations (TDZ 0.5 mg/l +2,4-D 1 mg/l +BAP 0.1 mg/l) and (2,4-D 2 mg/l +BAP 0.25 mg/l) were able to form different volatile oils components such as *m*-menthane, *p*-menthane, *p*-menth-2-ene, cumene, *dl*-limonene, *m*-cymene, *trans*-anethole, *cis*-methyl-isoeugenol, *beta*-pinene, *gamma*-terpinene, *alpha*-terpinolene and propiovanillone as shown in Tables 4 and 5. Although, the notable point was that these volatile oils components were formed in very low concentrations in different analysed calluses. This has given the green light for proceeding towards micropropagation to optimize formation of different valuable volatile oils components.

Analysis of extracts of micropropagated plants obtained either by organogenesis or somatic embryogenesis also showed different valuable volatile oils componenets such as *trans-p*-menthane, *cis-p*-menthane, *m*-menthane, 6,6-dimethyl-menth-2-ene, *cis*-asarone and apiole. The most interesting point was that apiole which is a valuable antimicrobial, insecticidal and colon anti-proliferative volatile oil which is a major constituent of parsely (23%) and Dill (16.8%), was formed in different extracts in highly different concentrations which may be in some extracts higher than parsely or Dill.

Apiole percentage in extracts of micropropagated plants obtained by organogenesis varied between those obtained from leaf, root and stem callus on the hormonal combination NAA 1 mg/l + BAP 0.1v as shown in Table 6.

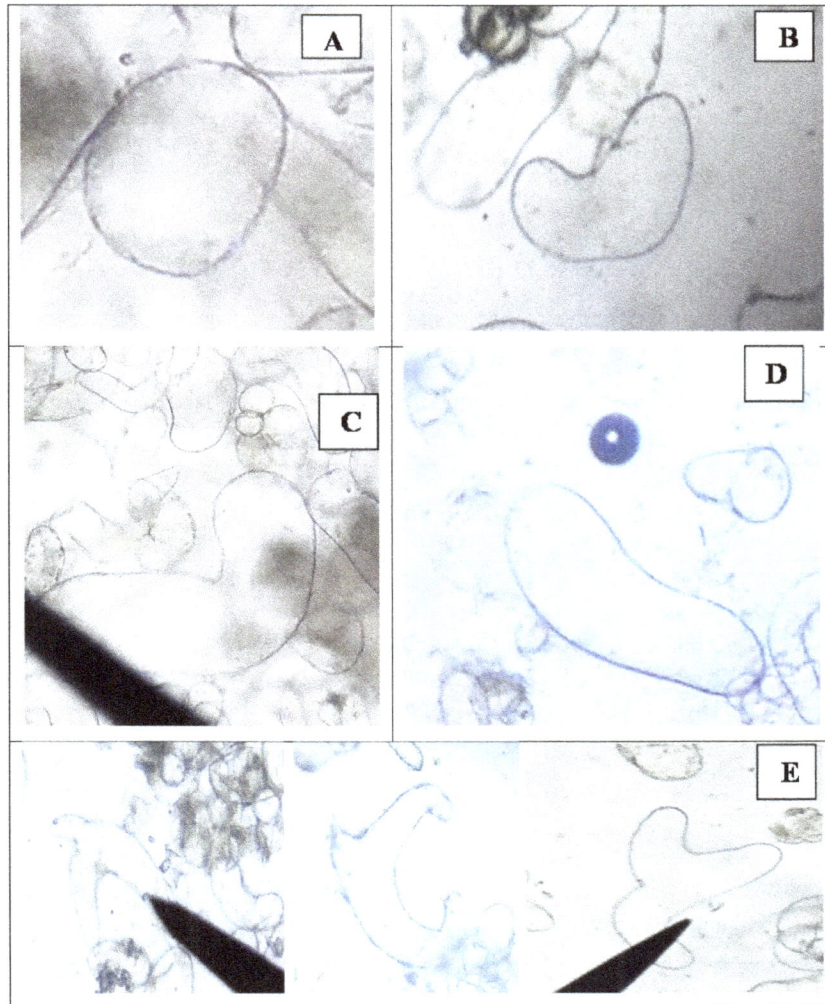

Figure 4. Different forms of somatic embryos of callus cells; A:globular form of embryo; B:early heart form of embryo; C:late heart form of embryo; D:torpedo form of embryo; E: dicotyledonary forms of embryo.

Figure 5. Hardened root system of *in vitro*micropropagated plants of wild fennel after placement in liquid media containing the auxin NAA (1 mg/l) for 30 days.

While apiole obtained by the solely hormonal combination (TDZ 0.5 mg/l +2,4-D 1 mg/l +BAP 0.1 mg/l) of stem callus that showed somatic embryogenesis, reached about 53% .

The analysis of hydrodistilled volatile oils of fruits of both cultivated and wild fennel as well as the hydrodistilled volatile oil of aerial parts of wild fennel showed different medicinally valuable volatile oils components such as *trans*-anethole, *l*-limonene, *beta*-myrcene, *alpha*- phellandrene, fenchone, *cis* and *trans*thujone. The most interesting point was that estragole which is a constituent responsible for most of the genotoxicity and carcinogenicity of the oil was present in lower percentages in the fruits of the wild plant (12.36%) as well as the aerial parts (38.61%) in comparison with the fruits of the cultivated plant (57.12%). Also, it was obviously clear that there is an inverse relationship between the percentage of estragole

Table 4. Different volatile oils identified by GC-MS analysis of stem explants callus of (TDZ 0.5 +2, 4-D 1+ BAP 0.1) hormonal combination.

No. at 60°c	M⁺	BP.	Fragments	R.T	Identification	% in TDZ stem Callus
1	140	97	123,112, 81,71,55	6.66	m-menthane	0.01
2	120	105	91,77,65,51	7.71	cumene	0.12
3	136	68	121,105,93,79, 53,39	7.84	dl-limonene	0.05
4	134	93	121,105,77,67,55	8.83	m-cymene	0.02
5	148	148	133,117,105,93,77,63,51	12.87	trans-anethole	0.01
6	178	178	163,147,135,115,107,91,77, 65,55	16.33	Cis-methyl- isoeugenol	0.42

Table 5. Different volatile oils identified by GC-MS analysis of different explants callus of (TDZ 0.5+2,4-D 1+BAP 0.1) and (2,4-D 2+BAP 0.25) hormonal combinations.

N0. At 40°c	M⁺	BP.	Fragments	R.T	Identification	% in TDZ ROOT	% in TDZ LEAF	% in DBAP ROOT	%in DBAP LEAF	% in DBAP STEM
1	140	97	123,112, 81, 71,55	9.19	*p*-Menthane	0.25	0.32	-	-	-
2	136	93	121, 69,53	9.35	*beta*-Pinene	-	-	-	0.07	-
3	138	97	125,111,83,69,55,41	9.56	*p*-Menth-2-ene	-	-	0.14	-	-
4	140	97	121,111,81,69,55	9.74	*m*-Menthane	0.17	0.2	0.2	-	-
5	136	68	119,93,51	10.55	*dl*-Limonene	-	-	-	0.34	-
6	136	93	121,105,77,55	11.23	*gamma*-Terpinene	-	-	-	0.07	-
7	136	121	105,93,79,67	11.88	*alpha*-Terpinolene	-	-	-	0.05	-
8	180	151	123,91,57	12.11	Propiovanillone	-	-	-	-	0.37

Table 6. Showing different volatile oils identified by GC-MS analysis of plantlets derived from indirect organogenesisof NAA1+BAP0.1 hormonal combination from leaf, stem and root explants callus.

No. at 40°C	M⁺	BP.	Fragments	R.T	Identification	% in ROOT	% in STEM	% in LEAF
1	140	97	123,112,81,71,55	9.19	*Trans-p*-Menthane	0.02	-	-
2	140	97	123,112,81,71,55	9.29	*Cis-p*-Menthane	-	-	0.15
3	140	97	121,111,81,69,55	9.74	*m*-Menthane	0.43	-	0.16
4	168	97	151,137,125,111,83,69,55	13.9	6,6-dimethyl-menth-2-ene	0.31	-	-
5	222	222	195,177,147,119,91,65	21.79	**Apiole**	13.72	39.1	41.55

Table 7. Showing different volatile oils identified by GC-MS analysis of plantlets of somatic embryogenesis derived from (TDZ 0.5+2,4-D 1+BAP 0.1) of stem explants callus.

No. at 40°C	M⁺	BP.	Fragments	R.T	Identification	% in TDZ stemplantlets
1	208	208	193,177, 165,148,133,121,105,91,77,57	17.22	*cis*-Asarone	2.93
2	222	222	195,177, 147,119,91,65	21.79	**Apiole**	53.02

and *trans*-anethole in the same part of the plant.

Conclusion

The wild fennel *Foeniculum vulgare var. vulgare*contains many valuable volatile oils. In this study, conditions for

seeds sterilization, plant germination, callus production and plant micropropagation were investigated. The best condition for seed sterilization was immersing in ethyl alcohol (70%) for 3 min along with immersing in sodium hypochlorite (5%) for 25 min showed complete germination with no contamination.

The best hormonal combinations for callus induction were

(TDZ 0.5+2,4-D 1+BAP 0.1) and (2,4-D 2+BAP 0.25) which also showed the best growth rate particularly from stem explants. Also, NAA 1+ BAP 0.1 hormonal combination was the best for organogenesis while TDZ 0.5 +2,4-D 1 +BAP 0.1 hormonal combination was the only one that showed somatic embryogenesis. Transplantation trials were ultimately conducted but never completed as wilting was occurring each time due to weakness of root system. Even after hardening of root system, plants remained vital for longer time but ended with wilting. GC-MS analysis of callus and micropropagated plants showed valuable volatile oils particularly apiole which showed significantly higher percentages in the micropropagated plants. GC-MS analysis of the volatile oils of fruits of cultivated, wild fennel as well as the aerial parts of the wild fennel showed different medicinally valuable volatile oils but the most notable point was the reduction of estragole percentage in the fruits of the wild fennel in comparison of the fruits of the cultivated fennel.

Conflict of interests

Theauthors didnot declare any conflict of interest.

REFERENCES

Alexandrovich I, Rakovitskaya O, Kolmo E, Sidorova T, Shushunov S (2003). The effect of fennel (Foeniculum vulgare) seed oil emulsion in infantile colic: a randomized, placebo-controlled study. Altern.Ther. Health Med. 9: 58-61.

Aprotosoaie AC, Hăncianu M, Poiată A, Tuchiluş C, Spac A, Cioană O, Gille E, Stănescu U (2007). In vitro antimicrobial activity and chemical composition of the essential oil of Foeniculum vulgare Mill.Rev. Med.Chir. Soc. Med. Nat. Iasi. 112: 832-836.

Chevallier A (1996). The Encyclopedia of Medicinal plants, A DK Pub. Book, 1sted., 210.

Egyptian Pharmacopoeia (1984).English Text, 3rd ed., University Press, Cairo.

Guang-shou T, Man-ling L, Feng-feng M, Yan H, Peng Y, Lei S, Min C (2011). Anti-inflammatory and Analgesic Effects of Volatile Oil Extracted from Foeniculum vulgare Mill.Seeds.Prog.Mod.Biomed. 2:044.

Gulfraz M, Mehmood S, Minhas N, Jabeen N, (2008). Composition and antimicrobial properties of essential oil of Foeniculum vulgare. Afr. J. Biotechnol. 7: 4364-4368.

Haze S, Sakai K, Gozu Y (2002). Effects of fragrance inhalation on sympathetic activity in normal adults.Jpn J.Pharmacol. 90: 247-253.

Kapai V, Kapoor P, Rao I (2010). In Vitro Propagation for Conservation of Rare and Threatened Plants of India-A Review. Int. J. Biol. Technol. 1:1-14.

Kishore RN, Anjaneyulu N, Ganesh MN,Sravya N (2012). Evaluation of anxiolytic activity of ethanolic extract of Foeniculum vulgare in mice model. Int. J. Pharm. Pharm. Sci. 4:584.

Leung AY, Foster S (1996). Encyclopedia of Common Natural Ingredients Used in Food, Drugs and Cosmetics, A Weily-Interscience Publication, John Wiley and Sons, INC. New York, Chichester, Briban, Toronto and Singapore, 2nd ed.p.240.

Mansour SA, Heikal TM, Refaie AA, Mossa A (2011). Antihepatotoxic activity of fennel (Foeniculum vulgare Mill.) essential oil against chlorpyrifos-induced liver injury in rats.Glob. J. Environ. Sci. Technol. 1: 1-10.

Mills S, Bone K, (2000). Principles and Practice of Phytotherapy, Modern Herbal Medicine, Churchill Livingstone, Edinburg, London, New York, Philadelphia, Sydney, Toronto. pp. 379-84.

Ozbek H (2005). The anti-inflammatory activity of the Foeniculum vulgare L. essential oil and investigation of its median lethal dose in rats and mice. Int. J.Pharmacol. 1:329-331.

Singh B, Kale R (2008). Chemomodulatory action of Foeniculum vulgare (Fennel) on skin and forestomachpapillomagenesis, enzymes associated with xenobiotic metabolism and antioxidant status in murine model system. Food Chem. Toxicol. 46: 3842-3850.

Tanira MO, Shah AH, Mohsin A, Ageel AM, Qureshi S (1996). Pharmacological and toxicological investigations on Foeniculum vulgare dried fruit extract in experimental animals. Phytother. Res. 10:33-36.

17

Healthcare workers uniforms (HCWU)

Salah A. Latief Mohammed

Textile Engineering Department, College of Engineering, Sudan University of Science and Technology, Khartoum North, Sudan. E-mail: sam_1112@yahoo.com.

There is a major concern for healthcare workers (HCW) regarding transmission of bacteria to and from their patients. Because of this potential contamination, protection is a major issue. Healthcare workers uniforms (HCWU) are often used as barriers to help eliminate or reduce the risk of infection for both the doctor and the patient. Without sufficient barriers, harmful bacteria can reach and penetrate the skin of surgeons and/or patients, with an associated potential for infection. In addition, when pathogens contaminate (HCWU), they can be transmitted to other persons beyond the initial wearer. For the prevention of surgical infection through contamination from aqueous liquids and bacteria, guidelines have been issued for surgical gowns by several organizations. This paper looks into this crucial issue in order to highlight the importance of this subject and discuss the parameters that govern it.

Key words: Healthcare workers (HCW), chitosan, AEGIS microbe shield (AMS), poly-hexamethylene biguanide (PHMB), reusable, disposable.

INTRODUCTION

During every hour of a major surgical operation, about 30,000 to 60,000 organisms are deposited on a three to four meter squared sterile field and about one –half of all surgical procedures resulted in an accident where at least one medical worker was contaminated with blood (Conn et al., 1986). During these operations, one of the primary sources of contamination of healthcare workers (HCW) is from open wounds.

However, a major concern for HCW is the problem of transmission of pathogens and bacteria from their patients to themselves and the reverse contamination. Any blood contamination could pose a risk of transmission of bacteria. Because of this potential contamination, protection is a major concern. Healthcare workers' uniforms (HCWU) which include surgical gowns, scrub suits, lab coats, and nurses' uniforms, are often used as barriers to help eliminate or reduce the risk of infection for both the doctor and the patient. Surgical gowns, which were used as early as the 1800s, are traditionally made from cotton fabric (Smith and Nichols, 1991). Although cotton gowns are comfortable for wearer, unfinished cotton fabrics does not protect against bacteria penetration, or the penetration of biological liquids (e.g. blood, body fluid) and associated bacteria (Laufman et al., 1975). Without sufficient barriers, harmful pathogens can reach and penetrate the skin of surgeons and/or patients, with an associated potential for infection. In addition, when pathogens contaminate (HCWU), they can be transmitted to other persons beyond the initial wearer.

The center for disease control (CDC) has estimated that 8.8 million people work in the healthcare industry, and 27 million surgical procedures are performed in the U.S. every year (Mangram et al., 1999). From these procedures, surgical wound infections can be transferred from worker to patient or vice versa (Hughes et al., 1983). A way to combat these infections occurring from penetration or transmission is for workers to wear proper HCWU.

STANDARDS FOR HCWU

Standards for (HCWU) are very important for the welfare of healthcare workers (HCW) as well as their patients. Several organizations have made recommendations or mandates on how to protect HCW as well as patients from exposure to blood borne pathogens and bacteria. For the prevention of surgical infection through contamination from aqueous liquids and bacteria, guidelines have been issued for surgical gowns by several organizations. The following organizations have provided detailed information concerning (HCWU).

Association of operating room nurses (AORN)

AORN is a professional organization of preoperative registered nurses. This organization promotes quality patient care through education, standards, services, and representation. AORN issued standards as early as 1975 for draping and gowning materials (AORN, 1975). It proposed that surgical drapes and gowns should be made of fabrics that form an effective barrier by eliminating the passage of bacteria between sterile and non-sterile areas. An effective barrier should be fluid resistant (e.g., blood and aqueous), abrasion resistant to eliminate bacteria penetration, and lint free to reduce the number of particles in the air. These guidelines emphasized that HCW have a serious concern for barrier protection clothing. AORN also recommended that surgical gowns need to be changed after becoming visibly soiled and then laundered in an approved facility, in order to maintain their barrier properties (AORN, 1993). Most importantly, HWCU manufacturers need to provide data to customers (e.g., HCW) regarding the bacteria and liquid barrier performances of their products (AORN, 1992).

Center for disease control (CDC)

Since 1946, the CDC is the leading federal agency for the protection of health and safety of U.S. citizens both in the United States and in their travel abroad. Today, the CDC is a vital force in protecting the U.S. public from most widespread diseases that could affect public health.

The guidelines of the CDC mandate that surgical gowns and drapes, either disposable or reusable, should be impermeable to liquids and viruses and be comfortable to wearer (Bolyard et al., 1998). If a HCWU (e.g., scrub suit) is soiled, contaminated, or penetrated by any infectious material, the CDC recommends that it be changed immediately.

Occupational safety and health administration (OSHA)

OSHA is a division of the Department of Labor and was established in 1971 to save lives, prevent injury, and protect workers health. OSHA recommends that appropriate protective clothing must be worn to form an effective barrier when an employee has a potential for exposure on the job (OSHA, 1989). The type of clothing needed depends upon the occupational task and the degree of potential exposure. If the clothes are potentially soiled from blood or other potentially infectious materials, protective clothing must be worn to prevent the

employees underlying clothing from contamination. Fluid-resistant clothing must be worn when workers could become contaminated through splashing or spraying of blood or other potentially infectious materials. Because a larger volume of blood and other potentially infectious materials are associated with the work of the HCW, a specific protective type of barrier clothing is needed.

OSHA further recommends that the contaminated uniform should be removed at the end of the work shift. A contaminated uniform should not be taken home but be left at the work area for cleaning, laundering, and/or disposing. Furthermore, Matthews et al. (1985) stated that HCWU should be comfortable, cheap, durable, non-toxic, and able to resist transfer of bacteria. Bacteria have different modes of transports (that is, air particles, blood, body fluids).

LIFE CYCLES OF TEXTILES FOR HCWU

Fabrics that are used for HCWU have two life cycles: Reusable and disposable. Reusable fabrics are usually made of woven fabric and often woven from cotton or polyester yarns or a blend of these two fibres. These fabrics are laundered and sterilized after use in order to remove stains and kill bacteria.

Based on Batra's (1992) report approximately 20% of surgical gowns are of the reusable type. In a cost study, reusable fabrics were found to be more cost-effective than disposable fabrics (DiGiacomo et al., 1992). The benefits of reusable fabric include less solid waste from limited disposal and more comfort to the wearer. In contrast, the problems associated with reusable fabrics include the loss of durability and the reduction of barrier protection after repeated washing (Laufman et al., 1975). If the barrier protection of the fabric is removed or weakened after repeated washing, the fabric becomes useless as protection for HCW.

On the other hand, disposable gowns are for single use only. Furthermore disposable fabrics are mainly used in surgical gowns but reusable fabrics are found in various (HCWU) (e.g., nurses. uniforms, lab coats, and scrub suits). They are generally made from a non-woven fabric and contain either wood pulp/polyester fibers or olefin (that is, polypropylene) fibers (Huang and Leonas, 1999). They are good in providing protection. However, the problems associated with disposable fabrics are high-risk contamination, environmental issues through waste and landfill, expense, and discomfort if they are reinforced with a plastic film (DiGiacomo et al., 1992; Hatch, 1993). Two benefits of disposable fabrics are that they do not need washing after use (that is, they are not reused), and they are already sterilized prior to use. By adding a plastic film to disposable fabrics, they can be made impermeable to bacteria. Reusable HCWU is usually more comfortable than disposable fabrics; however, reusable cotton fabric without a finish does not protect against bacterial penetration (Leonas, 1993).

Leonas (1993) studied disposable surgical gowns and found that improved repellency and reduced pore size of these gowns contributed to barrier protection. Some problems associated with disposable fabrics are expense, risk of contamination with disposal outside of the hospital setting, and other environmental issues related to disposal (DiGiacomo et al., 1992). In addition, although a plastic film added to disposable fabrics can increase protection, it could make the fabric bulky, hot, and uncomfortable to the wearer (Hatch, 1993), and increases the problems for disposal solutions.

BARRIER PROTECTION OF TEXTILES FOR HCWU

Both reusable and disposable HCWU have been used to provide barrier protection for HCW (Leonas, 1993, 1998; Leonas and Jinkins, 1997). Study results have shown that disposable HCWU) could provide better barrier protection if they were reinforced with a plastic film, and reusable HCWU could provide better protection if a textile finish such as a water-repellent finish or antibacterial finish was applied (Huang and Leonas, 1999; Laufman et al., 1975). A textile finish is defined as .the process of applying mechanical energy, thermal energy, or chemical materials to a textile product to alter its end-use performance (American Association of Textile Chemist and Colorists (AATCC), 2000: 397). One specific textile finish is the barrier protection finish. The barrier protection finish is usually a chemical finish, which is formed by bonding a chemical to the fiber or fabric. Such a finish forms a barrier or coating on the fabric and enhances the fabric's barrier protection properties. Examples of barrier protection finishes are oil/water-repellent and antibacterial finishes. Oil/water-repellent finishes cause oil/water to bead on the fabric surface, while allowing perspiration to pass through the spaces between the fabrics warp and filling yarns (Hatch, 1993). Fabrics with the oil/water-repellent finish can reduce the spread, wetting, and penetration of oil or water on and into the fabric.

Laufman et al. (1975) used the water-repellent finish, Quarpel, as a barrier protection finish against the bacteria *Serratia marcesens* and found that the finish inhibits bacterial penetration. Three types of mechanisms (that is, controlled-release, regeneration, barrier block) for antibacterial agents are used to control or inhibit bacteria. Those mechanisms are:

1. The controlled-release mechanism: It is the most commonly used among the antibacterial agents (Brumbelow, 1987). In the controlled-release finish, chemicals in the finish, are released from the fabric in enough quantities to kill or inhibit the growth of bacteria. The antibacterial agent, triclosan has been used as a controlled released mechanism on non-woven fabrics

(Huang and Leonas, 1999);

2. The regeneration model: It was first established by Gagliardi in 1962. In this model, an antibacterial chemical finish is applied to the fabric and is continually replenished by a bleaching agent during laundering. The antibacterial agent, monomethylol-5,5- dimethylhydantoin (MDMH) has been used as a regeneration mechanism on woven fabrics (Sun and Xu, 1999);

3. Barrier-block mechanism: It inhibits bacteria through direct surface contact. The antibacterial agent bonds (yhat is, covalent, ionic) to the fabric surface thus making the fabric an effective barrier against bacteria and remains durable during laundering.

The first two antibacterial finish methods have known problems in usage with HCWU. Problems with the controlled-release mechanism are its durability after laundering and leaching of the agents from the fabric. Leaching can often cause problems if the antibacterial agents come in contact with skin of HCW. These agents have the potential to affect the normal skin flora, which could lead to extreme skin irritation and cause dermatitis (Sun and Williams, 1999). In addition, leaching can make skin bacteria build a tolerance to the agent. Additional problems for HCWU also occur for fabrics using a regeneration mechanism. The agents that use the regeneration mechanism require chlorine bleaching to activate its antibacterial properties after laundering; however, over time chlorine can degrade natural fibers such as cotton, which is often used in reusable HCWU (Hatch, 1993).

Barrier-block mechanisms do not pose the problems currently found with the other two methods. The agent that uses the barrier-block mechanism does not leach on the fabric surface and does not need bleaching to continue its effectiveness. They are bonded on the fabric surface and remain fixed to the surface, thereby killing any bacteria that come in contact with the fabric (Malek and Speier, 1982). Chitosan, AEGIS Microbe Shield (AMS), and poly-hexamethylene biguanide (PHMB) are three agents that use the barrier-block mechanism and are currently available in the marketplace. Chitosan has been used in many applications such as dietary additives because of its biodegradability and non-toxicity to mammals (Kim et al., 1998). However, Lin et al. (2002) indicated that chitosan has water fastness problems after repeated laundering, and therefore, it is not appropriate to be used on HCWU.

AMS, in contrast to chitosan, is found in many antibacterial-containing products such as socks, bed linen, and camping materials (Burlington Industries and Dow Corning Corporation, 1985). Many of these personal use items are often washed. PHMB is found in swimming pool sanitizers, preservation, and personal care products (Payne amd Kudner, 1996). In the studies on the efficacy of AMS and PHMB, these two agents have been evaluated as antibacterial agents on the reduction of odor

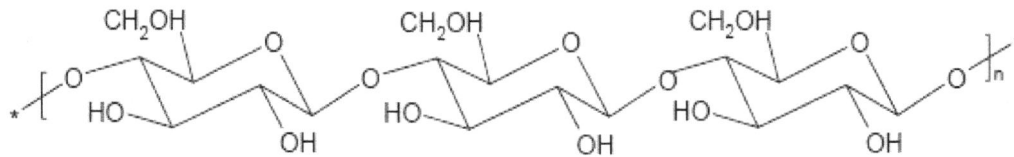

Figure 1. Polymer structure of cellulose.

(Malek and Speier, 1982; Payne and Kudner, 1996); however, their efficacy as antibacterial agents on the reduction of bacteria after laundering has been examined only in a limited arena. Malek and Speier (1982), in one study, examined the efficacy of AMS and found that it had significant antibacterial activity when used with a woven fabric. In addition, one study was found on the examination of antibacterial activities of PHMB combined with a fluoro-chemical compound, a water-repellent agent, on non-woven gowns before laundering (Huang and Leonas, 1999). The results showed that PHMB had significant antibacterial activity alone and when it was added to the fluoro-chemical compound. Payne and Kudner (1996) hypothesized that PHMB would show better durability than AMS due to its ability to bind at the different surfaces of cotton fabric. Their claim was supported by the information that AMS is bound to the fabric through one cationic group, but PHMB is bound to the fabric by multiple cationic groups. However, no study was found with the comparison of the antibacterial activity between AMS and PHMB on fabrics after repeated laundering.

The antibacterial agent, 3-trimethoxysily-propyldecyl-dimethyl ammonium chlorine (AEGIS Microbe Shield (AMS)) has been used as a barrier-block mechanism on cotton and cotton blended fabrics (Malek and Speier, 1982) as well as PHMB, which is commercially known as Reputex, has been used on woven and non-woven fabrics (Huang and Leonas, 1999; Wallace, 2001). Antibacterial finishes can be found on many products such as hosiery, shoe insoles, towels, underwear, bedding, and active wear (Thirty, 2001).

TYPES OF HEALTHCARE WORKERS UNIFORM (HCWU)

Healthcare workers uniform (HCWU) includes surgical gowns, scrub suits, lab coats, and nurses uniforms. They are categorized as reusable or disposable. Scrub suits, lab coats, and nurses uniforms are often made of reusable fabrics (Neely and Maley, 2000). However, surgical gowns are frequently made of either reusable or disposable fabrics (Granzow et al., 1998). The characteristics of reusable and disposable HCWU are dependent on fiber type, construction, and finishes to determine its optimal usage for protection. Reusable fabrics used for HCWU can be used over 50 times after laundering and sterilization (Sun and Xu, 1998); whereas, disposable fabrics for HCWU are used only once before being discarded.

Reusable HCWU

Reusable (HCWU) is used in many aspects of the healthcare industry such as in clinics, hospitals, and veterinary offices. Batra (1992) reported that reusable surgical gowns continue to represent 20% of the total number of (HCWU) being used. Reusable (HCWU) are often made of cotton, polyester, or cotton and polyester blend woven fabrics with a plain weave (Neely and Maley, 2000). In a plain weave, the warp yarn operates in an over-one and under-one pattern with the filling yarn throughout the fabric (Hatch, 1993). This weave pattern can provide a sturdy, comfortable fabric when made from cotton or a cotton/polyester blend fibers.

Cotton fabric is used for HCWU because of its properties of comfort, durability, and ease of care (Lee et al., 1999). The kidney bean shape permits the cotton fiber to contact skin randomly instead of continually, which is considered comfortable especially when the wearer perspires (Hatch, 1993).

The problem associated with cotton use for (HCWU) is its ineffectiveness in protection of HCW against bacterial penetration and transmission (Pissiotis et al., 1997). Cotton is hydrophilic due to its many hydroxyl (OH) groups (Figure 1). The OH groups make the fiber polar, which enables the fiber to attract water molecules. This property can increase the wearing comfort of HCWU containing cotton. Absorbency is important to comfort because cotton fibers can wick perspiration from the body of the wearer; however, the water molecules can discharge static electricity on the fiber, which accumulate and act as carriers for bacteria (Vigo, 1978). In addition, the hydrophilic nature of cotton allows for seepage and penetration when cotton HCWU is splashed with liquids (e.g., blood, body fluids).

Polyester is a synthetic fiber, which is usually a transparent white or off-white color. The longitudinal view of the polyester fiber reveals a smooth, rod-like shape, and its cross section is round or trilobal (Needles, 1981). The most common type of polyester is polyethylene terephthalate (PET), and it is composed of methylene groups, carbonyl groups, ester links, and benzene rings (Figure 2).

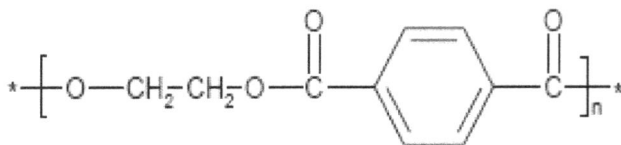

Figure 2. Polymer structure of polyester.

HCWU made of polyester are very durable due to the strength of the fibers. The well-aligned amorphous region of the polyester fiber makes the fiber very durable. The round, smooth, and flat shape of polyester can become uncomfortable because the fiber can directly stick to the skin of the wearer.

Polyester is a hydrophobic fiber, which means that it is non-polar and, therefore, does not attract water. The hydrophobicity of polyester can create a fabric environment that becomes uncomfortable if the wearer perspires. The polyester fibers would not be able to wick the perspiration or moisture away from the body, due to lack of hydrogen bonding in comparison to the structure and wicking properties of cotton. In addition, because of the hydrophobic characteristic of polyester, if the garment becomes contaminated, stains will become difficult to remove through laundering (Gohl and Vilensky, 1983).

A fabric with a polyester and cotton blend fiber content is the most common fabric type used in HCWU (Neely and Maley, 2000). Neely and Maley (2000) reported that polyester and cotton blended fabrics are used primarily for scrub suits, lab coats, and nurses uniforms. One of the reasons why the blending of polyester and cotton fibers is so successful for HCWU is their combined properties of comfort from cotton fibers and durability from polyester fibers (Hatch, 1993). Fabrics containing a polyester and cotton blend are stronger than fabrics made of 100% cotton and are more absorbent than fabrics made only of 100% polyester.

Comparison of various types of reusable gowns

The fiber content and bacterial transmission have been the focus of some studies using various fabrics found in HCWU. Laufman et al. (1975) conducted a study of bacterial transmission on various surgical gowns fabrics. One gown was made of a double layer of 100% regular cotton fabric, and the other gown was made of a single layer of tightly woven 100% Pima cotton fabric. Pima cotton has longer and more uniform staple fibers than regular cotton. No treatment was applied on the double layer regular cotton fabric. The Pima cotton fabric was evaluated in various conditions: (a) Before a water-repellent finish; (b) After a water-repellent finish but before washing, and (c) after a water-repellent finish and 2, 25, 55, and 75 launders cycles and sterilization. The tests for transmission were conducted after 5 and 30 s as well as after 1, 5, 15 and 30 min. Pressures were exerted

on the gowns with weights to simulate stresses that a surgeon exerts during surgical operations. The results showed that the untreated, double layer, regular cotton fabric and the untreated Pima cotton fabric did not prevent bacterial transmission. The treated Pima cotton fabric did not show any transmission even after 75 laundering cycles when the test was conducted after 15 min of contact. When the test was conducted after 30 min of contact, treated Pima cotton fabric that had been laundered for 75 cycles did show bacterial transmission. Comfort changes were not measured in this study.

Leonas (1998) conducted a study that examined the protection properties of several reusable fabrics after laundering. Three woven fabrics, containing one of three fiber contents - (a) cotton, (b) polyester, or (c) polyester and cotton blend were compared. The results showed that only the polyester fabric did not exhibit any penetration of Staphylococcus aureus (S. aureus) after laundering.

Contrasting results have been found in other studies, which also examined fiber content as a variable in preventing bacterial penetration and transmission. Smith and Nichols (1991) conducted a study on various gown fabrics. One gown was made of a single layer of 50/50% polyester and cotton blend fabric, and the other gown was made of a double layer of 100% polyester fabric. The researchers used an apparatus to simulate abdominal pressure that occurs during surgery. The pressures were evaluated from 0.25 to 2.0 psi between 1 s and 5 min. Both gowns allowed maximum 37 and 53% penetration, respectively after 5 min at pressures exceeding 1.0 psi. Another study was conducted by Leonas and Jinkins (1997) on three reusable surgical gowns. One gown was made of a single layer of 100% polyester fabric, a second gown was made of a double layer of 100% polyester fabric, and the third gown was from a fabric with a single layer of 50/50% polyester and cotton blend. The gowns were tested for liquid penetration and bacterial transmission against S. aureus and Escherichia coli (E. coli). The results showed that both the single and double layers of the 100% polyester gowns had liquid penetration in three of the six trials. The gown with the double layer of polyester allowed bacterial transmission of E. coli and the gown with a single layer of polyester allowed liquid penetration of S. aureus. The single layer, 50/50% polyester and cotton blend gown provided no resistance to either liquid penetration or bacterial transmission of S. aureus and E. coli.

Some results showed that a 100% polyester fabric resisted penetration better than a 50/50% polyester and cotton blend fabric (Smith and Nichols, 1991). In contrast, some results showed no difference among fabrics with varying fiber contents. Lastly, no difference in barrier protection was found in one study between reusable fabrics with a single layer and reusable fabrics with double layers of the same fiber type (Leonas and Jinkins, 1997).

Disposable HCWU

Disposable HCWU are mainly used for surgical applications. In most operating rooms, non-woven fabrics are the most commonly used disposable textiles and represent an expenditure of over $1.5 billion per year (Huang and Leonas, 1999). Non-woven fabrics are used in approximately 80% of all surgical procedures. An average of three billion square yards of non-woven fabrics is consumed for surgical textiles each year (Sun et al., 2000). Another disposable fabric used for HCWU is tissue, usually fiber or scrim reinforced (Laufman et al., 1975). Scrim reinforced tissue is strengthened by a polyester fiber web, and varies from fiber tissue which is tissue made from fibers (that is, cotton or polyester).

COMPARISON OF VARIOUS TYPES OF DISPOSABLE GOWNS

Laufman et al. (1975) tested various disposable surgical gown fabrics for bacterial penetration of *Serratia marcesens*. These fabrics came from different manufacturers and were made of a (a) single layer of spun-laced non-woven, (b) single layer of wet-laid non-woven, (c) scrim reinforced tissue, (d) fiber reinforced tissue, and (e) spread tow plastic film composite. A pressure of two kilograms was used to simulate a surgeon's elbow as he/she leans on the operating table. After five minutes of contact, the fiber reinforced tissue allowed bacterial transmission in most of the trials, and the wet-laid non-woven failed in one of six trials. After 15 min of contact, both the scrim reinforced tissue and the spun-laced non-woven allowed some bacterial transmission. After 30 min of contact, all of the tested surgical gown fabrics allowed bacterial transmission except one fabric. Only the spread tow plastic film composite fabric remained impermeable to bacterial transmission.

Smith and Nichols (1991) also studied various types of disposable gown fabrics. One was made of wood pulp/polyester spun-lace, and the other was an olefin SMS. The evaluated gowns were (a) a single layer of fabric, (b) a reinforced fabric with a layer of the same fabric, or (c) a fabric reinforced with an impervious material. The fabrics were tested with a pressure apparatus.

The single layer, wood pulp/polyester spun-laced gown fabric had a maximum of 92% liquid penetration. The double layer fabric of wood pulp/polyester spun-laced had a maximum penetration of 73%. The single and double layers of olefin SMS gown fabrics allowed 30 and 9% penetration, respectively. All of the gown fabrics that were reinforced with impervious fabrics had no (0%) penetration.

Leonas (1993) studied bacterial transmission on five disposable fabrics that were commercially available. Three of the fabrics were made of wood pulp/polyester, and two were made of olefin. Among the three wood pulp/polyester fabrics, two were a single layer composition but were manufactured by separate companies. The third wood pulp/polyester fabric was a double layer composition. The two olefin fabrics were either a single or double layer. The bacteria used in the test were *S. aureus* and *E. coli*. The results showed that all fabrics allowed no bacterial transmission, except one of the single layer wood pulp/polyester fabrics. The author indicated that this fabric allowed bacterial transmission because the pore size of this fabric was significantly larger than pore size of the other fabrics. Leonas and Jinkins (1997) conducted a similar study on disposable gowns from several manufacturers and found similar results to Leonas study. The gowns in the Leonas and Jinkins study were made of either wood pulp/polyester or olefins that were either single or double layers. The single and double layered fabrics of the wood/pulp polyester content gowns did not result in any liquid penetration; however, both the single and double layers of olefin content gowns had liquid penetration in one and two of the six trials, respectively. Although the olefin content gowns did allow some liquid penetration, none of the gowns allowed bacterial transmission of *S. aureus* and *E. coli*.

COMPARISON OF REUSABLE AND DISPOSABLE GOWNS PROTECTION

Garibaldi et al. (1986) study showed that there was no difference in barrier protection from reusable gowns made of polyester/cotton blend woven fabrics and disposable gowns made of polyester spun-laced non-woven fabrics, used with inter-operative and post-operative wound infections. From the data of 500 patients operations, this study revealed that the bacterium *S. aureus* was found on 13.1% of reusable and 15.5% of disposable gown fabrics. The authors concluded that the bacteria protection of reusable and disposable fabrics were similar. Laufman et al. (1975) studied various types of reusable and disposable gowns and found that after 30 min of contact, reusable Pima cotton fabrics treated with a water-repellent finish did not allow bacterial penetration even after 55 laundering cycles.

The disposable fabrics made of a spread tow plastic film composite also did not allow any bacterial transmission. In contrast, both untreated reusable gowns and non-reinforced disposable gowns allowed bacterial penetration after 15 min of contact. The study of Smith and Nichols (1991) showed that both single and double layers of wood pulp/polyester spun-lace disposable fabrics allowed a liquid penetration of 92 and 73%, respectively. The single layer of 50/50% polyester and cotton blend reusable gown fabric allowed a maximum penetration of 37%, while the double layer of 100%

Table 1. Internet search of antimicrobial uniforms.

Manufacturer	Antimicrobial finished uniforms (Y/Yes, N/No)
Crest	N
Peaches White Swan	N
Med Gear	N
White Cross	N
PL of California	N
Caduceus	N
Cherokee	N
Barco	N
Disney	N
Scrub by Design	N
Premier	N
AllHeart	N
G.A.L.S.of California	N
ScrubMate	N
Life Uniform	N
L. A. Rose	N
Jasco	N
Graves	N
Scrubs-R-Us	N

polyester gown fabric allowed a maximum penetration of 53%. The single and double layers of olefin SMS disposable fabric allowed only 30 and 9% penetration, respectively. All disposable gowns with an impervious fabric layer prevented penetration in all trials. Leonas and Jinkins (1997) also found that reusable fabrics allowed some liquid penetration and bacterial transmission, but disposable fabrics with an impervious layer prevented liquid penetration.

CONCLUSION

As stated before, there is a major concern for the healthcare workers HCW regarding transmission of bacteria to and from their patients. Bacteria have different modes of transports (that is, air particles, blood, body fluids) that aid in their transmission. The readily available presence of bacteria on healthcare workers uniform greatly increases the potential for penetration and transmission of these bacteria. To reduce this problem and to protect the workers is to have a proper barrier as part of the HCWU. This uniform should be comfortable, durable, non-toxic, cheap, and able to resist bacteria transport. Although cotton gowns are comfortable for wearer, unfinished cotton fabrics does not protect against bacteria penetration, or the penetration of biological liquids (e.g. blood, body fluid) and associated bacteria. Studies conducted in this field have shown that some water-repellent finish can reduce bacteria transmission, such finishes have had very limited commercial use on HCWU. According to a market survey conducted by the

researcher (Table 1) through the Internet, no oil/water repellent finishes were found on commercially available HCWU. Few soil-release finishes were found to be available on some reusable HCWU. However, soil-release finishes cannot provide barrier protection. This point may provide an option of using antibacterial agents to treat HCWU and create a niche for companies selling HCWU. The process of applying the antibacterial finish through padding and drying is easy and economical.

A study is needed to determine the minimum amount of finish add-on to the fabric and it would be beneficial in reducing the cost of using enough antibacterial agents to inhibit a maximum of bacteria for HCWU. More research, such as developing new agents or making derivatives from commercially available agents to enhance the properties is recommended.

Using chemicals such as fluorocarbons to create a more hydrophilic surface, soils and stains could be removed more easily from (HCWU) with a soil-release finish. In addition to a water repellent finish, researchers suggested that antibacterial fabrics could be used to create barrier protection by preventing harmful bacteria from penetrating through the fabric. However, antibacterial agents which are placed on the surface of the fabric to inhibit bacteria growth must remain effective after repeated laundering.

Clothing comfort is a state of an individual's satisfaction indicating physiological, psychological, and physical harmony between the person and their environment. The length of time worn, type of operation for which the uniform is used, and the fiber content and construction of the garment are important factors in determining comfort for the wearer. The comfort of HCWU is important for several reasons.

When doctors feel hot in their uniforms, their performance may be impaired in the operating room or in the office. In addition, when a protective garment is not comfortable, it is not worn. If not worn, the HCWU is not providing a protective barrier to the HCW.

In order to achieve comfort, a balance of heat produce by the body and the change in environmental conditions are needed. Moisture transmission, heat transmission resistance, and air permeability are the three factors that can mimic this balance. For a garment to be considered comfortable, water vapor transmission from the skin must occur. Cotton reusable HCWU are usually more comfortable than HCWU made from other fiber contents because of its better water vapor transmission, which enables water to wick from workers skin. The air permeability of a reusable gown is affected by yarn and fabric structure. The tighter the twist of the yarn and the closeness of the fabric, the less air will permeate through the fabric. The air permeability of a non-woven disposable gown is affected by the distribution of the fibers and the pore size in the fabric.

Disposable gowns reinforced with a plastic film are usually hotter than reusable gowns because no air can permeate through the plastic reinforcement. Studies have reported that if a worker is uncomfortable in their uniform, they are more likely not to wear it properly.

Generally, the length of contact of fluids on the gowns made a difference in the amount of transmission (that is, the longer the contact, the greater rate of bacterial transmission). Variations in fiber content and fabric construction provided varying degrees of protection against bacterial transmission. Olefin SMS non-woven was better than wood pulp/polyester spun-laced non-woven in protection against liquid penetration; however, regular olefin non-woven fabrics had similar results in bacterial transmission to the wood pulp/polyester non-woven fabric. In addition, contradictory results were found regarding the function of layers in bacterial protection. In one study, non-woven gowns with double layers of woven fabrics were superior to those with a single layer; however, two other studies showed that no differences in bacterial transmission were found between non-woven gowns with a single layer and double layers of the same non-woven fabric. One constant result was that non-woven gowns with plastic or some other impervious fabric did not allow any liquid penetration or bacterial transmission.

Reported results varied in the comparison of reusable and disposable gowns for barrier protection. One study showed that disposable gowns had better protection than reusable gowns and the other study showed no difference. To prevent bacterial penetration, a finish such as water-repellent finish possibly needs to be added to reusable fabrics, and an impervious layer needs to be added to disposable fabrics.

With regard to the different mechanisms mentioned, it is clear that the controlled-release mechanism and the regeneration model have known problems in usage with HCWU. Problems with the controlled-release mechanism are its durability after laundering and leaching of the agents from the fabric. Leaching can often cause problems if the antibacterial agents come in contact with skin of HCW. These agents have the potential to affect the normal skin flora, which could lead to extreme skin irritation and cause dermatitis. In addition, leaching can make skin bacteria build a tolerance to the agent. Additional problems for HCWU also occur for fabrics using a regeneration mechanism. The agents that use the regeneration mechanism require chlorine bleaching to activate its antibacterial properties after laundering; however, over time chlorine can degrade natural fibers such as cotton, which is often used in reusable HCWU.

Barrier-block mechanisms do not pose the problems currently found with the other two methods. The agent that uses the barrier-block mechanism does not leach on the fabric surface and does not need bleaching to continue its effectiveness

The costs of reusable and disposable HCWU are difficult to ascertain because the cost of a gown represents not only the manufacturing and retail cost but also the values of safety and comfort. In general, disposable gowns are considered to cost more because of the large storage space needed for fresh gowns and the continued disposal fees for used gowns. DiGiacomo et al. (1992) reported a study comparing the expenses of operation rooms in two hospitals. One hospital used disposable gowns and the other used reusable gowns. The hospital that used disposable gowns spent $155,664 per year compared to an expenditure of $35,680 in the hospital that used reusable gowns. The figure for the expense of disposable gowns included the disposal cost, and the figure for the reusable gowns included the long-term expense of reusable gowns such as cost of washing, sterilizing, and repackaging. However, these comparisons are not exact because data from surgical gown companies are not standardized. The Baxter Healthcare Corporation stated that disposable and reusable gowns cost $3.10 and 3.60 per use, respectively (Jinkins, 1994). Another surgical gown company, Medline, calculated that reusable gowns cost about $3 per use and disposable gowns were $4 per use (Anders, 1993). According to the market survey through the Internet by the researcher, it was found that in 2008, reusable gowns ranged between $15 and 25 per gown depending on brand and style with an expected lifetime of at least 25 times, and most disposable gowns cost between $40 and 100 for 30 to 50 pieces per case with an average per gown price of $2.

REFERENCES

American Association of Textile Chemist and Colorist (AATCC) (2000). Technical Manual of the American Association of Textile Chemist and Colorist, 75. Research Triangle Park, NC: AATCC.

Anders G (1993, April 2). Hospitals are returning to reusable surgical supplies. Wall St J., p. B4

Association of Operating Room Nurses (AORN) (1992). Recommended practices: Protective barrier materials for surgical gowns and drapes. Assoc. Oper. Room Nurs. (AORN) J., 55(3): 832-835.

Association Operating Room Nurses (AORN) (1993). Recommended practices: Universal precautions in the perioperative setting. Assoc. Oper. Room Nurs. (AORN) J., 57(2): 554-558.

Association of Operating Room Nurses (AORN) (1975). AORN Standards OR wearing apparel, draping and gowning materials. Assoc. Oper. Room Nurs. (AORN) J., 21(4): 594-596.

Batra SK (1992). The non-woven fabrics handbook. Cary, NC: International Non-wovens and Disposable Association (INDA).

Bolyard EA, Tablan OC, Williams WW, Pearson ML, Shapiro CN, Deitchman SD (1998). Guidelines for infection control in health care personnel, 1998. Am. J. Infect. Contr., 26(3): 289-354.

Brumbelow JB (1987). The effectiveness and cleanability of antimicrobial finishes on carpet tiles. Text. Chem. Colorist, 19 (4): 27-31.

Burlington Industries and Dow Corning Corporation (1985). A new and durable antimicrobial finish for textiles. Retrieved April 21, 2001, from http://www.microbeshield.com/txfinish.htm.

Conn J, Bornhoeft JW, Almgren C, Mucha DP, Olderman J, Patel K, Herring CM (1986). In vivo study of an antimicrobial surgical drape system. J. Clin. Microbiol., 24(5): 803-808.

DiGiacomo JC, Odom JW, Ritota PC, Swan KG (1992). Cost contaminants in the operating room: Use of reusable versus disposable clothing. Am. Surg., 58(10): 654-656.

Garibaldi RA, Maglio SM, Lerer T, Becker DB, Lyons R (1986). Comparison of non-woven and woven gown and drape fabric to prevent intraoperative wound contamination and postoperative infection. Am. J. Surg., 152(5): 505-507.

Gohl EPG, Vilensky LD (1983). Textile Science. (2nd ed.). Melbourne: Longman Cheshire.

Granzow JW, Smith JW, Nichols RL, Waterman RS, Muzik AC (1998). Evaluation of hospital gowns against blood strike-through and methicillin-resistant Staphylococcus aureus penetration. Am. J. Infect. Cont., 26(2): 85-93.

Hatch K (1993). Textile Science. San Diego, CA: Academic Press.

Huang W, Leonas KK (1999). One-bath application of repellent and antimicrobial finishes to non-woven surgical gown fabrics. Text. Chem. Colorist, 31(3): 11-16.

Jinkins RS (1994). Influence of crosslinked polyethylene glycol on barrier and antimicrobial properties of surgical gown fabrics (Doctoral dissertation, University of Georgia, 1994). Dissertation Abstract International, 55: 10B.

Kim YH, Choi H, Yoon JH (1998). Synthesis of a quaternary ammonium derivative of chitosan and its application to a cotton antimicrobial finish. Text. Res. J., 68(6): 428-434.

Laufman H, Eudy WW, Vandernoot AM, Liu D, Harris CA (1975). Strike-through of moist contamination by woven and non-woven surgical materials. Ann. Surg., 181(6): 857-862.

Lee S, Cho JS, Cho G (1999). Antimicrobial and blood repellent finishes for cotton and non-woven fabrics based on chitosan and fluoropolymers. Text. Res. J., 69(2): 104-112.

Leonas, K. K. (1998). Effect of laundering on the barrier properties of reusable surgical gown fabric. Am. J. Infect. Control, 26(5): 495-501.

Leonas KK (1993). Evaluation of five non-woven surgical gowns as barriers to liquid strike-through and bacterial transmission. Int. Non-woven Disposable Assoc. (INDA) J., 5(2): 22-26.

Leonas KK, Jinkins RS (1997). The relationship of selected fabric characteristics and the barrier effectiveness of surgical gown fabric. Am. J. Infect. Cont., 25(1): 16-23.

Lin J, Winkelmann C, Worley SD, Kim J, Wei CI, Cho U, Broughton RM, Santiago JI, Williams JF (2002). Biocidal polyester. J. Appl. Polym. Sci., 85(3): 177-182.

Malek JR, Speier JL (1982). Development of an organosilicone antimicrobial agent for the treatment of surfaces. J. Coated Fabrics, 12(7): 38-45.

Mangram AJ, Horan TC, Pearson ML, Silver LC, Jarvis MD (1999). Guideline for prevention of surgical site infection, 1999. Am. J. Infect. Cont, 27(2), 97-132.

Matthews J, Slater K, Newsom SW (1985). The effect of surgical gowns made with barrier cloth on bacterial dispersal. J. Hyg., 95(1), 123-130.

Needles HL (1981). Handbook of textile fibers, dyes, and finishes. New York: Garland Press.

Neely AN, Maley MP (2000). Survival of Enterococci and Staphylococci on hospital fabrics and plastic. J. Clin. Microbiol., 38(2): 724-726.

Occupational Safety and Health Administration (OSHA) (1989). Occupational exposure to blood borne Pathogens: Proposed rule and notice of hearing. (29 CFR Part 1910; FR Doc. 89-12470, pp. 23042-23139). Washington, DC: Department of Labor, U.S. Government Printing Office.

Payne JD, Kudner DW (1996). A new durable antimicrobial finish for cotton textiles. Am. Dyestuff Reporter, 28(5): 26-30.

Pissiotis CA, Komborozos V, Papoutsi C, Skrekas G (1997). Factors that influence the effectiveness of surgical gowns in the operating theatre. Eur. J. Surg., 163(8): 597-604.

Smith JW, Nichols RL (1991). Barrier efficiency of surgical gowns: Are we really protected from our patients. pathogens? Arch. Surg., 126: 756-763.

Sun G, Xu X (1999). Durable and regenerable antibacterial finishing of fabrics: Fabric properties. Text. Chem. Colorist, 31(1): 21-24.

Sun G, Williams JF (1999). Dressing to kill. Chemistry and Industry, 17: 658-661.

Sun G, Xu X (1998). Durable and regenerable antibacterial finishing of fabrics: Biocidal properties. Text. Chem. Colorist, 30(6): 26-30.

Thiry MC (2001). Small game hunting: Antimicrobials take the field. Am. Text. Chem. Colorist (AATCC) Rev., 1(11): 11-17.

Vigo TL (1978). Modified cellulosics. New York: Academic Press.

Wallace M (2001). Testing the efficacy of polyhexamethylene biguanide as an antimicrobial treatment for cotton fabric. AATCC Rev., 1(11): 18-20.

18

Mimicry of a natural, living intra-epidermal micro pattern used in guided tissue regeneration of the human epidermis

Denis E. Solomon

96 Standishgate, Wigan WN1 1XA, England. E-mail: denissolomon@yahoo.com.

The methodology for the isolation of the living, intra-epidermal micro pattern using a tissue culture method is described. The disassembly of the stripped off epidermis (after Dispase digestion at 37°C) into epidermal brown rosettes (their morphology under a phase contrast microscope) constitute intra-epidermal micro patterns. The stepwise reasoning in the recognition of epidermal basal layer rosettes and the difference between the attached upright and inverted brown rosette spreading their content of epidermal cells onto a prepared extracellular matrix is disclosed. Mimicking the circular shape of the brown rosette layers via the use of a microscopic sterile, cornstarch granule (with the shape of a donut under the microscope) as a biodegradable, nutritional scaffold is proposed. A lightweight antibiotic ointment and the cornstarch granules were used together and this mixture's significance in the guided tissue regeneration of the epidermis after Mohs surgery for basal cell carcinomas, as a treatment for second degree burns, and for possible healing of donor sites after skin biopsies is discussed.

Key words: Natural micro patterns, Mohs surgery, basal layer, tissue repair, epidermal-melanin unit, human epidermis, Cellular Potts Model, microscopic nutritional cell scaffold.

INTRODUCTION

The human skin, the largest organ in the human body, has little of the dramatic colouring of animals, butterfly wings or peacock feathers. This has led to skin envy on the part of many humans, and in turn to colourful skin tattoos. However a dermatological problem, allergic eczema against henna dye allergens can arise and in steps the dermatologist whose job it is to recognize skin patterns. These can be referred to as macro patterns, like the pattern of human male *androgenic alopecia*, more commonly known as, 'male pattern baldness'. Living natural micro patterns inside the human body have not been described *in vitro*. There is one example of cellular mimicry and intriguingly, it also involves the skin and its cells, but ones that had undergone some unknown kind of molecular perturbation. Aggressive melanoma cells can form part of tumour vasculature; a process described as 'vasculogenic mimicry'; thought to be driven by expression of endothelial specific genes in melanoma cells such as ESM-1 and VE-Cathedrin (Maniotis et al., 1999; Hendrix et al., 2001; Gaggioli and Sahai, 2007). The epidermal brown rosette, which I described in 2002, looks like a brown micro mass under the phase contrast microscope. If upright or inverted, only the top **or** bottom is being shown in cellular photomicrographs. Likewise, a stack of coins in its sleeve wrapper, if photographed from above, only shows a single coin (with an edging of wrapper), but we do know there are hidden 'layers' of coins out of sight. Hence, the top coin might show 'heads'. On inverting the stack of coins, does the possibility exist that we will see 'tails'? Yes, it does. I will explain that there are normal brown rosettes which do not preferentially attach to an extracellular matrix and those which do attach. If inverted, their cell content is exposed and can be seen to be morphologically different to upright attached ones. Basal layer cells, like melanocytes and the epidermal-melanin unit, which are important in diagnostic tests for basal cell carcinomas can be now morphologically recognized *in situ*. A full explanation of factual occurrences leading to the elucidation of these facts will be described in stepwise fashion. Using a similar shape and size, in the form of self-aggregating granules comprising a nutritional source and a microscopic cell scaffold, I will show how this simple set of 'imitation' principles can be guided towards tissue repair of the human epidermis by describing their use in postoperative treatments. Along the way, I will question the standard practices of both medical doctors and cell culture scientists, like myself.

METHODOLOGY

Methodology was previously described (Solomon, 2002). Briefly, it consisted of trimming the subcutaneous fat from a skin biopsy sample, washing the trimmed biopsy sample with Dulbecco's phosphate buffered saline (DPBS) pH 7.4 and antibiotics, laying epidermis side down in a small pond of 2% Dispase solution, incubating at 37°C, 5% CO_2/ 95% O_2 until separation of the epidermis from the dermis could be achieved. Rinsing in complete culture medium (Medium 199 or IMDM) Iscove's modified Dulbecco's medium plus 20% (FBS) foetal bovine serum, followed by DPBS was done. The epidermis was stripped using a pair of forceps. A cellular photomicrograph (phase contrast microscopy) of the epidermal-dermal junction is shown in Figure 5. The stripped epidermis was inverted in DPBS for 30 min until disassembly into brown rosettes.

Basal layer brown rosettes

If brown rosettes are transferred to an extracellular matrix substrate (ECM), either autologous dermal fibroblast (secreted by both papillary and reticular fibroblasts) or (HUVECs), human umbilical endothelial cells ECM, some preferentially attach. These are basal layer rosettes. The attached brown rosettes are fed initially with complete medium then only 'topped up' with culture medium until they shed their epidermal cells onto the underlying extracellular matrix substrate. If the brown rosettes are inverted, the spilled cells will display morphologically, recognizable melanocytes and melanosomes and the epidermal-melanin unit (Figure 4). Specific cell markers will identify other basal layer epidermal cells, for example, the Merkel cells.

First Aid antibiotic ointment - Net weight ½ oz. (14g).

Bacitracin zinc 400 units, Neomycin 3.5 mg, Polymyxin B sulfate, 5000 units. No name of manufacturer was listed on tube...just Distributor....Walgreen Company, Deerfield, Illinois 60015-4616.

Sterile cornstarch granules

A cornstarch granule under the microscope has the morphology of a donut. Aggregates of cornstarch occur in solution or in a cell

culture medium allowing multi-directional cell attachment (Type of cell could vary). Wound fluid will determine type and size of aggregates occurring. Particle size was a consideration (see www.engineeringtoolbox.com). Red blood cells are 5-10 microns in size, whereas cornstarch is of the order of 0.1-0.8 microns, but the latter do form aggregates of different sizes. Published reports of the invasive use of cornstarch granules as a lone tool in wound healing or as glove powder inadvertently contaminating wound beds resulted in tissue inflammation. A Medline search for published material containing the concomitant use of cornstarch granules and an antibiotic (or antibiotics) yielded no results. These granules are being put forward for use as biodegradable, nutritional, cell scaffolds with the lightweight triple antibiotic ointment.

Guided tissue regeneration

The primary scaffold of wound healing is the fibrin blood clot and the cascade of subsequent events results in scar tissue (Broughton et al., 2006). Interference with the formation and amount of the fibrin clot by application of the nutritional cell scaffold and triple antibiotic ointment does cause subcutaneous tissue repair with little or no scar tissue. Three 'case reports' described below will serve to illustrate this tentative conclusion:

(1) An 88 year-old man had undergone a procedure for skin cancer on his right cheek. Post-operatively, he was medically advised to use Vaseline and Hydrogen peroxide for home wound care management. He and his elderly wife were upset with the lack of post-operative care and distressed with the amount of dribbling seepage. Within 6 weeks, with use of the cornstarch granules/antibiotic mixture applied after every two days with a change of normal dressing by his wife on home premises, the dermis was rebuilt flush with his cheek, the epidermis was also regenerated and only close examination could reveal two thin scalpel lines. There was no skin crater or scarring.
(2) A 60 year old man, obese and diabetic developed a basal cell carcinoma about an inch and a half distal to his left shin bone. After Mohs surgery, wherein the blood vessels were cauterized, he was prescribed Mupirocin (Bactobran) and Gentamicin sulphate ointment. Use of the nutritional scaffold and antibiotic ointment resulted in tissue repair after three weeks. I was advised that there had been no inflammation and looking at the skin site now, no one could tell it had been the focal point of a medical procedure.
(3) A second degree burn on the first web space of the dorsal surface of the author's right hand was obtained through accidental contact with the pre-heated filament of a domestic kitchen oven. No scar tissue resulted after concomitant repeated use of the nutritional scaffold and the triple antibiotic ointment with a simple gauze dressing. Too much of the cornstarch granules were used. A basket weave pattern of raised healing epidermis, similar to that described by Hoath and Leahy (2003) in their Figure 1 (a photomicrograph of human skin following transplantation to a mouse with severe combined immunodeficiency) was observed. This was naturally discarded eventually to leave an unblemished surface area. Since 1999, untreated burns on the dorsal surface of my right hand leave behind an area of perturbed discoloured skin, unlike the hard-to-perceive cosmetic result after treatment with cornstarch granules and the lightweight antibiotic ointment.

RESULTS AND DISCUSSION

Pattern formation in nature is best thought of as a

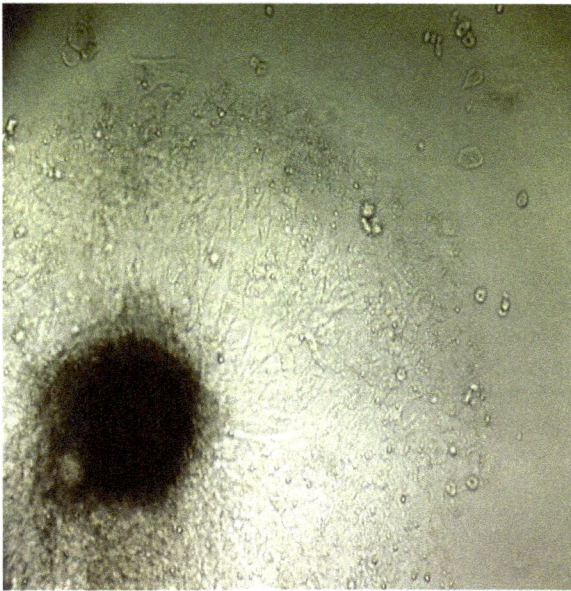

Figure 1. An *upright* attached brown rosette in an upright position displaying its intact micro pattern while shedding its cell load onto a prepared extracellular matrix and creating another micro pattern Magnification: x 100.

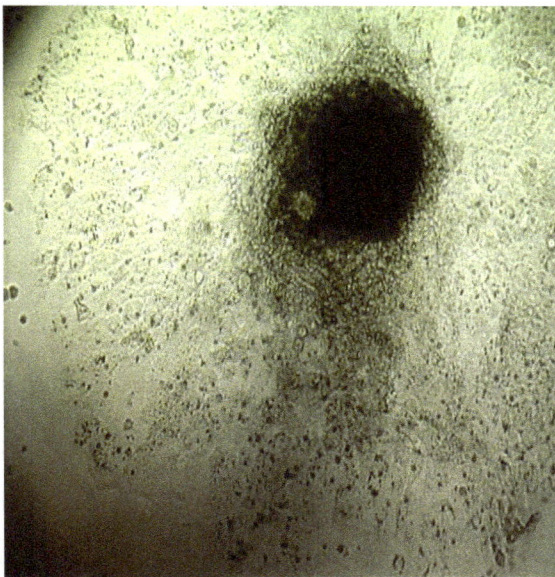

Figure 2. An *inverted* attached brown rosette shedding its cell load of epidermal cells. Magnification: x 200

process of symmetry breaking, that is, an initially homogeneous system becomes spatially, and sometimes temporally, inhomogeneous. Examples include the wind-dependent generation of sand dunes (Chuong et al.,

2006). The essential reference book for understanding patterns is 'The Self-Made Tapestry: Pattern Formation in Nature', by Philip Ball, Oxford University Press, 1999. There is no reference material on natural 'live' micro patterns in the human body's living tissues. Five are presented here; the attached basal epidermal brown rosette spilling its cell content, being upright or inverted and the epidermal-dermal junction.

Published after my 2002 report, Hoath and Leahy (2003) theorized the concept of functional epidermal units centred around a phi (1.618034) proportionality, providing a central organizing principle. The melanocyte: keratinocyte ratio was given as 1:36 (Frenck and Schellhorn, 1969) and the Langerhans cell: epidermal cell as 1:53 (Bauer, et al., 2001). These strikingly constant ratios allied with the fact that the epidermal brown rosette is circular do give this theory some credence.

Fifty years after Turing (1952) proposed a reaction-diffusion mechanism for biological pattern formation providing a biological explanation for animal coats, feather buds and fish skin patterns, an attempt was made to expand his framework by proposing a developmental mechanism. The two-dimensional (CPM) Cellular Potts Model (Zeng et al., 2004) speaks to enhanced local cell-cell adhesion and preferential cell-extracellular matrix (ECM) adhesion in 'condensation' (*in vitro* biological cell clustering), depending only on biological mechanisms and chemicals shown experimentally to be significant during patterning. Their experimental images at *low* density condensations, in their Figure 2, particularly (E) and (F), show a passing pattern resemblance to the epidermal brown rosette, which under the phase contrast microscope appears to be a wholly amorphous micro mass.

Of note, is their commentary that condensing precartilage cells employ transmembrane adhesion molecules, N-CAMs and N-Cathedrins (see my earlier comment on vasculogenic mimicry) and the time of maximal expression of N-Cathedrin (a calcium-dependent integrin of neural origin) corresponds to the period of active precartilage mesenchymal 'condensation' (also see Crosby et al., 2005).

Cross talk between the integrins (Monier-Gavell and Duband, 1997) was not considered in the CPM simulations, for example, control of N-Cadherin activity by intracellular signals elicited by beta1 and beta3 integrins in migrating neural crest cells. Different migrating cells, e.g. basal keratinocytes and neural crest cells have the bet1 integrins as a common denominator. Interestingly, Merkel cells of the basal layer of the epidermis are believed to function as sensory mechanoreceptors and are thought to be derived from either neural crest cells or basal keratinocytes. Whereas the basal keratinocytes may use the epithelial beta1 integrins as ECM receptors, the possibility exists that the neural crest cells may use

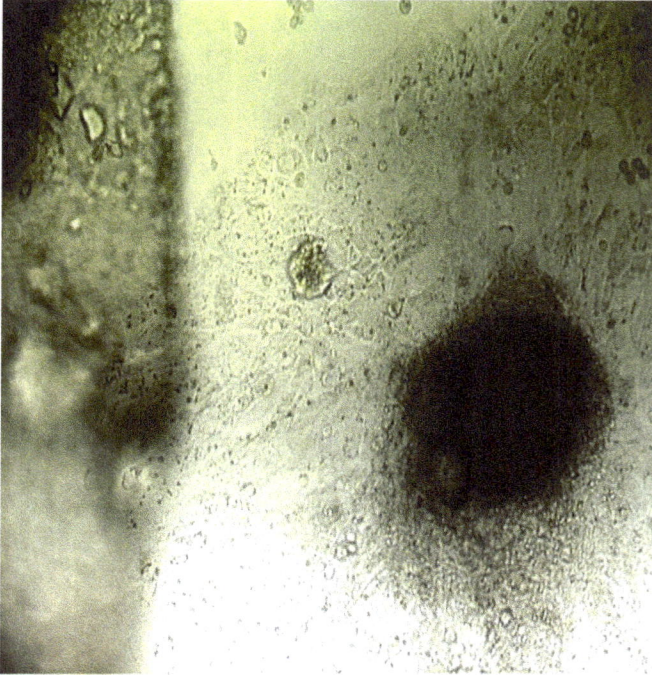

Figure 3. An *inverted* attached basal brown rosette. There appears to be an encapsulated basal cell lesion showing a cellular configuration within. To the left are the edges of the plastic tissue culture dish. Magnification: x 200.

them for embryonic neural cell motility exhibiting the versatility of integrin useage.

The CPM can be adapted to explain basal brown rosette parameters using the preferential attachment to an ECM which can be of embryonic origin, human umbilical vein endothelial cells (HUVECs) or an ECM secreted by dermal fibroblasts, thus exhibiting a dual capability in terms of their ECM requirement; neonatal or mesenchymal. Interestingly, the majority of cells in the epidermis, the keratinocytes, are affected by calcium concentration which could be a reflection of inherent integrin needs for proper functioning.

In published reports, epidermis is separated from the underlying dermis after enzymatic digestion with 0.25 % Trypsin- 1 mM EDTA (Ethylene diamine tetra-acetic acid) (Sorrell et al., 2004) or Thermolysin (Germain et al., 1993) or Dispase (Stenn et al., 1989). Trypsin continues to be used in spite of Barton and Marks' (1981) report of changes (invagination of desmosomes, vacuolation, and redistribution of tonofibrils) in suspensions of human keratinocytes directly attributable to use of this enzyme. Dispase, on the other hand, is a neutral protease which is both a fibronectinase and type IV collagenase (Stenn et al., 1989), dissolving the attachments between the basal keratinocytes and the basement membrane, without

disturbing the desmosomal intercellular junctions between adjacent cells (Green, 1991). It had been reported that Dispase causes the internalisation of basal cell adhesion dependent domains containing the $\alpha6\beta4$ integrin (Poumay et al., 1992), a receptor for laminin-5, an ECM component. This finding was based on Dispase-detached *cultured* human keratinocytes. It must be emphasized that caution should be exercised in extrapolating previous research reports on individual epidermal cells to the brown rosettes. 2% Dispase was used not at 12°C (Normand and Karasek, 1995), but at an incubator temperature of 37°C because it was privately thought that both cadaver and fresh skin tissue segments would better keep their normal constitutive properties at an ambient temperature that approximated *in vivo* conditions of the human body. The incubation time of 16 h can be shortened, depending on type of skin tissue sample (thin skin, wrinkly skin from knees and elbows et cetera). To neutralise Dispase, the enzyme manufacturers in their catalogues recommend using 5-10 mM EDTA. It has not been realised that leaving the enzyme Dispase in contact with human epidermis layer on an overnight basis has two consequences. The enzyme will lose its potency; hence no need to use 5-10 mM EDTA to neutralise and furthermore, the stripped epidermis layer disassembles (after immersion in DPBS) into brown micro masses (under a phase contrast microscope), I had christened as 'brown rosettes' in my 2002 paper. These cellular structures have not been previously described because in so- called 'established techniques' researchers have persistently used a double enzyme digestion (e.g. Dispase at various temperatures other than 37°C, followed by Trypsin-EDTA on the stripped off layer of epidermis) which yields single keratinocytes.

My experience with the burn on the dorsal surface of the first web space of my right hand provided nagging thoughts that I was overlooking a scientific concept. I could not understand or explain scientifically why there was no scar tissue or darkening of the burn area and why a spur of the moment remedy had worked so well. After 2002, I guessed that the appropriate cells had filled in the hole in the donut (the cornstarch granule) and that it represented a microscopic cell scaffold. I also realised that all cultured skin substitutes and (CEAs), cultured epithelial autografts (Green, 1991) had an inbuilt flaw. There was no inbuilt source of nutrition. Why had a lightweight triple antibiotic ointment sold over the counter at an American drugstore done its job so proficiently?

Writing a Scientific American article (March, 2008) 'Regrowing limbs: Can people regenerate body parts?' Professor Ken Muneoka had made the point that medical treatment *inhibits* tissue regeneration because of its focus on preventing infection. Medical management of subcutaneous wound healing has not been practised by placing a quantity of a microscopic absorbent scaffold composed of a nutritional binding agent together with an antibiotic ointment within the wound bed to allow epithelial-mesenchymal crosstalk to dictate the architecture and regeneration of damaged tissue at its own pace. In essence, the microscopic scaffold sets up the sub sequential evolution of its own two-dimensional coordinates within the wound bed. Wound seepage will cause the manifestation of the binding, aggregation and absorbent properties of the scaffold. Body heat will cause the antibiotic ointment to degrade to a semi-liquid form and bind to the scaffold resulting in something approaching, a 'powdery filler'. Cells within the damaged subcutaneous tissue will be directly fed by the carbohydrate nutritional agent.

The question then arose as a result of my friends' post operative experiences whether a full-strength prescriptive medical antibiotic ointment was really necessary to help regenerate an avascular tissue layer that had a thickness measured in millimetres. The thickness of human epidermis and the papillary dermis layer is 0.3 mm; a further 0.7 mm down lies the reticular dermis (Sorrell et al., 2004), shows a 'keratinocyte mass' formed in the presence of papillary fibroblasts. It was not fully understood that the well being of the epidermal layer of human skin was influenced by both papillary and reticular fibroblasts (contained in the living dermis) acting in consent with the dermal-epidermal junction possibly acting as a traffic policeman directing and/or monitoring the cross-talk between the epithelial and mesenchymal cell layers with their inherent autocrine, paracrine loops plus the release of diffusible growth factors and cytokines.

Looking back at my own paper with the benefit of hindsight and with a much clearer understanding of these matters, I can now fully understand why the brown rosettes when co-cultured with autologous human dermal microvascular endothelial cells and dermal fibroblasts (papillary and reticular fibroblasts) lifted off from all substrates. A major epidermal player was completely out of its normal, accustomed milieu. Back then, I (and others) thought a proliferation limit had been reached and I reported it in those terms.

It was simply not understood at the time that (1) the attached brown rosettes originated from the basal layer of the epidermis and (2) the spread of cells emanating from an attached brown rosette on the dermal ECM was composed of the *full complement* of basal epidermal cells. They were misidentified. Also not properly recognised was that the cellular photomicrograph showed only the *top layer* of a possible series of epidermal layers hidden and hence out of view, underneath or within. Secondly, there was no published identity test or specific cell marker for human basal keratinocytes. A single report appeared years after (Spichkina et al., 2006) describing the selective adhesion of human basal keratinocytes to ECM proteins. A simple observation made while idly fingering the smooth surface of a green leaf and inverting it to see a configuration of veins provided a 'Eureka' moment. Chlorophyll makes the green leaves, green but its porphyrin rings which act as photoreceptors are hidden within.

I dashed off to find and examine old cellular photomicrographs to scrutinise the now understood, attached 'inverted' brown rosettes and their spread of epidermal cells onto the ECM. This led me to a literature search for integrin receptors (in basal epidermal cells) with a similarity in apical and basal polarity. Only one report (Bishop et al., 1998) was found, stating that keratinocyte beta1 integrins in the basal layer of the epidermis did not display an intrinsic polarity with regard to their ligand-binding capacity. Therefore, it was concluded that upright or *inverted* 'brown rosettes' would adhere to the dermal fibroblast or HUVECs ECM in identical fashion and the *full complement of epidermal cells* would spread out from the attached brown rosette. If the attached *inverted* 'brown rosettes' were truly basal layer epidermal cells, the cell-spread, away from the micro mass, should expose resident keratinocytes together with other *in situ* epidermal cells within the micro pattern, for example, those described by Quevedo (1969, 1972) as *the epidermal-melanin unit*. Quevedo (1972) described this concept wherein 'a structural and functional organization of melanocytes and keratinocytes exists at levels of biological organisation that transcend those characterizing the individual component cells'. The refractive index of the melanosomes provided persuasive signposts in the positive identification of the melanocyte, which are not found as contiguous cells in the stratum basale of human epidermis, but within the epidermal-melanin unit. Hindsight is the only exact science.

The analogy I drew in my Introduction, of a stack of coins in its wrapper does have some relevance. It will be noted that the perimeter of an attached, *inverted* brown rosette (Figure 2 to 4) shows a lighter colour to the main body of the brown micro mass, indicative perhaps of the migration of a 'concentric circle layer of epidermal cells' away from the main body onto the ECM. On close examination of Figure 2 to 4, tiny circles are seen migrating away in a 'V' shape from the main body of the micro mass, reminiscent of nascent tiny buds arrayed within a floral pattern. Perhaps my fingering of a leaf was not far off the mark; in other words, a resonant botanical pattern/symmetry might be applicable here. The realisation that a cornstarch granule with its inherent attributes of being a microscopic cell scaffold was providing a form of mimicry was verified by me, and my friends' medical experiences with postoperative tissue regeneration.

Taken together, these facts question the standard

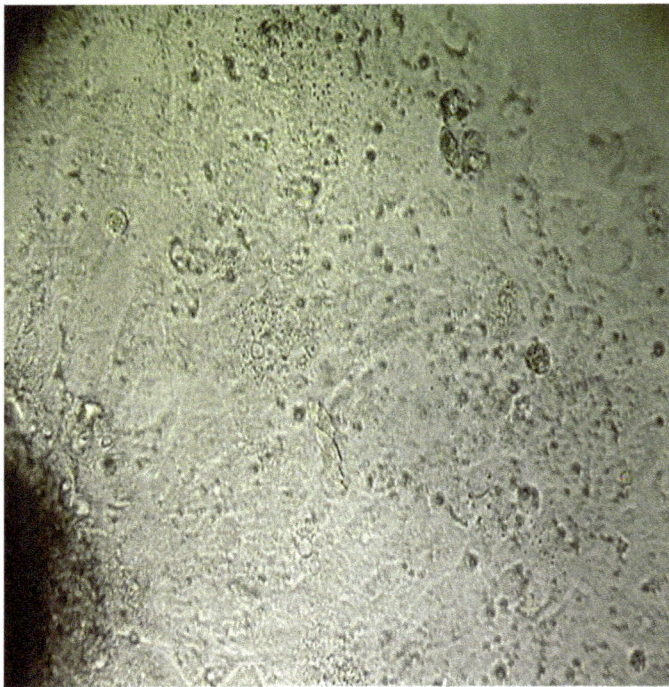

Figure 4. An *inverted* attached basal brown rosette shedding its cell load of epidermal cells. The *epidermal-melanin unit* can be seen. The refractive index of the melanosomes provided persuasive signposts in the positive identity of the melanocyte, which are not found as contiguous cells in the stratum basale of human epidermis. Magnification: x 100.   .

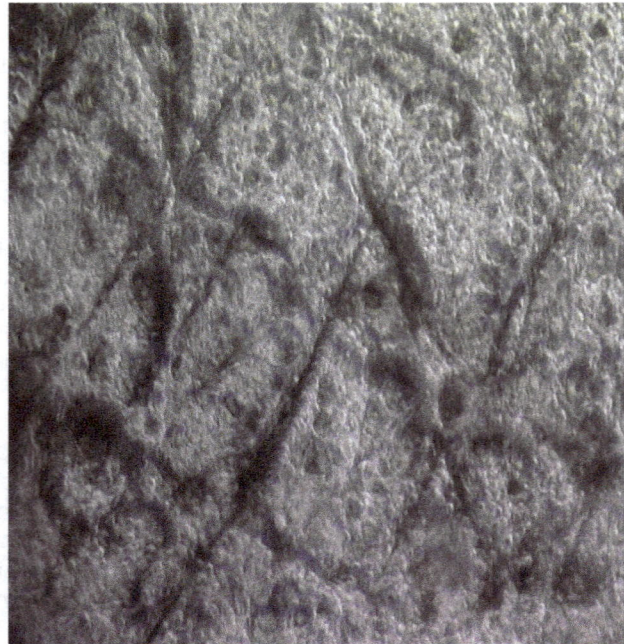

Figure 5. Another micro pattern. The epidermal-dermal junction (after stripping off the Dispase-digested epidermis) displaying dermal papillae and sweat ducts (the 'round holes'). Magnification: x 200.

medical practice of covering wounds with CEAs, a sheet material, only containing one type of epidermal cell, the keratinocyte. This also might explain why on occasion, they do not 'take', that is become vascularized and in certain instances are rejected. This is the routine 'Outside-In' of medical science. Perhaps, this needs to be reconsidered and as I am describing in this paper, an 'Inside-Out' technique might have its merits with respect to a better cosmetic outcome.

Surgical treatment for basal cell carcinomas (Rubin et al., 2005) includes curettage and electrodessication, cryosurgery, surgical excision, and Mohs micrographic surgery, for which there is no standardized post-operative treatment.

My friend's father deemed the medical advice of hydrogen peroxide and Vaseline, not fit for purpose. With my other friend, the medical prescription of Gentamicin ointment was curious and puzzling. I had stopped using this antibiotic in the tissue culture laboratory, after a report (Goetz et al., 1979) that it inhibited proliferation of human diploid fibroblasts, had become widely accepted. What was the explanation behind the prescription? I could not pursue this matter because of medical records

confidentiality. My own medical doctor told me I probably had a skin quality with a propensity for normal healing, when eventually I sought him out, on my return home. The intricacies of inter-cellular tissue repair are more or less unknown to family doctors.

It would be easy to prove or disprove this methodology by clinically employing it to treat donor sites after skin biopsies are taken. The whole dermis would now become subject to this tissue repair method. A similar principle would apply to treatment of second degree burns. Further work is needed to determine the cellular architecture of the epidermal brown rosettes and what role they play in melanomas. If it can be shown that allogenic brown rosettes are non-antigenic like dermal fibroblasts, it will provide an improvement to the techniques described by Gerlach et al. (2011).

Finally, little do the tattoo artists know that they are superimposing their colourful patterns over an invisible micro pattern that Nature has so brilliantly devised.

ACKNOWLEDGEMENTS

The author is grateful to the School of Medicine, University of Miami for all its assistance in this work. The author also wishes to thank Professor Malinin for graciously offering him his private laboratory with its cell culture facilities on a 24/7 basis. W.E. Buck MD (a Senior Pathologist in Tissue Bank) and D. Eton MD, FACS (Head of Vascular Surgery) are also acknowledged for supplying cadaver skin tissue and fresh skin tissue samples respectively.

REFERENCES

Barton SP, Marks R (1981). Changes in suspensions of human keratinocytes due to trypsin. Arch. Dermatol. Res., 271: 245-257.

Bauer J, Bahmer FA, Worl J, Neuhuber W, Schuler G, Fartasch M (2001). A strikingly constant ratio exists between Langerhans cells and other epidermal cells in human skin: A stereologic study using the optical dissector method and the confocal laser scanning microscope. J. Invest. Dermatol., 116: 313-318.

Bishop LA, Kee WJ, Zhu AJ, Watt FM (1998). Lack of intrinsic polarity in the ligand-binding ability of keratinocyte beta1 integrins. Exp. Dermatol., 7: 350-61.

Broughton G, Janis JE, Attinger CE (2006). Basic science of Wound healing. Plastic. Reconstr. Surgery. 117 (Suppl.): 12S-34S.

Chuong CM, Dhouailly D, Gilmore S, Forest L, Shelley WB, Stenn KS, Maini P, Michon F, Parimoo S, Cadau S, Demongeot J, Zheng Y, Paus R, Happle R (2006). What is the biological basis of pattern formation of skin lesions? Exp. Dermatol., 15: 547–564.

Crosby CV, Fleming PA, Argraves WS, Corada M, Zanetta L, Dejana E, Drake CJ (2005). VE-cadherin is not required for the formation of nascent blood vessels but acts to prevent their disassembly. Blood 105: 2771-2776.

Frenk E, Schellhorn JP (1969). Morphology of the epidermal melanin unit. Dermatologica, 139: 271-277.

Gaggioli C, Sahai E (2007). Melanoma invasion- current knowledge and future directions. Pigment. Cell. Res., 20: 161-172.

Gerlach JC, Johnen C, Ottoman C, Bräutigam K, Plettig J, Belfekroun C, Münch S, Hartmann B (2011). Method for autologous single skin cell isolation for regenerative cell spray transplantation with non-cultured cells. Int. J. Artif. Organs., 34: 271-279.

Germain L, Rouabhia M, Guignard R, Carrier L, Bouvard V, Auger FA (1993). Improvement of human keratinocyte isolation and culture using Thermolysin. Burns 19: 99-104.

Goetz IE, Moklebust R, Warren CJ (1979). Effects of some antibiotics on the growth of human diploid fibroblasts in cell culture. In Vitro 15: 114-119.

Green H (1991). Cultured cells for the treatment of disease. Sci. Am., (Nov.) pp. 96-102.

Hendrix MJ, Seftor EA, Meltzer PS, Gardner LM, Hess AR, Kirschmann DA, Schatteman GC, Seftor RE (2001). Expression and functional significance of VE-cadherin in aggressive human melanoma cells: role in vasculogenic mimicry. Proc. Natl. Acad. Sci. USA., 98: 8018-8023.

Hoath SB, Leahy DG (2003). The organization of human epidermis: functional epidermal units and phi proportionality. J. Invest. Dermatol., 121: 1440-1446.

Maniotis AJ, Folberg R, Hess A, Seftor EA, Gardner LM, Pe'er J, Trent JM, Meltzer PS, Hendrix MJ (1999). Vascular channel formation by human melanoma cells in vivo and in vitro: vasculogenic mimicry. Am. J. Pathol., 155: 739-752.

Monier-Gavell F, Duband JL (1997). Cross talk between adhesion molecules: control of N-cadherin activity by intracellular signals elicited by beta1 and beta3 integrins in migrating neural crest cells. J. Cell Biol., 137: 1663–1681.

Normand J, Karasek MA (1995). A method for the isolation and serial propagation of keratinocytes, endothelial cells, and fibroblasts from a single punch biopsy of human skin. In Vitro Cell. Dev. Biol.-Animal., 31: 447-455.

Poumay Y, Leclercq-Smekens M, Grailly SM, Degen A, Leloup R (1992). Specific interaction of basal membrane domains containing the integrin $\alpha6\beta4$ in dispase-detached cultured human keratinocytes. Eur. J. Cell Biol., 60: 12-20.

Quevedo WCJr (1969). The control of color in mammals. Amer. Zool., 9: 531-40.

Quevedo WCJr (1972). Epidermal Melanin Units: Melanocyte-Keratinocyte Interactions. Am. Zool., 12: 35-41.

Rubin AI, Chen EH, Ratner D (2005). Basal-cell carcinoma. N. Engl. J. Med., 353: 2262-2269.

Solomon DE (2002). An in vitro examination of an extracellular matrix scaffold for use in wound healing. Int. J. Exp. Path., 83: 209-216.

Sorrell JM, Baber MA and Caplan AI (2004). Site-matched papillary and reticular human dermal fibroblasts differ in their release of specific growth factors/cytokines and in their interaction with keratinocytes. J. Cell Physiol., 200: 134-45.

Spichkina OG, Kalmykova NV, Kukhareva LV, Voronkina IV, Blinova MI, Pinaev GP (2006). Isolation of human basal keratinocytes by selective adhesion to extracellular matrix proteins. Tsitologiia 48: 841-7.

Stenn KS, Link R, Moellmann G, Madri J, Kuklinska E (1989). Dispase, a neutral protease from Bacillus polymxa, is a powerful fibronectinase and type IV collagenase. J. Invest. Dermatol., 93: 287-290.

Turing AM (1952). The chemical basis of morphogenesis. Phil. Trans. Roy. Soc. B., 237: 37–72.

Zeng W, Thomas GL, Glazier JA (2004). Non-Turing stripes and spots: a novel mechanism for biological cell clustering. Physica A., 341: 482–494.

19

Morphology and histology of the alimentary tract of adult palm weevil, *Rhynchophorus phoenicis* Fabricius (Coleoptera: Curculionidae)

Omotoso Olumuyiwa Temitope

Department of Zoology, Faculty of Science, Ekiti State University, P.M.B. 5363, Ado-Ekiti, Ekiti, Nigeria.
E-mail: topeomoth@yahoo.co.uk.

Palm weevil (*Rhynchophorus phoenicis*) is a notorious pest of oil palm trees, worldwide. Investigations were conducted on the alimentary tract of the adult insect. Results of the morphology and histology of the alimentary tract revealed that the alimentary tract consisted of a foregut, which was made up of a buccal cavity, oesophagus, crop and proventriculus. The crop big or small and the proventriculus had bristle teeth which the insect used in the mechanical breakdown of ingested food. The midgut is the longest part of the alimentary tract and most prominent features there are goblet cells, columnar cells, villi, microvilli and intestinal glands. The midgut was well adapted for the digestion and assimilation of food. The hindgut was responsible for the re-absorption of water from undigested food (feaces) because of the presence of rectal pads. The results of these investigations revealed that the alimentary tract of the weevil was structurally and functionally adapted to digest and extract its nutritional requirements from its food.

Key words: Morphology, histology, alimentary tract, adult palm weevil, *Rhynchophorus phoenicis*.

INTRODUCTION

In all living animals the alimentary tract is one of the most important systems in the body because it deals with both the intake of food and its digestion. The alimentary tract of insects generally consist of three primary parts namely, the foregut, midgut and hindgut, however, differences exist among species due to differences in dietary habits. Palm weevil, *Rhynchophorus phoenicis* F. is a notorious pest of oil palm trees, in the tropics and subtropical regions of the world where it causes great economic losses. The insects feed on the trunk of palms thus creating empty cavities in the palm and causing the breaking of the palm trees. The physiology and the structure of the alimentary tract of this insect have not been reported. However, the physiology and the structure of the alimentary systems as well as the midguts of a variety of insects have been documented (Ferreira et al., 1981; Gartner, 1985; Dimitradis, 1991; Lee et al., 1998; Hung et al., 2000). Information on the morphology and histology of the alimentary tract of the adult palm weevil is not available. The purpose of the present study was to provide information on the morphology and histology of the alimentary tract of the insect.

MATERIALS AND METHODS

Collection of insects and the laboratory maintenance of the weevils for this study

The adult palm weevils (50) used for this work were obtained from palm wine tappers at Igbokoda (Latitude 6° 21° N and Longitude 4° 48° E) in Ondo State, Nigeria. The weevils were kept in plastic containers, which were filled to two-third with raphia palm chippings and covered with iron mesh. The containers were transported to the laboratory where the insects were allowed to acclimatize for 48 h before being used.

Morphology of the alimentary system

The guts of the adults palm weevils were prepared according to the methods described by Adedire (2002). The insects were placed in a deep freezer for 30 min to immobilize them before clipping their

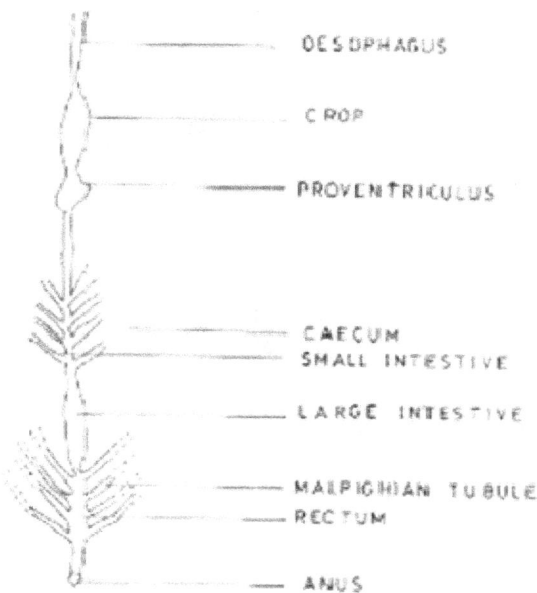

Figure 1. Morphology of the whole gut of adult palm weevil, *Rhynchophorus phoenicis*. X=1/4.

wings and dissect-ting them to remove the alimentary tract. The guts were fixed in Bouins solution for 24 h and then dehydrated for 3 h in 50, 60, 70, 80, 90 and 100% ethanol. The guts were stained with haematoxylin for 5 min and later washed with water for 2 min. The guts were differentiated in acid alcohol for 30 s and rinsed in water for 3 min. The guts were counter-stained with eosin for 3 min and dehy-drated through 50, 60, 70, 80, 90 and 100% ethanol. The guts were cleared in xylene for 30 min and mounted with DPX mountant on glass slides. Morphology of the alimen-tary system was observed under a hand lens and drawing made.

Histology of the sections of the alimentary system

The method described by Adedire (2002) was used in preparing the histological sections of the alimentary tract of the weevil. Adult weevils were immobilized by placing them in a deep freezer for 30 min; their guts were carefully dissected in 0.1 M NaCl and were preserved in Bouins solution for 2 days. The guts were thereafter cut into three distinct sections: foregut, midgut and hindgut. Each section was cleared in xylene for 30 min and infiltrated for 30 min in three changes of paraffin wax. The gut sections were embedded in paraffin wax and allowed to solidify and sectioned serially at 6 microns, using a microtome. The sections were placed on slides, washed with tap water and transferred into warm water (45°C) to allow the wax and the sectioned tissues to spread evenly. The slides were air-dried and transferred into xylene for 5 min to remove wax from the tissues. They were dehydrated in 50, 60, 70, 80, 90 and 100% ethanol for a period of 1 min each. The slides were stained with haematoxylin for 10 min and differentiated in acid alcohol for 30 seconds. The slides were rinsed in tap water for 3 min and counterstained with eosin for 3 min. The slides were rinsed in tap water and the tissues were dehydrated for 1 min each in 50, 60, 70, 80, 90 and 100% ethanol. The slides were cleared in xylene for 5 min and mounted with Canada Balsam and observed on a light microscope. Photomicrographs of the gut sections were taken at ×200 magnification with Leitz camera PM-C attached to Leitz (model ×40) microscope.

RESULTS

Foregut

The foregut of the adult consists of the buccal cavity, oesophagus, crop and the proventriculus. The oesophagus is long and it terminates in a small crop at the base of which is the strong proventriculus where food is pulverized (Figure 1).

Transverse section of the oesophagus showed that it consisted of stratified squamous epithelium, exhibiting a characteristic structure and arrangement. The cells are thin, flat and were irregularly shaped. The oesophagus has polyhedral cells, arteries, capillaries, oesophageal glands and squamous cells. Nerves and mucous alveoli are present in the submucosa (Plate 1). The proventriculus was made up of thick walls of circular muscles, longitudinal muscles, capillaries and nerves and it looks more like a star (Plate 2).

Midgut

The midgut was the longest part of the alimentary tract. It was about two times longer than the whole insect length and was situated between the foregut and the hindgut. Its anterior region was straight cylindrical structure located in the thorax and the posterior region was coiled shaped and lied in the abdominal segment. The direction of the coiling was downward and it then turns left where it finally runs downwards and terminates in the hindgut. The distinctive features of the midgut were the villi. The villi were the finger-like projections emanating from the midgut and they are used for food absorption. The transverse section of the midgut showed that the midgut had numerous villi which were well supplied with blood vessels (Plate 3).

The midgut had goblet cells, interstitial glands and columnar cells which form the bulk of the epithelium. The villi had brushlike border of apical microvilli projecting into the lumen. Numerous goblet cells were present while interstitial cells occupy the spaces between the goblet cells. Circular and longitudinal muscles were also present. The villus has central lacteal and capillaries. The villi differ in shape and length.

Hindgut

The hindgut formed the second longest part of the alimentary tract and was one and halve longer than the whole length of the insect. The hindgut, which was found at the posterior part of the alimentary tract, consisted of the rectum and anus. The rectum was the place where undigested food (faeces) stayed while water molecules were being removed from it before it is passed out through the anus. Malpighian tubules which were used to carry out excretion were present towards the posterior part of the hindgut. The same layers are present in the

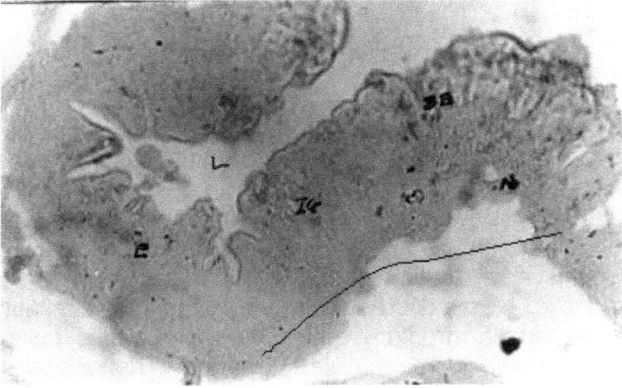

Plate 1. Transverse section of the oesophagus of adult palm weevil E = Epithelium, IG = Intestinal Gland, N = Nerve, SE = Squamous Epithelium, L = Lumen. X=240.

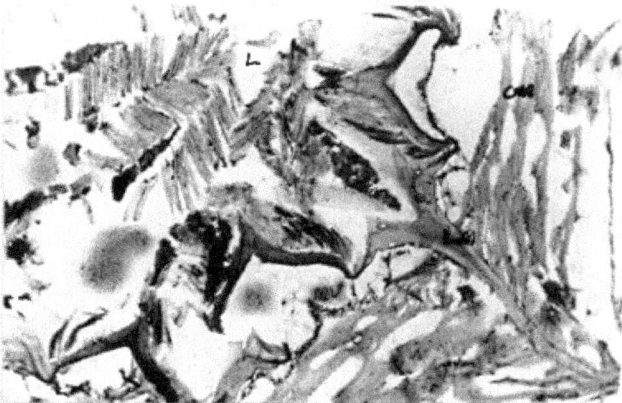

Plate 2. Transverse section of the proventriculus of adult palm weevil CM = Circular Muscle, LM = Longitudinal Muscle, L = Lumen. X=250.

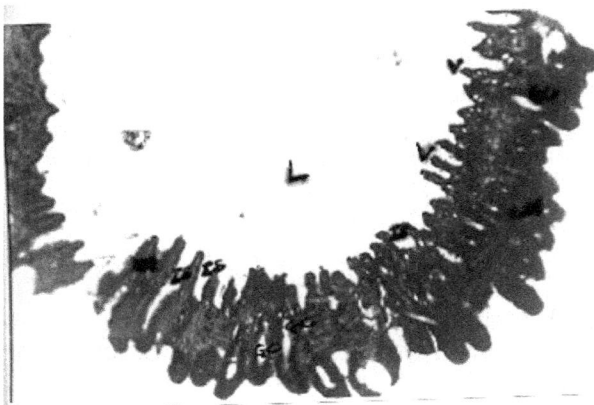

Plate 3. Transverse section of the midgut of adult palm weevil.IS = Inter villus Space, CM = Circular Muscle, GC = Goblet Cell, L = Lumen, V = Villus. X=250.

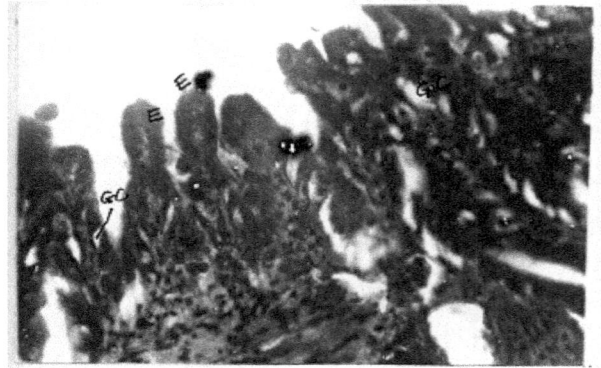

Plate 4. Transverse section of the ileum of adult palm weevil.E = Epithelium, GC = Goblet Cell. X=250.

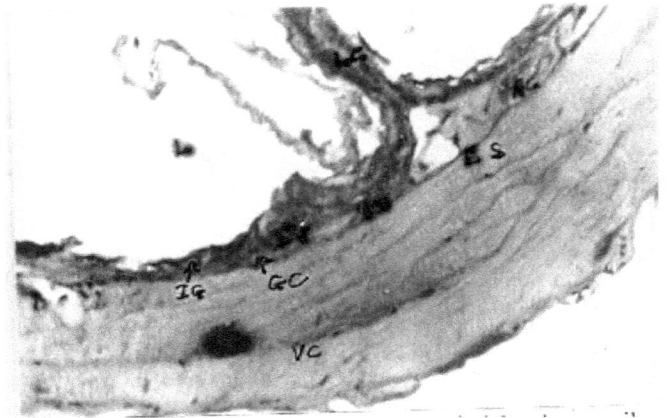

Plate 5. Transverse section of the rectum of adult palm weevilCM= Circular Muscle, GC = Goblet Cell, AC = Adipose Cell, L = Lumen,LM = Longitudinal Muscle, S = Submucosa, IG = Intestinal Gland. X=250.

(Plates 4 and 5).

The rectum had no villi but temporary folds of submucosa and mucosa layers. The rectum consisted of intestinal glands, lamina propria, surface epithelium which was lined by columnar cells, goblet cells, arteries and veins. There were adipose cells, intestinal glands, circular muscles and longitudinal muscles. Longitudinal folds were present and they had a core of submucosa which was covered by mucosa. The longitudinal folds reduce the lumen to small star-shaped spaces.

DISCUSSION

The alimentary tract of palm weevil consisted structurally of three distinct regions from the mouth to the anus. The gut consisted of the foregut, midgut and hindgut. Berridge (1970) reported that the foregut and the hindgut of insects had cuticle and that they were of ectodermal origin while the midgut, which had no cuticular intima,

wall and the same components are found in each layer

was developed from the endoderm. The foregut of insect begins at the base of the mandible and maxillary and extends to the oesophagus and proventriculus (Tsai and Perrier 1996; Lee et al., 1998). The arrangement of the oesophagus of palm weevil was the same as those reported for oriental fruit fly by Chun-Nu et al. (2000). The larva had a big crop where it stores all its food. Yoloye (1988) and Adedire (2002) reported similar occurrences in the alimentary tract of cockroach, *Periplaneta americana* and kolanut weevil, *Sophrorhinus insperatus* respectively. Backus (1985) had described the ciberial and the oesophageal regions of the mouthparts of several leafhoppers.

The function of the oesophagus of the insect is to pass food downwards to the midgut while the crop serves as a storage depot for ingested food (Wigglesworth, 1965). The oesophageal glands secrete mucous which lubricates food and makes swallowing easy. The oesophagus and the crop of palm weevil must have contained strong elastic muscles which allows peristaltic movement of the food in the oesophegus and makes the distention of the crop possible. Lee et al. (1998) reported similar observation in the foregut of *Bactrocera dorsalis*. The oesophagus of the insect was made up of stratified squamous epithelium. Smith (1968) reported that there may be little or no absorption of nutrients into the haemolymph across the crop wall of insects. The proventriculus plays a very prominent role (grinding) during food digestion in palm weevil. The proventriculus is muscularly built in this insect to withstand the pressure associated with the mechanical grinding of food in the chamber. Chapman (1985) reported that proventriculus was absent in fluid feeders.

The architecture of the midgut of this insect was the same as those reported for other insects (Gartner, 1985; Adedire, 2002). The midgut was the region for both digestion of food and absorption of nutrients. The epithetlium was responsible for both the production of many digestive enzymes and the uptake and transfer of nutrients to the haemolymph (Wigglosworth, 1965; Billen and Buschinger, 2000). Adedire (2002) reported that the larva of *S. insperatus* eats more food because it uses more energy for eclosion and the periodic moulting of its cuticle.

Columnar cells were concerned with the roles of enzymes secretion and the absorption of the products of digestion (Chapman, 1985). Smith (1968) stated that the cytoplasm of columnar in the midguts of insects was richly supplied with cisternae of the rough endoplasmic reticulum as well as the granular cisternae of the golgi complexes. Dimitradis (1991) reported that the anterior and posterior midgut regions of the adult *Drosophila auraria* are probably involved in the absorption of nutrients from the gut lumen to the haemolymph. Both of these regions displayed cells with well-developed villi and microvilli. The midgut epithelium of insects lack permanent uniform cuticle and the food in this part of the digestive system is separated from the peritrophic membranes (Chapman, 1985). There are two types of peritrophic membranes in insects according to their modes of formation. Chapman (1984) reported that the first type of peritrophic membrane occurred in Dipterans in which the membrane is secreted and formed at the anterior end of the midgut while the second type is formed by delamination from the entire surface of the midgut and it occurs in many orders of insects, including Orthoptera and Coleoptera. This structure generally plays a role similar to that of mucous in the vertebrates gut by protecting the midgut cells from mechanical damage caused by abrasive food particles (Richards and Richards, 1971). The peritrophic membrane in palm weevil midgut contained two layers like that reported for oriental fruit fly and *D. auraria* by Chun-Nu et al. (2000) and Dimitriadis (1991) respectively. The presence of endocrine cells around the regenerative cells had been reported in the midgut of *Meliopona quadrifasciata* by Nerves et al. (2003).

The role of the hindgut was very prominent in the alimentary tract of palm weevil. In the rectum more water is removed from the faeces and a somewhat semi-solid material is ejected through the anus. The presence of more goblet cells and intestinal glands in the rectum and anal region lend credence to the fact that more mucous are produced for the lubrication and easy egestion of feacal materials through the anus. These regions are muscularly built to withstand the pressure of defecating. In addition, the presence of more blood vessels in this region may account for the re-absorption of more water and the conveyance of this (that is, water) by the blood.

REFERENCES

Adedire CO (2002). Functional morphology of the alimentary canal of the kolanut weevil, *Sophrorhinus insperatus* Faust (Coleoptera: Curculionidae). Nig. J. Exp. Appl. Biol. 3:137-147.

Backus EA (1985). Anatomical and sensory mechanisms of leafhopper and planthopper feeding behavior. In: L. R. Nault and J. G. Rodriguez [eds.], The leafhoppers and planthoppers. John Wiley and Sons, New York. pp. 163-194.

Berridge MJ (1970). A structural analysis of intestinal absorption. In AC Neville, ed. Insect ultrastructure. London: Sym. Roy. Entomol. Soc. 5:135-151.

Billen J, Buschinger A (2000). Morphology and ultrastructure of a specialized bacteria pouch in the digestive tract of *Tetraponera* ants (Foramicidae: Pseudomyrmecinae). Arthropods Struct. Dev. 29:259-266.

Chapman RF (1985). The insects: structure and function. 3rd ed. Location ECBS. pp 54-56.

Chun-Nu H, Tai-lung L, Wen-Yung L (2000). Morphology and ultrastructure of the alimentary canal of the oriental fruit fly, *Bactrousa dorsalis* (Hendel) (Doptera: Tephritidae): the structure of the midgut. Zool. Stud. 39:387-394.

Dimitradis VK (1991). Fine structure of adult *Drosophila auraria* and its relationship to the sites of acidophilic secretions. J. Insect Physiol. 37:167-177.

Ferreira C, Ribeiro AF, Tera WR (1981). Fine structure of the larval midgut of the fly *Rhynchgosciara americana* and its physiological implications. J. Insect Physiol. 27:559-570.

Garthner LP (1985). The fine structural morphology of the midgut of adult Drosophila: a morphometric analysis. Tiss. Cell 883-888.

Hung CN, Lin TL, Lee WY (2000). Morphology and ultrastructure of the alimentary canal of the oriental fruit fly, *Bactrousa dorsalis* (Hendel) (Doptera: Tephritidae): the structure of the midgut. Zool. Stud. 39: 387-394.

Lee W, Chen M, Lin T (1998). Morphology and ultrastructure of the alimentary canal of oriental fruit fly *Bactrocera dorsalis* (Hendel) (Diptera: Tephritidae) (I): The Structure of the Foregut and Cardia. Zool. Stud. 37:95-101.

Neves CA, Gitirana LB, Serrao JE (2003). Ultrastructure of the midgut endocrine cells in *Melipona quadrifasciata anthidioides* (Hymenoptera: Apidae). Braz. J. Biol. 63:683-690.

Richards AG, Richards PA (1971). Origin and composition of the peritrophic membranes of the mosquito, *Aedes aegypti*. J. Insect Physiol. 17: 2257-2275.

Smith DS (1968). Insect cells: their structure and functions. Oliver and Boyd Press. Edinburgh, UK. pp. 223-266.

Tsai JH, Perrier JL (1996). Morphology of the digestive and reproductive systems of *Dalbulus maidis* and *Graminella nigrifrons* (Homoptera: Cicadellidae). Florida Entomol. 79:563-577.

Wigglosworth UB (1965). The principle of insect physiology. Methuen Press, London. pp. 273-299.

Yoloye VL (1988). Basic Invertebrate Zoology. 1st edt. University of Ilorin Press, Ilorin. pp. 242-249.

20

Achievement of balanced oncogenes and tumor-suppressor genes activity in normal and malignant cells *in vitro* and *in vivo*

Iskra Ventseslavova Sainova[1]*, Ilina Vavrek[1], Velichka Pavlova[1], Teodora Daneva[2], Ivan Iliev[1], Lilija Yossifova[1], Elena Gardeva[1] and Elena Nikolova[1]

[1]Department of Experimental Morphology, Institute of Experimental Morphology, Pathology and Anthropology with Museum, Bulgarian Academy of Sciences, "Acad. G. Bonchev" Str., 1113 Sofia, Bulgaria.
[2]Institute of Biology and Immunology of Reproduction, Bulgarian Academy of Sciences, Sofia, Bulgaria

The main goal is connected with providing, on the one hand, of active tumor-suppressor genes for prevention of eventual malignant transformations, and, on the other hand, of functionally active oncogenes for prevention of early aging and death, both *in vitro* and *in vivo*. Modulation of an adequate immune control was also necessary, and in this way any eventual unwished side effects from the genetic manipulations applied, could be escaped. Gene transfer in laboratory-cultivated mouse embryonic stem cells (mESCs) was made by use of appropriate recombinant DNA-constructs, which contained the promoter for gene, coding Elongation Factor 1-alpha (EF1-α), isolated from adeno-associated virus (AAV) (Parvoviridae); gene Dcn1, isolated from 3T3 fibroblasts of laboratory mice Balb/c, as well as gene for neomycin resistance, isolated from bacterial DNA-plasmid. Besides those indicated in the scientific literature inactivation of oncogene Dcn1 in the process of normal cell differentiation, its presence in the genome was supported and confirmed by our results from electrophorhesis of genomic DNA from normal mature epithelial cells of adult Balb/c laboratory mice. Furthermore, electrophorhetic profiles of genetic material from wild type (WT) on oncogene Dcn1 and "knock-down" (KD) on it inbred lines experimental mice differed not only on this oncogene, but also on the tumor-suppressor gene HACE1 in both categories of laboratory rodents. Similarly transfected Hela and RIN-5F malignant cells were then *in vitro*-co-cultivated with myeloid cell precursors, derived from populations of non-transfected laboratory-cultivated mESCs, in the presence of Doxyciclin, known from many literature data as activator of tumor-suppressor genes from STAT-family expression. Our results were also confirmed by the noticed differences in the degree of myeloid differentiation of derived precursor cells in their *in vitro*-co-cultivation with containing additional copies of tumor-suppressor genes malignant cells from both lines described, in comparison with the data, obtained in their laboratory co-cultivation with non-treated human cervical carcinoma Hela cells. Differences were also observed in *in vitro*-co-cultivation with the derived by us normal mESCs, containing additional copy of oncogene Dcn1 by the described above transfection with recombinant DNA-constructs. On the other hand, the derived normal cells with inserted additional copy of oncogene Dcn1 have indicated good safety and immunogenity. These cells have also indicated preserved normal cell characteristics, as well as eventual over-expression of the experimentally-activated oncogene Dcn1 in them.

Key words: Oncogenes, tumor-suppressor genes, myeloid cell precursors, recombinant gene constructs, cell transfection.

INTRODUCTION

The importance of co-ordinated oncogenes and tumor-suppressor genes action in the regulation and prevention of malignant transformation has widely been investigated as important in the regulation and prevention of malignant transformations and of many age-related disorders (Wood et al., 2000; Zhang et al., 2007). On the other

---

*Corresponding author. E-mail: isainova@yahoo.com.

hand, according many literature findings genetic interactions between oncogene and tumor-suppressor genes (Bellosta et al., 2005; Etard et al., 2005), as well as influence of the protein product on the one or two genes on the structure and functions of the other of both genes (Bauer et al., 2000; Vogelstein and Kinzler, 2004), have been proven as possible. In this way, correlations of gene p53 with gene NUMB has recently been indicated, characterized as a cell fate determinant because of its role in the asymmetric cell division in the mitosis process, as well as with gene Oct4, (known as regulator of the processes of stem cell self-renewal and differentiation and gene variation Cdk2ap1), by a mechanism of Oct2/4 promoter methylation.

In the last years of XX century, gene therapy and tissue engineering have been found as the most important approaches, which exploit the current knowledge in molecular biology and biomaterial science in order to direct stem cells to in vivo-differentiation to desired lineages and tissues (Barrette et al., 2000; Borysiewicz et al., 1996; Brachmann et al., 1998; Chen et al., 2003; Domi and Moss, 2005). In this aspect, widely studied is the ability for in vitro-cultivation of viruses in cell cultures, with the aim for development of both viral recombinants for malignant immunotherapy and of products for therapy of these disorders. As such tools can be used DNA- and RNA-viruses (Barrette et al., 2000; Borysiewicz et al., 1996), as well as bacterial plasmids and yeasts (Chen et al., 2003; Domi and Moss, 2005). For this aim, an intact gene tk, coding the enzyme timidinkinase (TK), has been found to be necessary, but, on the other hand – the integration of the searched gene(s) out of tk locus of the virus genome, as well as virus promoter, which could provide the expression of the inserted gene(s). Modifications by changes of the promoter and/or in the insertion site, as well as in the target vector repeats in fragments, expressing proteins with immunomodulator functions, have been proven to be possible. In this way have been inserted genes, coding cell receptors, cytokines, enzymes, complement activators, apoptosis activators and/or inhibitors, surface antigens, tumor markers. Besides the respective inserted gene(s), a marker gene has also been found to be necessary, but both gene types are controlled by appropriate promoter sequences. As a next step has been carried out polymerase chain reaction (PCR) of the received construction, by use of oligonucleotide primers for insurance of respective restriction sites – SfiI-site on the 5'-end and, respectively, RsrII-restriction site on the 3'-end of the PCR-product, which is obtained as a result of digestion by respective restriction enzymes (bacterial restrictases, which are particularly endonucleases), connected with respective early or late promoter in the virus genome or plasmid DNA.

Taking these data in consideration, the main goal was connected with providing of active tumor-suppressor genes for prevention of eventual malignant trans-formations, and, on the other hand, of active oncogenes for

prevention of early aging and death, both in vitro and in vivo.

MATERIALS AND METHODS

Stem cells, isolated from mouse Balb/c embryos, were cultivated for 48 to 72 h on previously formed monolayers of feeder primary MEFs after their previously treatment by Mitomycin-c (mm-c) (Sigma-Aldrich) and/or 3T3 fibroblasts. After tripsinization, they were transfected by electroporation (5×10^6 cells/ml). For this aim, recombinant DNA-genome from adeno-associated virus (AAV) (Parvoviridae) (Chen et al., 2003), containing promoter for gene, coding Elongation Factor 1-alpha (EF1-α); gene Dcn1, isolated from 3T3 fibroblasts of laboratory mice Balb/c, as well as gene for neomycin resistance, isolated from bacterial DNA-plasmid, were used. For gene transfer, electroporator for cell transfection (BioRad) was used. Separate sub-populations of non-transfected mESCs were cultivated in the presence of 2 µg/ml Doxicyclin (Sigma-Aldrich) for suppression of cell proliferation and eventual stimulation of myeloid cell differentiation, by the indicated in the scientific literature activation of genes from STAT-family on its influence. At the same time, malignant cells from cell lines Hela and RIN-5F of human cervical carcinoma and rat insulinoma, containing additional copies of tumor-suppressor genes HACE1 and Secretagogin gene, respectively, inserted by their transfection with recombinant gene constructs, were also cultivated and supported in analogical conditions. On the other hand, non-transfected cell cultures of the derived from human cervical carcinoma cell line Hela were also prepared. All cells were incubated at 37°C in incubator with 5% CO_2 and 95% air humidification, in Dulbecco's Modified Minimal Essential Medium (DMEM) (Sigma-Aldrich), supplemented with 10% Fetal Calf Serum (FCS) (Sigma-Aldrich), 100 U/ml penicillin (Sigma-Aldrich) and 100 µg/ml streptomycin (Sigma-Aldrich), and they were observed by inverted light microscope (Leica).

After tripsinization of the transfected cells and their consequent treatment with mixture of phenol-chlorophorm-isoamil alcohol (PCI) (Sigma-Aldrich), the isolated nuclear genetic material was treated with lysis buffer (Sigma-Aldrich) for isolation of genomic DNA. The last was subjected on standart Polymerase Chain Reaction (PCR) of previously isolated nuclear DNA and its consequent 1% agarose gel (Sigma-Aldrich) electrophorhesis, in the presence of DNA-primers against the inserted DNA-fragment (Sigma-Aldrich), mixture of the four types deoxy-nucleosid-tri-phosphates (dNTP - Sigma-Aldrich), enzyme Taq-polymerase (Sigma-Aldrich).

For differentiation in myeloid and lymphoid precursors, populations of non-transfected mESCs were further cultivated in medium, containing GM-CSF (Sigma-Aldrich) and complement proteins, respectively, by addition of 10% non-inactivated FCS (Sigma-Aldrich) in the last case.

Consequently, to the cell sub-populations of both non-transfected cell populations, malignant antigens were added. The last were derived by cultivation of Hela cells in serum-free DMEM (Sigma-Aldrich) for 24 h, its consequent centrifugation and filtration. Fixed light microscopic preparations were prepared by their consequent fixation with 95% ethanol (Sigma-Aldrich) or paraphormaldehyde (Sigma-Aldrich), washing with 1:9 diluted PBS (Sigma-Aldrich) and Giemsa-staining (Sigma-Aldrich).

RESULTS

In our experiments 9 transfected by electroporation cell clones were received and derived (Figures 1a and 1b).

According to the genomic assays results, 2 of the cell clones derived were positive on the additionally inserted

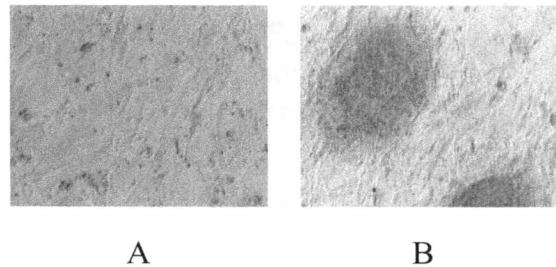

Figure 1. Native light-microscopy preparations from transfected mESCs negative (a) and positive on additionally-inserted copy of the oncogene *Dcn1*, respectively (b).,

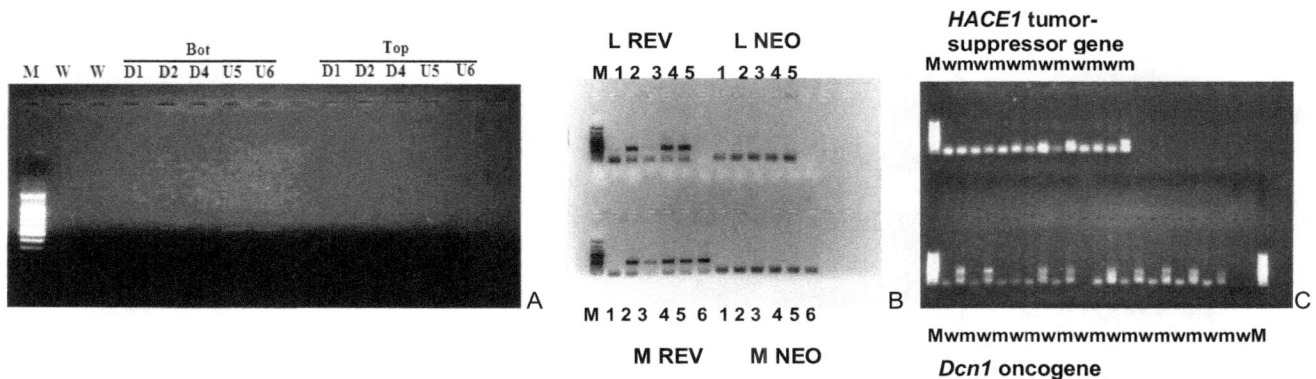

Figure 2. Agarose gel electrophorhesis for prove of the presence and/or the absence of additionally-inserted copy of the oncogene *Dcn1* in cell clones, derived from transfected by electroporation *in vitro*-cultivated mESCs (a) and in the used for cell transfection recombinant gene constructs (b); in mature epithelial cells, isolated from tail skin of adult experimental mice Balb/c with its normal expression (c), as well as from wild type (WT) and partially knock-down mutant (MT) on the same oncogene aboratory mice (d). Differences in the electrophorhetic profiles both of the oncogene *Dcn1* and the opposite tumor-suppressor gene *HACE1* could also be noted.

copy of the oncogene *Dcn1* and the other 7 cell clones - negative on it (Figure 2a). These results were confirmed by the data, obtained by electrophorhesis of the used recombinant DNA-constructs in the same conditions (Figure 2 – b). In genetic assay of nuclear DNA-material from high differentiated epithelial cells of homozygous on the oncogene *Dcn1* wild type (WT - *Dcn1+/Dcn1+*) and partially "knocked-down" on the same gene mutant heterozygous (MT - *Dcn1+/Dcn1-*) inbred lines of adult experimental mice, besides on it, differences in the electrophorhetic profiles of it, but also of the tumor-suppressor gene *HACE1*, were noticed (Figure 2 c).

Decreased cell proliferation level on the one hand and active myeloid differentiation on the other was established in cultivation of cell sub-populations in the presence of 2 µg/ml Doxicyclin (Sigma-Aldrich) (Figure 3).

These results could be confirmed by the observed signs of early myeloid and lymphoid differentiation in the presence of respective external factors (Figures 4 and 5).

According the results, the tendency for *in vitro*-differentiation in both myeloid and lymphoid precursors is stronger in the presence of transfected mESCs, containing additional copy of the oncogene *Dcn1*. Hence, the obtained data have also suggested that the so derived cells are safe enough both, and, on the other hand they have good immunogenic potential. The results obtained were compared with data, received from malignant rat insulinoma RIN-5F cells, containing additional copy of the Secretagogin gene, inserted by their transfection with recombinant gene construct *pGEX-1λT* (Amersham Pharmacia Biotech) of bacterial *Escherichia coli* strains, where a decreased malignant potential of the so transfected cells as a result of eventual induced Secretagogin over-expression, was supposed. As a proof about that could be accepted the observed effects of early myeloid differentiation and suppression on the cell proliferation in the presence of Doxicyclin could be explained with its activation effect on the tumor-suppressor genes of *STAT*-family. The data obtained

Figure 3. Decreased levels of cell *in vitro*-proliferation and activated *in vitro*-differentiation, in particular in myeloid precursor cells, by activation of the tumor-suppressor genes from *STAT*-family by cultivation in the presence of Doxicyclin (2 µg/ml - Sigma-Aldrich). Immune progenitor cells in different phases of immune differentiation, in particular, in different types myeloid precursors, most of which contain different types of granules in their cytoplasm, small cytoplasm amount with basophilic and/or eosinophilic granules, could be seen: *In vitro*-differentiation of non-transfected mESCs in the presence Doxicyclin, but in the absence of malignant cells (a); *In vitro*-differentiation in the presence of Doxicyclin and malignant cells Hela. A lot of cytoplasmic excrescences and cell-cell contacts are seen (b); *In vitro*-differentiation in the presence of GM-CSF and malignant cells Hela, containing tumor-suppressor gene *HACE1* (c); *In vitro*-differentiation in the presence of Doxicyclin and malignant cells RIN-5F, containing tumor-suppressor gene for Secretagogin (d); *In vitro*-differentiation in the presence of Doxicyclin and normal transfected mESCs, positive on additional copy of the oncogene *Dcn1* (e).

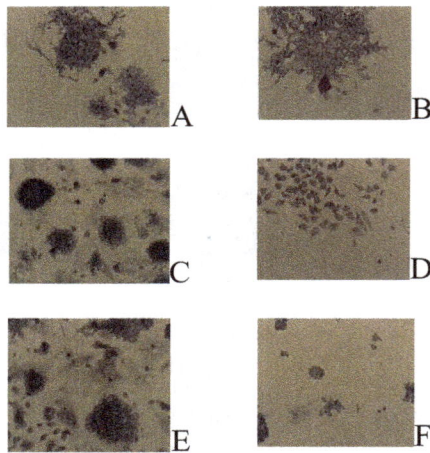

Figure 4. Early stages of myeloid *in vitro*-differentiation of non-transfected mESCs in different conditions: in the absence of differentiation factors and transfected cells (a); in the presence of GM-CSF, but in the absence of Hela-antigens and transfected cells (b); in the presence of Hela-antigens and in the absence of GM-CSF and transfected cells (c); in the presence of both GM-CSF and Hela-antigens, but in the absence of transfected cells (d); in the presence of GM-CSF, Hela-antigens and transfected cells, negative by additionally-inserted copy of oncogene *Dcn1* (e); in the presence of GM-CSF, Hela-antigens and transfected cells, positive on additionally-inserted copy of oncogene *Dcn1* (f).

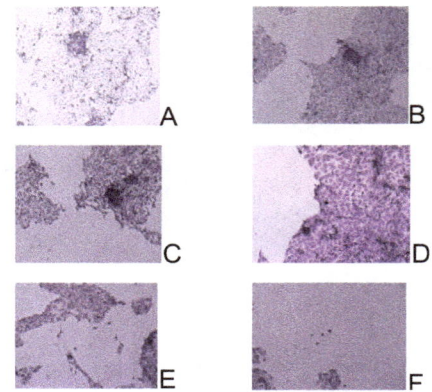

Figure 5. Early stages of lymphoid *in vitro*-differentiation of mESCs indifferent conditions: in the absence of differentiation factors and transfected cells (a); in the presence of complement components and absence of Hela-antigens and transfected cells (b); in the presence of Hela-antigens, but in the absence of complement proteins and transfected cells (c); in the presence of both complement components and Hela-antigens, but in the absence of transfected cells (d); in the presence of complement proteins, Hela-antigens and transfected cells, negative by additionally-inserted copy of oncogene *Dcn1* (e); in the presence of complement proteins, Hela-antigens and transfected cells, positive on additionally-inserted copy of oncogene *Dcn1* (f).

also slightly differed of the results, obtained in their laboratory co-cultivation with human cervical carcinoma non-transfected Hela cells in the same conditions, as well as in the absence of malignant cells, mainly in the number of formed cytoplasmic excrescences and contacts between cells by the so formed structures

(Figure 4).

DISCUSSION

Despite of the fact that the expression of oncogene

Dcn1has been proof to be inhibited in the mature normal cells (Ma et al., 2008; O-charoenrat et al., 2008), the results obtained have supported its presence in their genomes (Figure 1, Figure 2 and Figure 3). These results were confirmed by previously published data, obtained from PCR and subsequent electrophoresis of the used recombinant vector constructs in the same conditions (Figure 2 - b), as well as of genomic DNA, isolated from mature epithelial cells from skin of adult Balb/c experimental mice (Figure 3 - a), and from normal wild type (WT) on the oncogene Dcn1 and partially knock-down on it mutant (MT) adult laboratory rodents (Figure 3 – b). Taking in consideration literature data about the importance of coordinated oncogenes and tumor-suppressor genes action in the regulation and prevention of malignant transformation (Wood et al., 2000; Zhang et al., 2007), in both WT and knock-down MT on oncogene Dcn1 in inbred experimental mice lines, respective electrophorhetic profiles of this gene, as well as of tumor-suppressor gene HACE1 were made and compared, and the results obtained have indicated certain differences in both genes between the separated categories of laboratory rodents (Figure 3). The indicated high self-renewal potential of the stem cells in in vitro-conditions makes them strong candidates for delivering of genes, as well as for restoring organ systems function have been found to be included in these processes (Liang and Van Zant, 2003; Rubin, 1997; Vaziri and Benchimol, 1998; Vogelstein and Kinzler, 2004). This understanding could be applied toward the ultimate goal of using stem cells not just for various forms of therapy, but rather as a tool to discover the mechanisms and means to bring, reconstituting them from old and young individuals has exhibited indistinguishable progenitor activities both in vivo and in vitro (Smith, 2001; Vaziri and Benchimol, 1998). The properties of "malignant stem cells", have outlined initial therapeutic strategies against them (Smith and Boulanger, 2002; Vogelstein and Kinzler, 2004). A broad expression of oncogene Dcn1, characterized as a regulator of gene p53, has been detected in many tumor tissues and cultivated cell lines (Colaluca et al., 2008; Kurz et al., 2008; Ma et al., 2008; O-charoenrat et al., 2008). Function of this gene has also been found to be sufficient for cullin neddylation in a purified recombinant system, as well as, on the other hand – contribution of its over-expression to malignant disorders, as well as a potential marker for metastatic progression (Bowerman, 2007; Colaluca et al., 2008; Eferl et al., 2003; Gartner et al., 2007; Zhang et al., 2007). Links between DNA-replication, chromatin and proteolysis has been confirmed by the newly discovered cullin-RING E3-ubiquitin ligases, assembled on the CUL4 platform (Jin et al., 2006). In this aspect, a conserved component of CUL4-Dbd1 E3-ligase has been found as essential for the replication factor Cdt1 destruction and thus – for ensure proper cell cycle regulation of the DNA-replication process. Cullin-based E3-ligases, have recently been

proven as crucial regulators of mitosis. A key role of the enzyme CUL7 E3-ubiquitin ligase in the proteolytic targeting insulin receptor substrate-1, which has been proven as a critical mediator for insulin/IGF1-signalling, has been demonstrated (Jin et al., 2006). On the other hand, both positive and negative roles of ubiquitin-mediated proteolysis in the regulation of longevity in the eukaryotic organism Cenorhabditis elegans by insulin/IGFs–signaling pathways, have been established (Bowerman, 2007). Studies on the biology of the stem cells are often focused on their self-renewal and differen-tiation (Amit et al., 2000; Coulombel, 2005; Cumano et al., 1992; Keller, 1995; Liang and Van Zant, 2003; Molofsky et al., 2004; Rubin, 1997; Smith, 2001; Vaziri and Benchimol, 1998; Vogelstein and Kinzler, 2004). On the other hand, a rapid lymphoid-restricted (T-, B-, and NK) reconstitution capacity in vivo, as well as completely lacked myeloid differentiation potential both in vivo and/or in vitro, has been reported in stem cells from bone marrow material of adult laboratory mice (Kobari et al., 2000). The observed effects of early myeloid differen-tiation and suppression on the cell proliferation in the presence of Doxicyclin could be explained with the described in many literature sources activation effect of this substance on the tumor-suppressor genes of STAT-family (Figure 4) (Fitzgerald et al., 2008; 2009; 2005; Kyba et al., 2003; Poehlmann et al., 2005; Suman et al., 2009). According to other literature findings, a calcium-dependent SCGN–TAU interaction, as well as co-appearance of both proteins is shown (Gartner et al., 2007; Maj et al., 2010; Wagner et al., 2000), despite the fact that two different genes code them. The noticed by us cytoplasmic excrescences and cell-cell-contacts in co-cultivation with non-transfected malignant Hela cells with no induced tumor-suppressor gene over-expression, known as signs of phagocyte cell differentiation, could be accepted as a proof for eventual decrease of the oncogene potential in malignant cells, containing additional copy of tumor-suppressor gene, in in vitro-conditions, as well as an indication about eventual over-expression of the experimentally-activated oncogene in genetically-manipulated normal cells. The absence of the mentioned above features in the process of myeloid differentiation in the presence of the received positive on additional copy of the oncogene Dcn1 normal transfected cells could be accepted as a proof for the safety and immunogenity of these so derived transfected cells, which have preserved their non-tumorigenic/normal cell characteristics in vitro.

Conclusion

The noticed cytoplasmic excrescences and cell-cell-contacts in co-cultivation with malignant cells are known as signs of phagocyte cell differentiation from the scientific literature. In this connection, the observed highest degree of the formed structures in the presence of malignant Hela cells with no induced tumor-suppressor gene over-expression, could be accepted as a proof for eventual decrease of the oncogene potential in malignant cells, containing additional copy of tumor-suppressor gene, in *in vitro*-conditions. The absence of the mentioned above features in the process of myeloid differentiation in the presence of the received positive on additional copy of the oncogene *Dcn1* normal transected cells could be accepted as a proof for the safety and immunogenity of these so derived cells, which have preserved their non-tumorigenic/normal cell characteristics *in vitro*, as well as for eventual over-expression of the experimentally-activated oncogene in genetically-manipulated normal cells.

REFERENCES

Amit M, Carpenter MK, Inokuma MS, Chiu CP, Harris CP, Waknitz MA, Itskovitz-Eldor J, Thomson JA (2000). Clonally derived human embryonic stem cell lines maintain pluripotency and proliferative potential for prolonged periods of culture. Dev. Biol., 227: 271-278.

Barrette S, Douglas JL, Seidel NE, Bodine DM (2000). *Lentivirus*-based vectors transduce mouse hematopoietic stem cells with similar efficiency to *moloney murine leukemia virus*-based vectors. Blood 96: 3385-3391.

Bauer A, Chauvet S, Huber O, Usseglio F, Rothbacher U, Aragnol D, Kemler R, Pradel J (2000). Pontin52 and reptin52 function as antagonistic regulators of beta-catenin signalling activity. EMBO J., 19: 6121-6130.

Bellosta P, Hulf T, Diop SB, Usseglio F, Pradel J, Aragnol D, Gallant P (2005). *Myc* interacts genetically with Tip48/Reptin and Tip49/Pontin to control growth and proliferation during *Drosophila* development. Proc. Nat. Acad. Sci. USA., 102: 11799-11804.

Borysiewicz LK, Flander A, Numako M, Man S, Wilkinson GW, Westmoreland D, Evans AS, Adams M, Stacey SN, Boursnell ME, Rutherford E, Hickling JK, Inglis SC (1996). A recombinant *vaccinia virus* encoding *human papillomavirus* types 16 and 18, E6 and E7 proteins as immunotherapy for cervical cancer. Lancet 347: 1523-1527.

Bowerman B (2007). *C. elegans* aging: proteolysis cuts both ways. Curr. Biol., 17: R514-R516.

Brachmann CB, Davies A, Cost GJ, Caputo E, Li J, Hieter P, Boeke JD (1998). Designer deletion strains derived from *Saccharomyces cerevisiae S288C*: a useful set of strains and plasmids for PCR-mediated gene disruption and other applications. Yeast 14: 115-132.

Chen S, Agarwal A, Glushakova OY, Jorgensen MS, Salgar SK, Poirier A, Flotte TR, Croker BP, Madsen KM, Atkinson MA, Hauswirth WW, Berns KI, Tisher CC (2003). Gene delivery in renal tubular epithelial cells using recombinant *adeno-associated viral* vectors. J. Am. Soc. Nephrol., 14: 947-958.

Colaluca IN, Tosoni D, Nuciforo P, Senic-Matuglia F, Galimberti V, Viale G, Pece S, Di Fiore PP (2008). NUMB controls *p53* tumor suppression activity. Nat. Lett., 451: 76-80.

Coulombel L (2005). Adult stem cells: who are they, what do they do? Bull. Acad. Natl. Med., 189: 589-602.

Cumano A, Paige CJ, Iskove NN, Brady G (1992). Bipotential precursors of B-cells and macrophages in murine fetal liver. Nat., 356: 612-615.

Domi A, Moss B (2005). Engineering of a *vaccinia virus* bacterial artificial chromosome in *Escherichia coli* by *bacteriophage* λ-based recombination. Nat. Meth., 2: 95-97.

Eferl R, Ricci R, Kenner L, Zenz R, David JP, Rath M, Wagner EF (2003). Liver tumor development. *c-Jun* antagonizes the proapoptotic activity of *p53*. Cell, 112: 181-192.

Etard C, Gradl D, Kunz M, Eilers M, Wedlich D (2005) Pontin andReptin regulate cell proliferation in early *Xenopus* embryos in collaboration

with *c-Myc* and *Miz-1*. Mech. Dev., 122: 545-556.

Fitzgerald JS, Poehlmann TG, Schleussner E, Markert UR (2008) Trophoblast invasion: the role of intracellular cytokine signaling via signal transducer and activator of transcription 3 (STAT3). Trophoblast invasion: the role of intracellular cytokine signalling via signal transducer and activator of transcription 3 (STAT3) Human Reproduction Update, 14: 335-344.

Fitzgerald JS, Todd B, Jeschke U, Schleussner E, Markert UR (2009). Knocking off the suppressors of cytokine signaling (SOCS): their roles in mammalian pregnancy. J. Reprod. Immunol., 83: 117-123.

Fitzgerald JS, Tsareva SA, Poehlmann TG, Berod L, Meissner A, Corvinus FM, Wiederanders B, Pfitzner E, Markert UR, Friedrich K (2005). Leukemia inhibitory factor triggers activation of signal transducer and activator of transcription 3, proliferation, invasiveness, and altered protease expression in choriocarcinoma cells. Int. J. Biochem. Cell Biol., 37: 2284-2296.

Gartner W, Vila G, Daneva T, Nabokikh A, Koc-Saral F, Ilhan A, Majdic O, Luger A, Wagner L (2007). New functional aspects of the neuroendocrine marker secretagogin based on the characterization of its rat homolog. Am. J. Physiol. Endocrinol. Metab., 293: E347-E354.

Jin J, Arias EE, Chen J, Harper JW, Walter JC (2006). A family of diverse Cul4-Dbd1-interacting proteins includes Cdt2, which is required for S phase destruction of the replication factor Cdt1. Mol. Cell, 23: 709-721.

Keller GM (1995). *In vitro*-differentiation of embryonic stem cells. Curr. Opin. Cell. Biol., 7: 862-869.

Kobari L, Pflumio F, Giarratana M, Li X, Titeux M, Izac B, Leteurtre F, Coulombel L, Douay L (2000). *In vitro*- and *in vivo*-evidence for the long-term multilineage (myeloid, B-, NK, and T-) reconstitution capacity of *ex vivo*-expanded human CD34(+) cord blood cells. Exp. Hematol., 28: 1470-1480.

Kurz T, Chou YC, Willems AR, Meyer-Schaller N, Hecht ML, Tyers M, Peter M, Sicheri F (2008). Dcn1 functions as a scaffold-type E3 ligase for cullin neddylation. Mol. Cell, 29: 23-35.

Kyba M, Perlingeiro RCR, Hoover RR, Lu C-W, Pierce J, Daley GQ (2003). Enhanced hematopoietic differentiation of embryonic stem cells conditionally expressing Stat5. Proc. Nat. Aca. Sci. USA., 100: 11904-11910.

Liang Y, Van Zant G (2003). Genetic control of stem-cell properties and stem cells in aging. Curr. Opin. Hematol., 10: 195-202.

Ma T, Shi T, Huang J, Wu L, Hu F, He PF, Deng WW, Gao P, Zhang Y, Song Q, Ma D, Qiu X (2008). *DCUN1D3*, a novel UVC-responsive gene that is involved in cell cycle progression and cell growth. Mol. Cell, 99: 2128-2135.

Maj M, Gartner W, Ilhan A, Neziri D, Attems J, Wagner L (2010). Expression of TAU in insulin-secreting cells and its interaction with the calcium-binding protein secretagogin.J. Endocrinol., 205: 25-36.

Molofsky AV, Pardal R, Morrison SJ (2004). Diverse mechanisms regulate stem cell self-renewal. Curr. Opin. Cell. Biol., 16: 700-707.

O-charoenrat P, Sarkaria I, Talbot SG, Reddy P, Dao S, Ngai I, Shaha A, Kraus D, Shah J, Rusch V, Ramanathan Y, Singh B (2008). SCCRO (DCUN1D1) induces extracellular matrix invasion by activating matrix metalloproteinase 2. Clin. Cancer Res., 14: 6780-6789.

Poehlmann TG, Fitzgerald JS, Meissner A, Wengenmayer T, Schleussner E, Friedrich K, Markert UR (2005) Trophoblast invasion: tuning through LIF, signalling via *Stat3*. Placenta 26: S37-S41.

Rubin H (1997). Cell aging *in vivo* and *in vitro*. Mech. Ageing Dev., 98: 1-35.

Smith AG (2001). Embryo-derived stem cells: of mice and men. Annu. Rev. Cell. Dev. Biol., 17: 435-462.

Smith GH, Boulanger CA (2002). Mammary stem cell repertoire: new insights in aging epithelial populations. Mech. Ageing Dev., 123: 1505-1519.

Suman P, Poehlmann TG, Prakash GJ, Markert UR, Gupta SK (2009) Interleukin-11 increases invasiveness of JEG-3 choriocarcinoma cells by modulating *STAT3* expression. J. Reprod. Immunol., 82: 1-11.

Vaziri H, Benchimol S (1998). Reconstitution of telomerase activity in normal human cells leads to elongation of telomeres and extended replicative life span. Curr. Biol., 8: 279-282.

Vogelstein B, Kinzler KW (2004). Cancer genes and pathways they control. Nat. Med. 10: 789–799.

Wagner L, Oliarnyc O, Gatner W, Nowotny P, Groeger M, Kaserer K, Waldhausl W, Paternack MS (2000). Cloning and expression of secretagogin, a novel neuroendocrine- and pancreatic islets of Langerhans– specific Ca^{2+} binding protein. J. Biol. Chem., 275: 24740–24751.

Wood MA, McMahon SB and Cole MD (2000). An ATPase/helicase complex is an essential cofactor for oncogenic transformation by c-Myc. Mol. Cell, 5: 321-330.

Zhang L, Anglesio MS, O'sullivan M, Zhang F, Yang G, Sarao R, Ngheim MP, Cronin S, Hara H, Melnyk N, Li L, Wada T, Liu PP, Farrar J, Arceci RJ, Sorensen PH, Penninger JM (2007). The E3 ligase HACE1 is a critical chromosome *6q21* tumor suppressor involved in multiple cancers. Nat. Med., 13: 1060-1069.

21

Nucleofection an efficient non-viral transfection technique for IFN-t-EGFP gene expression study in Sahiwal cattle fibroblast cells

Anand Laxmi N.[1], Gunjan G.[1] and Prateesh M.[2]*

[1]DCP, NDRI, Karnal, India.
[2]IVRI, Barrielly, India.

Viral based techniques were considered to be the most efficient systems to deliver DNA into fibroblast cells as they show high transgene expression in many cellular models. Viral approaches are complicated by immune response, intracellular trafficking potential mutations and genetic alterations due to integration. The nucleofector TM technology is electroporation based gene transfer technique which has been proved to be an efficient tool for transfecting primary cells and hard to transfect cell lines. The present study was designed to examine Sahiwal fibroblast cell to act as competent donor cell in gene expression studies. Using a green fluorescent protein reporter vector, a high transgene expression level was obtained using U-12 and U-23 pulsing programs: 45 and 70% respectively. Cell recoveries and viabilities were 90.5 and 95% respectively for U3-23 program. Overall transfection efficiency was 60% as observed on evaluation by flowcytometry. Further, the cells confirmed to be positive for gene expression when subjected to PCR, RT- PCR and, flow cytometry analysis using pGFP repoter other than supplied by Amaxa system. Hence, the cells were transfected with bIFN-t---GFP reporter gene construct by nucleofection technique and similarly, cells were evaluated for expression of gene by fluorescence microscopy. A sixty percent of cells were positive for fuorescence on enumeration of transfected cells by fluorescence microscopy 72 h post transfection. Further it was confirmed by PCR technique using primers for IFN-t gene. The cell cultures when analysed for IFN-t protein expression, the protein could be detected by silver staining of SDS-PAGE gels. Reports are available where this technology has been successfully applied for the transfection of other cells like monocytic cell lines and neural stem cells. This technique can be further utilised for transfection of genes of economic importance in primary cell lines which be in turn can be utilised for production of transgenic embryos. Through out the study the transfected cell cultures were negative for apoptosis. The Sahiwal fibroblast cells culture system can be utilized for transfection and gene expression studies. Green Fluorescent Protein acts as a useful visual indication system for gene insertion or targeting studies, when conjugated to the gene of interest

Key words: Transfection, gene, fibroblast cell.

INTRODUCTION

Gene transfer technology has many potential applications in gene expression studies. Usually most commonly used systems to deliver DNA in to cells and study its expression is based on viral based techniques. This is mainly useful for stable integration in to host cell genome

(Blesch, 2004). The non-viral techniques available are lipofection microinjection, electroporation and gene gun technique which are associated with low gene expression but associated with either mortality or toxicity. The nucleofector technology is a non-viral technology which can be used for transfecting primary cell lines. Transfection of cells is not dependent on cell division, and still yields high transfection efficiency (Naito et al., 2007). This further paves way for studying gene

*Corresponding author. E-mail: dnana44@yahoo.co.in.

expression for longer duration of time. Gene transfection allows, introduction of normal genes in animals with genetic dysfunctions. Growth factors and cytokines also deliver biological agents in considerable amounts. In the present study, we evaluated the efficacy of Nucleofector technology as a gene delivery vehicle in to primary fibroblast cells of Sahiwal cattle and local production or expression of recombinant bIFN-t- EGFP protien. Nucleofection is a transfection method which enables transfer of nucleic acids such as DNA, RNA, Small interfering RNA molecules into cells. Nucleofection, also referred to as Nucleofector Technology, was invented by the biotechnology company *amaxa*. "Nucleofector" and "nucleofection" are trademarks, owned by Lonza Cologne AG. The reported pulsing programs are patented programs set by Amaxa systems. All EGFP positive cell lines were observed by fluorescence microscopy confirmed by PCR, flow cytometry and SDS-PAGE analysis. The objective of this study was to obtain transgenic cell lines available for cloning. Aiming at this p bIFN-t-GFP gene construct was introduced by nucleofection into Sahiwal cattle skin fibroblasts. GFP gene enhanced version EGFP was used in the present study. Studies on expression of GFP in cattle have been carried out by Chan et al. (2002); Rosochacki et al. (2001). Studies have been conducted on expression of IFN-t (Interferon-tau) in bacterial and yeast cell cultures (Fang-Fang et al., 2008).

In the present study, long term culture *in vitro* did not have any effect on the proliferation rate of the cells and also on expression. The bIFN-t-GFP expression could be observed even after six passages post transfection. This is the first study in which expression of recombinant IFN-t protein in fibroblast cells of Sahiwal cattle has been reported for successive passages. The protein fraction extracted from the transfected cells exhibited antiproliferative activity on lymphocyte cell cultures.

METHODOLOGY

Harvesting of cell culture

Skin piece was collected from ear pinna of Sahiwal cattle, transported to the lab in PBS supplemented with streptomycin – pencillin antibiotics. Tissue pieces were cultured in petridishes under sterile conditions. When the cells started extruding out of the tissue and became confluent, piece was removed, medium was replaced. The culture was subjected to trysinization and was transferred to the 75 mm^3 culture flasks.

On attaining 70 to 80% confluency, cultures in the flasks were washed with 2 ml of Ca^{++} Mg^{++} free D-PBS buffer. One ml. of cold trypsin EDTA solution (trypsin-0.125% and EDTA-0.0125% in D-PBS) was added to each flask. This dose was standardized for Sahiwal cattle. They were incubated at 37°C for eight minutes. On trypsinization, cells appeared spherical in shape which could be observed under inverted microscope. Further 2 ml of complete medium (DMEM+Ham' s F12 with 10% FBS) was added to each flask. The suspended cells were immediately transferred to centrifuge tubes and were centrifuged at 1000 rpm for 8 min. The supernatant was discarded and cells were suspended in 1 ml of

medium and cell counts were taken on haemocytometer. After harvesting of primary cell cultures and further on seeding they were designated as passage number one.

Cell transfection

Nucleofection of fibroblast cells was performed according to optimized protocol provided by the manufacturer (Amaxa Biosystems, Cologne, Germany, www.amaxa.com). Briefly cells were suspended in 100 μl of nucleofector solution (Amaxa Biosystem), mixed with either pEGFP or pIFN-t-EGFP construct [Source Promoter (Toshihiko et al., 1999), and coding regions were generously gifted by Dr. R.M. Roberts Lab, and Dr. Alan D. Ealy Missouri] and were subjected to nucleofection with the U-23 or U-12 program respectively (Figure 1). Immediately after transfection cells were transferred in to prewarmed fresh medium in six well plates. Cells were analysed for viability and GFP expression respectively 36h post nucleofection. The fluorescence of the cells was viewed at different passages post transfection.

The vector construct

1. The promoter region of IFN-t was PCR amplified using sequence specific primers designed based on cattle genome sequence data. The promoter region was amplified, the amplicon was gel purified and subsequently restriction digested and cloned within a cloning vector (pUC18). A few clones were sequenced to get a clone with no mutations.
2. The coding region of the IFN-t was PCR amplified using sequence specific primers designed based on cattle genome sequence data. The primers had suitable restriction sites for assembling the gene into the final construct. The gene was subcloned into pUC18 for sequence confirmation. A few clones were sequenced to get one with no mutations. The gene was PCR amplified using primers designed in such a way that the gene would be in-frame after it's cloned within the vector backbone (Appendix – Figure 1).

RNA/DNA / protein extraction

RNA/DNA/protein fractions were extracted from cell culture lysates using Norgen's RNA/DNA/Protein purification kit. Purification is based on spin column chromatography using Norgen's proprietory resin.

The transfected cell culture samples were centrifuged and to the pellet containing 2 x 10^6 no. of cells, 350 μl of lysis solution was added and vortexed for 15 s for complete dissolution. To the lysate 200 μl of 95% ethanol was added and vortexed gently for 10 min.

With the help of DNA, RNA elution solutions and according to the kit's manufacturer's instructions, the RNA was eluted and fraction was stored at -70°C and similarly the extracted DNA fraction was stored at -20°C.

The flow through collected after binding of RNA to the column, the pH of the solution was adjusted and applied on to the column. The flow through obtained was discarded and protein fraction was eluted using protein elution buffer. The fraction was neutralized with the neutralizer.

PCR Protocol for EGFP gene

Genomic DNA from 2 x 10^6 no. of transfected and non-transfected cells which served as control was extracted with the help of DNA extraction kit. Approximately 1 μg of DNA was used as a template in PCR reactions containing forward and reverse primers.

Template DNA was denatured at 94 °C for 1 min. The PCR cycle consisting of denaturation, annealing and extension steps were as follows:

Denaturation at 94 °C for 30 s annealing at 56 °C for 30 s, extension at 72 °C for 1 min. Final extension step was performed at 72 °C for 15 min. The PCR amplification was subjected to 30 cycles. The resulting amplified PCR products were analysed on 1.8% agarose gels. A100 bp ladder (Chromous Co., Bangalore) was used as marker and 100 ng of pEGFP served as standard.

ATGGTGAG CAAGGGGCGAGGAGCT F- primer:
: GTACCGTCGACTGCAGAATTCGAAGCT R- primer

PCR protocol for IFN –t gene

Template DNA was denatured at 94 °C for 2 min. The PCR cycle consisting of denaturation, annealing and extension steps were as follows:

Denaturation at 94 °C for 30 s, annealing at 56 °C for 30 s, extension at 72 °C for 30 s Final extension step was performed at 94 °C for 2 min. The cycle was repeated 35 times.
The resulting amplified PCR products were analysed on 1.8% agarose gel with 100 bp ladder as marker. 100 ng of pIFN-t-GFP served as a control.

Primers for IFN-t

*CAGTGACATATGGCCTTCGTGCTCTCTCTACTGATG-F primer
*CAGTGACTCGAGAAGTGAGTTCAGATCTCCACCCATCT-R primer.

Analysis of EGFP expression

The EGFP fluorescent cells were visualized under fluorescence microscope and analyzed by flowcytometry. For control, cells transfected without vector were considered. Briefly cells were detached from the flaks by incubation with 0.125% Trypsin and 0.012% EDTA for 5-6 minutes, recovered by centrifugation at 800g for 6 minutes. The pellet was suspended in DMEM: Ham's F12 medium. An aliquot of the suspension was visualized under fluorescence microscope. For flow cytometry, the cells were washed with PBS buffer containing 1% BSA. Cells were collected by centrifugation and resuspended in PBS buffer containing BSA and 2% paraformaldehyde before analysis. Both control and pIFN-t-EGFP transfected cells were respectively analyzed on Bector-Dickinson instrument after fixation of cells for analysis by flowcytometry as follows.

Approximately 1×10^6 transfected cells were washed with PBS buffer. Samples were centrifuged at 1500 rpm for 5 min. Supernatant was discarded and 500 µl of cold PBS was added and vortexed gently. Further, 500 µl of cold buffered 2% formaldehyde solution was added. They were incubated for 1h at 2-8°C. The samples were again centrifuted at 1500 rpm for 5 min. and washed once more with cold PBS. To the pellet 1 ml of 40° ethanol chilled at -20°C was added dropwise. The suspension was incubated overnight at 2 to 8°C. The next day the samples were centrifuged and supernatant was discarded. They were incubated with 20 µl RNAse at 37°C for 30 min in the dark. The cells were subjected to flowcytometry for measurement of green fluorescence. Fluorescence data was collected using 488 nm excitation and filter combination consisting of filters with 395 and 475 long pass filters. Dot plot scatter for green fluorescence was created using Cell Quest Program.

SDS-PAGE

Pellets enriched with cellular proteins were suspended in 500 µl of water and 500 µl of sample buffer. The preparations were heated for 10 min at 98°C and stored frozen at -20°C. Proteins were visualized by staining with Coosamie brilliant blue R-250 or by silver staining. The apparent molecular weights of the proteins detected were estimated by comparison with a broad-range standard protein marker obtained from Genei Co. (Bangalore).

Separating gel mixture was prepared containing acrylamide monomer, Tris HCl (pH 8.8) and 10% SDS solution. After degassing the gel mixture ammonium per sulphate and TEMED were added to the solution. The resulting acrylamide solution was transferred to the casting chamber with Pasteur pipette. The solution was allowed to run down along the side of the spacers. Stacking gel was also prepared containing acrylamide monomer, Tris buffer (pH 6.8) and 10% SDS. Ammonium per sulfate and TEMED were added just before loading the gel chamber. One ml of freshly prepared stacking gel was poured in to the gel chamber and Teflon comb was placed to form sample wells. The cells and chamber were filled with Tris-Glycine-SDS buffer. Protein standard/markers or extracted protein fractions mixed with sample buffer were placed in to bottom of different wells respectively. The gel was removed from the casting stand and assembled in to appropriate slab for electrophoresis. The proteins were resolved by electrophoresis till the tracking dye reached the bottom of the separating gel. After confirming; the gel was stained with 0.25% Comassie Brilliant Blue R250 in methanol: water: glacial acetic acid (5:5:1).

Silver staining technique

To detect the presence of IFN-t-GFP, monoclonal antibodies to EGFP+ were used, and the antigen antibody complex was immunoprecipitated with protein A beads (Chromous Co. Bangalore). The antigen eluted out was separated by SDS-PAGE and further silver stained. Similarly, EGFP was immunoprecipitated from EGFP+ transfected cells and the purified protein was subjected to SDS-PAGE and also subjected to silver staining (Rabilloud, 1992).

Reverse transcriptase-PCR

Reverse transcriptase – PCR

The gene expression of the recombinant, pIFN-t-GFP construct was monitored by RT-PCR, using BioRT, two step RT-PCR kit (Taurus Scientific, USA). In the first step, cDNA was synthesized from total RNA by reverse transcription using AMV reverse transcriptase and gene specific down stream primers for transfected IFN-t gene, dNTP mixture (10 mM) and Rnase inhibitor. For confirmation the cDNA obtained was subjected to DNAse treatment, upon incubation with DNAse for 30 min, cDNA band was not observed on 1.8% agarose gel. The program for cDNA synthesis is as follows:

RT reaction

60°C - 45 min
95°C - 5 min
Ice bath - 5 min

The cDNA obtained in the first step was subjected to amplification by PCR.

Apoptosis assay

The genomic DNA was extracted from the fibroblast cell cultures and was subjected to DNA ladder assay on 0.8% agarose gel. A 1 kbp ladder was used as marker (Genei Co., Bangalore). Briefly the cell cultures were transpsinized and DNA was extracted from the cells using Norgen's kit (Canada) based on column extraction method. Approx. 500 ng of DNA ladder and 1 µg of DNA extracted from transfected cells with the loading dye were loaded in to separate wells on 0.8% agarose gels. The gels were subjected to electrophoresis for 1h and were further viewed under UV transilluminator (System and Control, US A).

Cell culture of lymphocytes and antiproliferative test

The lymphocyte cell suspension was diluted with RPMI 1640 culture media containing 10% FCS, so that final concentration was 50×10^5 lymphocytes/ml. The viability of the lymphocytes was evaluated by trypan blue exclusion method. A 200 µl of the medium containing approx. 10×10^5 cells were pippeted out into each well in a 96 well plate. For stimulation of mitotic activity phytohemoagglutinin (PHA-P) was added at a concentration of 5 µg/ml in a total volume of 200 µl /well. The cell cultures were propagated/ incubated for 24 h in a CO_2 incubator at 37°C and 5% CO_2. The samples with and without mitogen were run in duplicates. The proliferative response of the lymphocytes was estimated using colorimetric MTT test (Mossman, 1983).

MTT assay

MTT[3-(4,5-dimethylthiazol-2-yl)-2,5-diphenyltetrazolium bromide] assay, is based on the ability of a mitochondrial dehydrogenase enzyme from viable cells to cleave the tetrazolium rings of pale yellow and form a dark blue formazan crystals. The accumulated formazan in the viable cells is released and solubilized by the addition of DMSO. The lymphocyte cultures were incubated for 48 h with mitogen and immuno-precipitated protein fraction of fibroblast cell culture lysate (1 µg of total protein/well) in DPBS and further with MTT at a conc. of 5 µg/well was also added. The plates were further incubated for 4 h in a CO_2 incubator under controlled conditions. A 100 µl of DMSO was added to each well at the end of 4 h of incubation. The solution in the wells was mixed thoroughly and incubated at room temperature for 15 min. The optical density or absorbance of the solution was recorded at 503 nm with reference to 630 nm.

Lymphocyte stimulation in index = O.D. of the complex formed from stimulated cells/O.D. of the complex formed from unstimulated cells The statistical analysis was done by two way ANOVA program.

RESULTS

Fibroblast cell culture

The Sahiwal fibroblast cell cultures were successfully propagated, till sixth passage post transfection. A morphologically homogenous population of confluent fibroblast cell cultures was obtained by 5 to 6 days post harvesting and seeding of cells. At first passage post transfection growing cells were uniformly positive for EGFP fluorescence expression. The Sahiwal fibroblast cells were transfected by the Nucleofector technology following the protocol provided by the manufacturer. The expression of transgene pEGFP and pIFN-t-EGFP

respectively was evaluated by fluorescence microscopy and by fluorescence activated cell sorter (FACS). At 36 h post nucleofection more than 50% were EGFP+ cells which increased to 70% after first passage post-nucleofection (Figure 1a and b). The genomic DNA extracted from cell cultures pre and post transfection respectively when subjected to gel electrophoresis, they were nonapoptotic (Figure 2). Similar were the results obtained with pIFN-t-EGFP gene constructs. But the percentage of bIFNt-EGFP+ cells, at first passage post nucleofection, was 60% (Figure 1c)and thereafter the percentage of EGFP/IFN-t-EGFP expressing nucleofected cells remained constant till sixth passage post nucleofection. Conversely U-12 program, resulted in 40 to 43% EGFP+ cells. U-23 program was found to be superior to U-12 program (Figure 1a and b). The percentage of viable cells estimated by Trypan blue exclusion method was observed to be 90 to 95% at first passage post nucleofection. The results obtained for EGFP+ cells on nucleofection with pIFN-t-EGFP construct was significantly less (60 vs. 70%), ($P < 0.05$) (Figure 2) with U-23 program when compared with pEGFP+ cells. To assess the stability of nucleofection based transfection, the cells expressing EGFP was determined at second, fourth and sixth passage post pIFNt-EGFP nucleofection respectively. It was observed that cells were expressing EGFP fluorescence at all the passages confirming the expression of IFN-t also. On analysis by flowcytometry it was observed that 50% of the cells were expressing fluorescence for the IFN-t transfected cell cultures (Figure 3a and b) when compared with that of the control.

PCR and RT PCR studies

The primers specific for EGFP and IFN-t were used for amplification of genomic DNA extracted from EGFP andIFN-t-EGFP gene transfected cell cultures respectively. After amplification of genomic DNA, PCR products were observed at 885 and 585 bp region for the EGFP+ and IFN-t-EGFP respectively confirming presence of the gene [Figure 4a and b]. Though the EGFP fluorescence could be observed post nucleofection, with IFN-t-EGFP gene, the expression of the gene was monitored at transcriptional level using specific primers for IFN-t gene against the exon coding region for the IFN-t transfected cell cultures at second, fourth and sixth passage respectively.

RT-PCR was performed using AMV reverse transcriptase. On amplification of the C-DNA a band was observed on 1.8% agarose gel, at 585 bp region against100 bp ladder used as marker, which confirmed the expression of gene at transcriptional level and was positive for culture at second, fourth and sixth passage, respectively (Figure 5). Real time PCR studies were conducted using forward and reverse primers for IFN-t

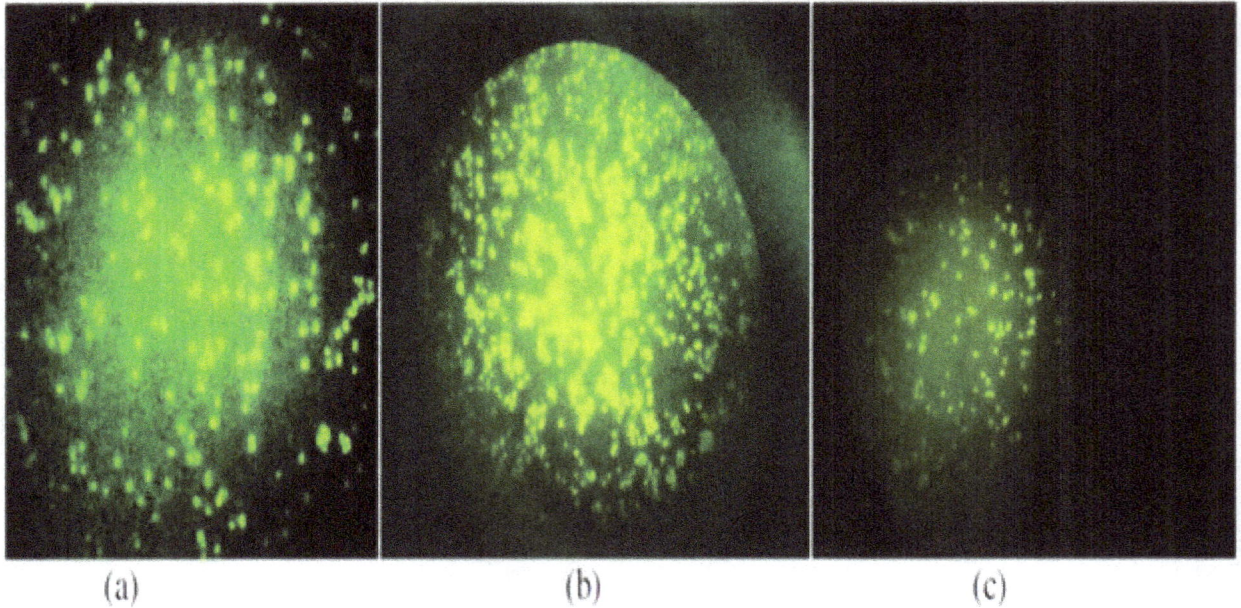

(a)   (b)   (c)

Figure 1. 45 and 70% of the cells were observed to be pEGFP⁺ with U12; (a) and U23; (b) nucleofector programs respectively; (c) Cells transfected with pIFN-t-EGFP- L-100 bp ladder, 60% fluorescent cells as observed under fluorescence microscope.

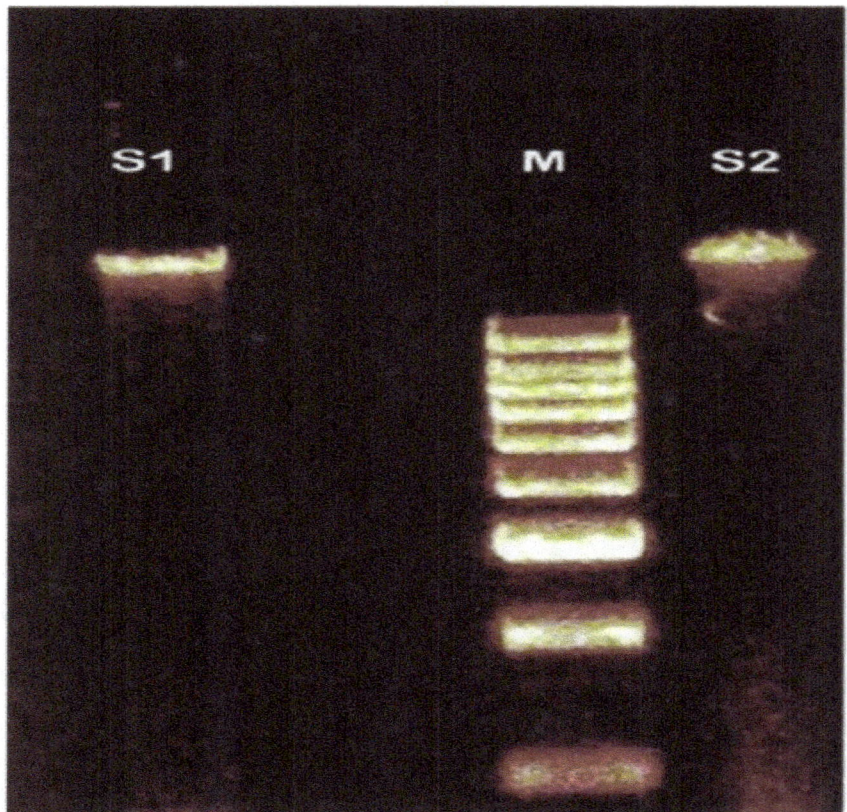

Figure 2. S1,S2- Genomic DNA of cell cultures pre and post transfection section respectively exhibiting non apoptosis M-Marker in Kbps.

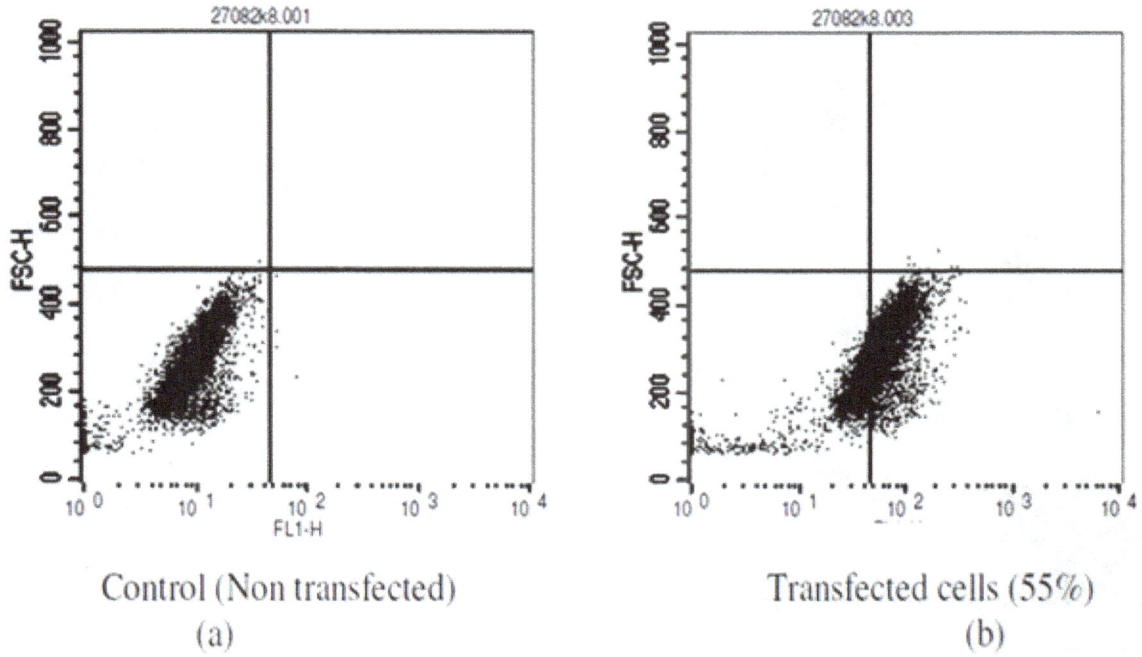

Figure 3. On analysis by flowcytometry.

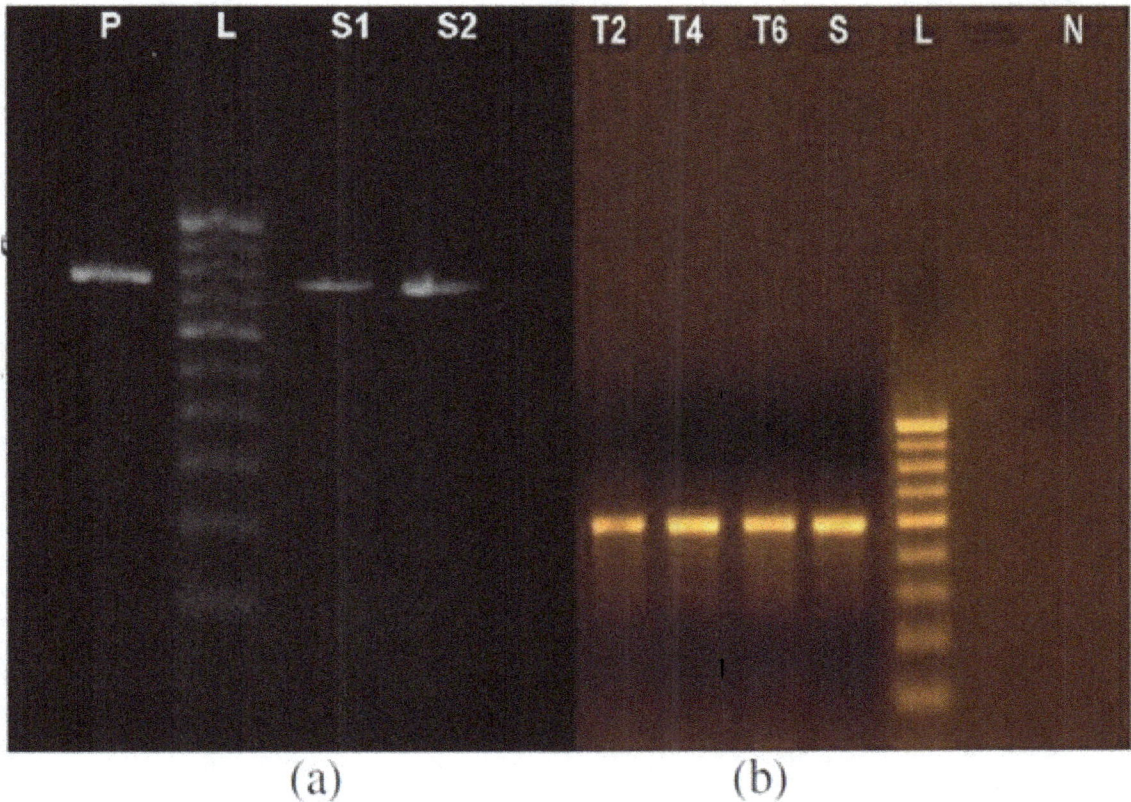

Figure 4. S1,S2 -PCR product of EGFP gene (885 bp) from cells post nucleofection P– Positive marker; L– 100 bp ladder, (b) passages post transfection N- PCR product not observed for DNA from non transfected cells product of IFN-t gene (585 bp) at different.

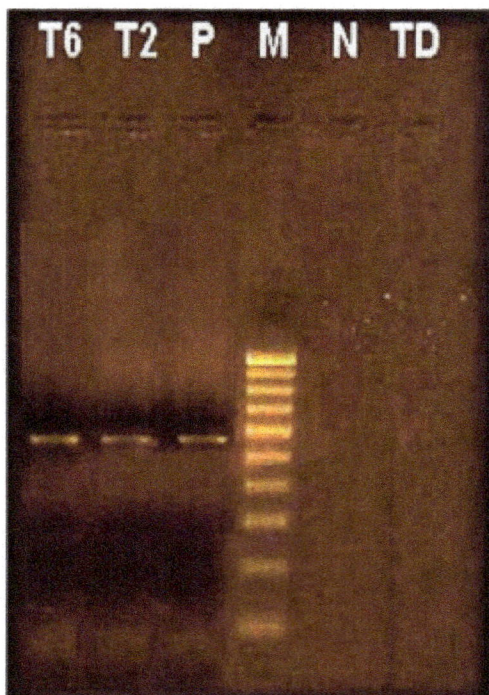

Figure 5. RT-PCR of gRNA obtained from transfected cells-cDNA amplified by PCR M – 100 bp marker, T2, T6- Passage no. post transfection P- Positive marker for IFN-t (585 bp), N– Product negative for non transfected cells TD- treatment with DNAse.

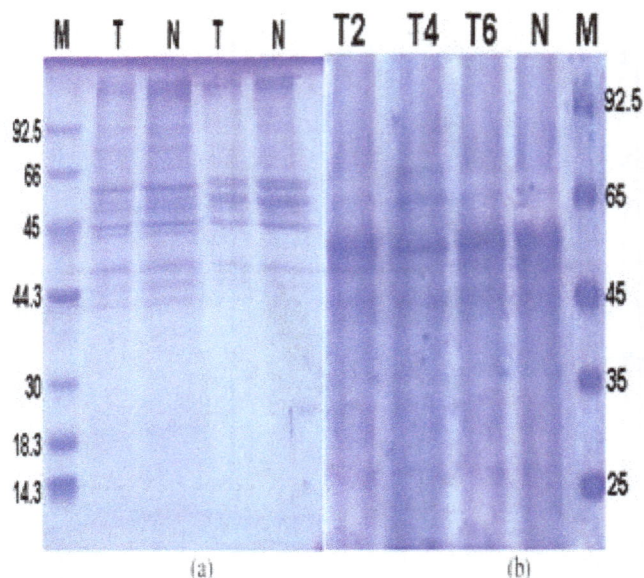

Figure 6. Protein fraction obtained from cell lysate subjected to 12% SDS– PAGE and Coomassie blue staining. (a) M – Marker T– transfected protein M- marker (kD) N– Non transfected protein, (b) T2-T6– passage no. for cultures post transfection.

gene, using SYBR green as fluorochrome using C-DNA obtained from transfected cell cultures, at different passages respectively. No significant amplification for expression of gene from second to sixth passage was observed (Data not given). Hence, transfection of the gene, and expression of the gene at transcriptional and translational level was confirmed.

SDS PAGE

Protein fraction extracted from the cell cultures containing EGFP/ recombinant fusion protein IFN-t-EGFP and also from non-transfected cell cultures when analysed on SDS-PAGE and further subjected to Coomasie Blue staining, there was no significant difference in the pattern of protein bands, between transfected and non-transfected cell cultures and also between the protein extracts obtained from transfected cell cultures at different passages post transfection [Figure 6a and b]. On silver staining of the SDS gels the protein was detected on SDS-PAGE at 53 and 31 KD region, former for IFN-t-EGFP fusion protein and the latter protein band for EGFP+ respectively which confirms the expression of the transfected gene (lane 1 and 3). In contrast, 31 KD protein

band was not detected from the protein extracted from the cultures transfected with the recombinant IFN-t-EGFP gene. A number of protein bands were observed for the supernatant fraction obtained during the immuno precipitation of proteins. Lane 2 represents the positions of the molecular mass standards (Figure 7).

Effect on lymphocyte proliferation

The bovine lymphocytes were separated from blood and subjected to mitotic stimulation with phytohaemoagglutinin (PHA). Lymphocytes devoid of PHA served as unstimulated cells. The culture medium alone served as blank. These cultures were subjected to MTT test and lymphocyte stimulation index (LSI) was calculated as ratio of the absorbance of the stimulated lymphocytes to the unstimulated lymphocyte cultures. It was observed that LSI of lymphocyte culture treated with immunoprecipitated protein and with elution buffer was 0.745 and 0.945 respectively. The LSI of lymphocyte cultures without addition of any protein was 1.15. The stimulation index with immunoprecipitated protein had significantly decreased ($P < 0.05$) when compared with that of untreated cultures Similarly LSI of lymphocyte cultures treated with protein fraction of the crude cell lysate obtained from transfected cultures significantly decreased the stimulation index when compared with the protein fraction from the untransfected cells ($P < 0.05$) (Figure 8).

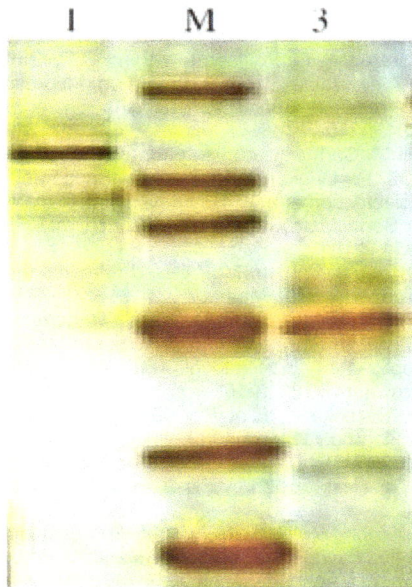

Figure 7. Lane-1 Recombinant IFN-tGFP Protein (53 kD) Lane-3 EGFP (31 kD)–Green fluorescent protein Lane-2 M- Standard protein markers from bottom14,18,31,44.3,45,66 (kD). Protein fraction obtained from cell lysate subjected to 12% SDS – PAGE and silver staining.

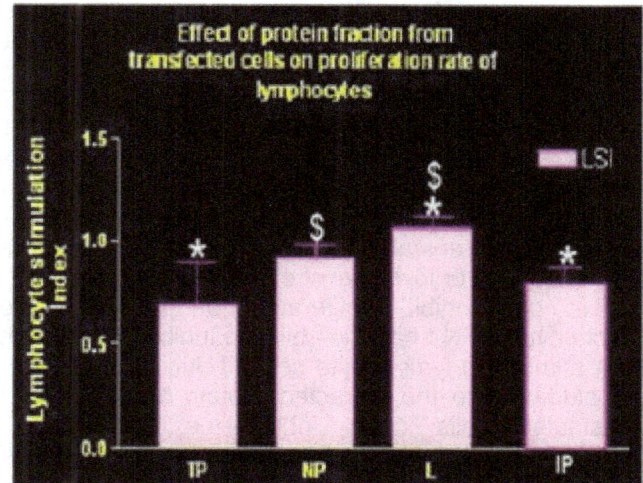

Figure 8. Lymphocyte stimulation index of cell cultures (LSI) When cultures were incubated with protein fraction from Transfected cell culture (TP) (P < 0.05), with immune - precipitated protein (IP) (P < 0.05), with protein fraction from non transfected cell culture (NP). Control (L) - lymphocyte culture treated with PHAw/o addition of any protein.

DISCUSSION

In this study, we demonstrated the expression of transfected gene IFN-t-GFP.in fibroblast cell culture systems. The recombinant IFNt-EGFP protein could be detected or resolved on SDS-PAGE only after silver staining of the gels.

In this study, we have applied a novel non-viral transfection system, nucleofection to induce high recombinant gene expression in primary fibroblast cell lines. In preliminary experiments, it was tested with different conditions of electrical pulsing to determine the program, inducing the highest transgene expression. By using a IFN-t-GFP and GFP as reporter, plasmid under the transcriptional control of a eukaryotic promoter (Materials and Methods), the highest fraction of GFP+ cattle fibroblast cells was obtained after transfection with U-23 program. Similar were the results on transfection with p EGFP only our results on cell survival and recovery are apparently in confirmation with the results obtained by Aluigi et al. (2006). However, even after six weeks of culture, significant percentage of GFP+ cells could be detected. Nucleofected fibroblast cells could be used in several types of studies where expression of transgene protein is required. There are some reports where low cell recovery was observed (Maasho et al., 2005; Laakshmipathy et al., 2004).

Available evidence indicates that IFN-t inhibits transcription of estrogen receptor α and oxytocin receptor genes to block development of the uterine luteolytic mechanism (Spencer and bazer, 1996; Spencer et al., 1998). Use of green fluorescent protein (GFP) from the Jellyfish *Acquorea victoria* is a powerful method for non-desctructive *in situ* monitoring, since expression of green fluorescence does not require any substrate addition (Anderson et al., 1998). The GFP gene may be transferred to and expressed in a wide range of organisms, e.g., mammals (Ludin et al., 1996).

Because of the central role of IFN-t in gestation, this protein has excited the interest of reproductive physiologists (Demmers et al., 2001). Supplementation of inseminated animals with exogenous IFN-t during the pregnancy recognition period has been considered as a means to compensate for inadequate secretion of IFN-t by conceptuses that are delayed in their development and enhance their chances for survival (Spencer et al., 1999). Hence in the present experiment we attempted to develop culture system for expression of pbIFN-t-EGFP in Sahiwal fibroblast cells. It proves and paves way for studies on transfection of other genes and their functions.

Reports are available, where efficient transfer of DNA expression vectors and SiRNA oligonucleotides into a variety of primary cell types from different species including human B-CLL cells, human CD34+ cells, human lymphocytes, rat cardiomycytes, human bovine chondrocytes (Greseh et al., 2004). Transfection has been achieved in mural stem cells also (Marchenko and Flanagan, 2007). The versatility and resolving capacity of polyacrylamide gel electrophoresis (PAGE) has resulted in the most popular method of protein identification and purification post Coomasie blue staining. The sensitive method used by Hochstrasser et al. (1998) and Rabilloud (1992) was used for silver staining procedure. The products of recombinant bIFN-t, expressed in fibroblast cells were analysed by SDS-PAGE both by Coomasie Blue and Silver staining procedure, where we were successful in analyzing protein by silver staining only, this may be due to unstable nature of the protein during and after extraction or low level of expression of protein.

IFN-t can inhibit proliferation of peripheral blood lymphocytes. IFN-t had the ability to inhibit proliferation of PHA stimulated lymphocyte cells (Tekin et al., 2000). In our studies also the extracted protein fraction from the transfected cells could inhibit the proliferation of lymphocytes *in vitro* indicating immunosuppressive activity. This may be due to IFN-t or some upregulated protein which is immunosuppressive in nature.

The current study describes for the first time expression of IFN-t-GFP genes in cattle fibroblast cells at different passages post transfection under the control of eurkaryotic promoter specific for IFN-t-expression (Toshihiko et al., 1999), it is necessary for the proliferated transfected cells to show an increase in number of transfected cells. To overcome the above mentioned limitations, the delivery of DNA encoding reporter genes such as EGFP is necessary to promote the expansions capacity and survival potential. As the fluorescence of the protein could be visualized, which was present at the downstream of the recombinant gene construct. It shows that protein IFN-t was expressed in cell cultures.

The detection of protein on silver staining demonstrates the tendency of cell cultures towards less production or less stability of the fusion protein and further recommends improvement in techniques or possibility to enhance the stability of the extracted recombinant protein has to be explored.

ACKNOWLEDGEMENTS

The authors are grateful to Department of Biotechnology and Indian Council of Agricultural Research, New Delhi, for providing funds tocarry out this research work. Acknowledgements are also due to Defence Research Development Organization (DRDO), New Delhi for providing flow cytometer facility. Thanks are due to Dr. B. S. Prakash, Head, Dairy Cattle Physiology Division, NDRI, Karnal for providing necessary laboratory facilities. The eukaryotic promoter for IFN-t was generously gifted by Dr. R. M. Robert's laboratory, and the coding region by Dr. Alan D. Ealy, Coloumbia University Missouri.

REFERENCES

Andersen J, Sternberg C, Poulsen LK, Bjørn S P, Givskov M, Molin S (1998). New Unstable Variants of Green Fluorescent Protein for Studies of Transient Gene Expression in Bacteria. Appl. Environ. Microbiol., 64: 2240-2246.

Aluigi M, Fogli M, Curti A, Isidori A, Gruppion E, Chiodoni C, Colombo MP, Versura P, Grigioni AD, Ferri E, Baccarani M, Lemoli RM (2006). Nucleofection Is an Efficient Nonviral Transfection Technique for Human Bone Marrow–Derived Mesenchymal Stem Cells. Stem Cells, 24: 454–461.

Blesh A (2004). Lentiviral and MLV based retroviral vectors for ex vivo and *in vivo* gene transfer. Methods,33: 164–172.

Chan AWS, Chong KY, Schatten G (2002). Transgenic bovine embryo selection using green fluorescent protein. In: Methods in Molecular Biology: Applications and Protocols (B.W. Hicks, Ed.). Human Press Inc. NJ, Totowa, NJ, 183: 3-15.

Demmers KJ, Derecka K, Flint A (2001). Trophoblast interferon and pregnancy. Reprod., 121: 41-49.

Toshihiko E, Ealy AD., Ostrowski MC, Roberts RM (1999). Control of interferon- gene expression by Ets-2. Proc. Natl. Acad. Sci. USA, 95: 7882–7887.

Fang-Fang G, Zhong-Yi W, Shen-Ming Z (2008). Bovine interferon-tau expression in *Escherichia coli* and identification of its biological activities. Chin. J. Agric. Biotechnol., 5: 189-195.

Gresch O, Engel FB, Nesic D, Tran TT, England HM, Hickman ES, Körner I, Gan L, Chen S, Castro-Obregon S, Hammermann R, Wolf J, Müller-Hartmann H, Nix M, Siebenkotten G, Kraus G, Lun K (2004). New non-viral method for gene transfer into primary cells. Methods, 33: 151–163.

Hochstrasser DF, Merril CR (1998). Catalysts' for polyacrylamide gel polymerization and detection of proteins by silver staining. Appl. Theor. Electrophoresis, 1: 35–40.

Lakshmypathy U, Pelacho B, Sudo K, Linehan JL, Coucouvanis E, Kaufman DS, Verfaillie CM (2004). Efficient Transfection of Embryonic and Adult Stem Cells. Stem Cells, 22: 531-543.

Ludin B, Doll T, Meili R, Kaech S, Matus A (1996). Application of novel vectors for GFP-tagging of proteins to study microtubule-associated proteins. Gene, 173: 107-111.

Marchenko S, Flanagan L (2007). Transfecting human neural stem cells with the Amaxa Nucleofector. J. Vis. Exp., 6: 240.

Maasho K, Opoku-Anane J, Marusina AI, Coligan JE, Borrego F (2005). NKG2D is a costimulatory receptor for human naive CD8+ T cells. J. Immunol., 174: 4480-4484.

Mossman J (1983). Rapid colorimetric assay for cellular growth and survival; application to proliferation and cytotoxicity assays. J. Immun. Methods, 65: 55-63.

Naito M, Minematsu T, Harumi TI, Kuwana T (2007). Intense Expression of GFP Gene in Gonads of Chicken Embryos by Transfecting Circulating Primordial Germ Cells *in vitro* and *in vivo*. J. Poultry Sci., 44: 416-425.

Rabilloud T (1992). A comparison between low background silver diamine and silver nitrate protein stains. Electrophoresis, 13: 429–439.

Rosochacki SJ, Duszweska AM, Kozikova LV, Olszewski R (2001). Expression of GFP in microinjected bovine embryos. Anim. Sci. Pap. Reports, 19: 193-202.

Spencer TE, Stagg AG, Ott TL, Johnson AG, Ramsey WS, Bazer FW (1998). Differential Effects of Intrauterine and Subcutaneous Administration of Recombinant Ovine Interferon Tau on the endometrium of Cyclic Ewes. Biol. Reprod., 61: 464-470.

Tekin SA, Ealy AD, Wang S, Hansen PJ (2000). Differences in Lymphocyte-Regulatory Activity AmongVariants of Ovine IFN-t. J. Interferon Cytokine Res., 20: 1001–1005.

APPENDIX

The vector construct

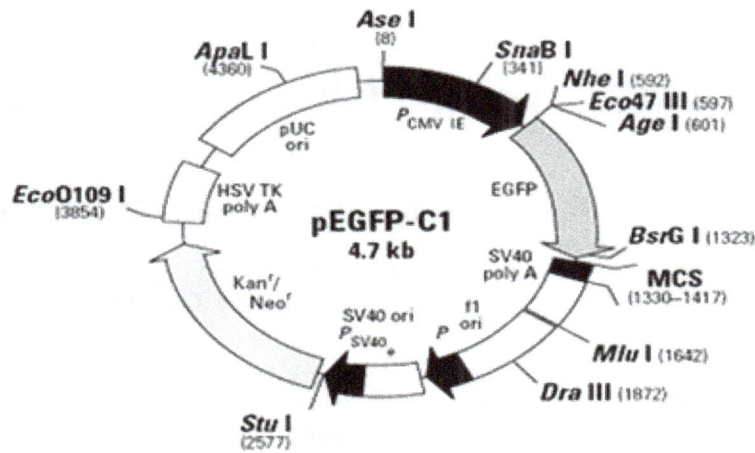

The original vector

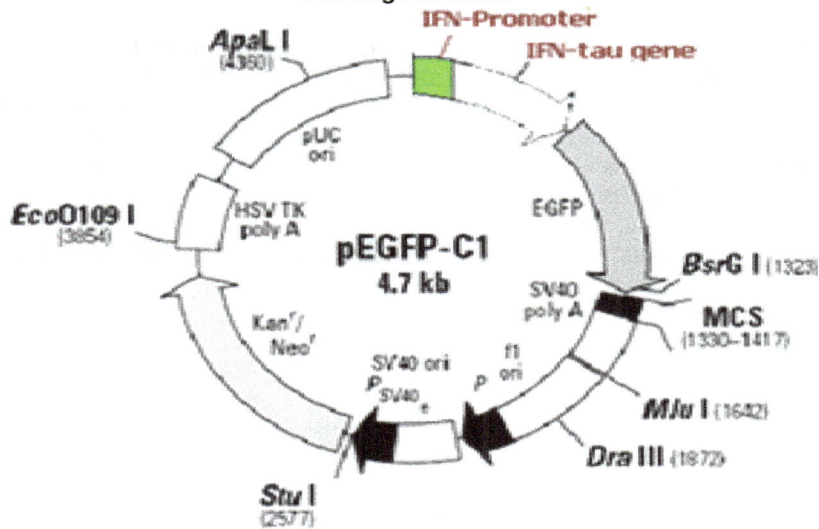

Figure 1. The cloned construct with IFN-t and EGFP construct at the downstream of IFN-t promoter.

22

3-D osteoblast culture for biomaterials testing

Jörg Neunzehn*, Sascha Heinemann and Hans Peter Wiesmann[1]

Max Bergmann Center of Biomaterials and Institute of Material Science, Technische Universität Dresden, Budapester Str. 27, D-01069 Dresden, Germany.

Micromass cell cultures, osteomicrospheres, were formed by aggregation of primary osteoblasts. Cell differentiation during sphere formation and the integrity of the osteomicrospheres was evaluated by analyzing the immunohistochemical expression of osteonectin, osteocalcin, and collagen type I. Transmission electron microscopy facilitated the proof of the tissue-like microstructure inside the osteomicrospheres and the arrangement of collagen fibers similar to early stages of natural bone formation. The *in vivo* situation of osteomicrospheres transferred to host tissue was simulated by embedding mature osteomicrospheres in a fibrin matrix. Additionally, cell spreading out of the compartments was investigated by electron microscopy of microspheres cultivated on the surface of three-dimensional fiber-like scaffolds. Concluding, the properties of the osteomicrospheres represented in this work demonstrate the potential as a tissue-specific *in vitro* test method to replace early small animal tests in the future.

Key words: Biomaterials testing, micromass culture, osteoblast, osteomicrospheres, ultrastructure.

INTRODUCTION

Tissue equivalent, three-dimensional cell agglomerates so-called micromass cultures show similar cell proliferation and differentiation like cells *in vivo*. In different studies micro mass cultures of osteoblasts, osteoblast-like cells and other cell types like stem cells (Langenbach et al., 2011, Zhang et al., 2010) were investigated intensively which affirms this tissue equivalent cell behavior.

New therapeutic approaches for the regeneration of bone require comprehensive understanding of the cellular and tissue level mechanisms that underlie bone development and bone healing. To improve bone tissue engineering strategies it seems suitable to mimic the native developmental mechanisms of bone growth and differentiation. Therefore, new insights are expected from 3-D spheroid cell culture systems (Bates et al., 2000). Numerous *in vitro* studies were performed to evaluate the cell behavior in various three-dimensional artificial scaffold materials (Meyer et al., 2004; Wiesmann et al.,

2004). Spheroids are developed from cells by a variety of methods: i) plating cells in gyratory shakers, roller flasks, or spinner flasks that continuously rotate preventing cellular adherence to the walls; ii) coating tissue culture surfaces with non adhesive substance (Iwasaki et al., 2009); iii) using the hanging drop culture method; iv) by centrifuge compression of cells and v) by the use of appropriate growth factors.

Increasing evidence has clearly stated that changes in cell shape as present under 3-D culture conditions can actually affect the cell fate. Cell morphology was shown to influence proliferation (Chen et al., 1997, Rossi et al., 2005), differentiation and gene expression (Rossi et al., 2005).The accumulation of experimental results indicates that in the absence of an anchoring material, intercellular adhesion may provide signals which promote cell activity. The aim of the present study was to generate 3-D micro-mass tissues to characterize tissue formation by histochemical, immunhistochemical, and microscopically techniques.Emphasis is given to the feature of mineral formation in the newly formed extracellular matrix (ECM).Furthermore the potential of the osteomicrospheres as an *in vitro* test system was evaluated.

*Corresponding author. E-mail: joerg.neunzehn@tu-dresden.de.

MATERIALS AND METHODS

Harvesting of osteoblast-like cells and sphere formation

The periosteal layer of calf metacarpus was aseptically stripped off the bone, cut into small pieces and plated into polystyrene culture dishes with their osteogenic side down-facing. High growth enhancement medium (HGEM)(MP Biomedicals GmbH) supplemented with 10 % fetal calf serum, 250 µg/ml amphotericin B, 10000 IU/ml penicillin, 10000 IU/ml streptomycin and 200 mmol/l L-glutamine (Biochrom KG seromed) was used for cultivation and was replaced once a week. Incubation was carried out at 37°C in a humidified atmosphere of 95% air and 5% CO_2.

Bovine primary osteoblast-like cells were harvested after three weeks by collagenase incubation (Biochrom KG seromed), counted using a coulter counting system (CASY I Model TT, Schaerfe System GmbH) and transferred to a non-attachment environment. Therefore agarose (Biozym Scientific GmbH) coated chambers of 96-well plates were prepared by applying 50 µl of a warm mixture of 20 mg/ml agarose (Biozym Scientific GmbH) in DMEM (Biochrom)/HGEM per well. A population of 200,000 cells per well was seeded and incubated in HGEM as mentioned above. The medium was changed twice weekly. After seven days of cultivation, osteomicrospheres were analyzed or used for biomaterial testing.

Cultivating osteomicrospheres on biological environments

In order to evaluate the potency of the osteomicrospheres to interact with biological environments, seven days old microspheres were aspirated with a pipette and transferred to be cultured in a fibrin matrix or on a fibroid silica-gel biomaterial scaffold (Bayer Innovation GmbH). The sphere-seeded materials were evaluated after three days by histological techniques or scanning electron microscope (SEM), respectively. As a reference, cell microspheres were cultivated on glass chamber slides (Nunc) and evaluated by light microscopy after seven and 28 days after hematoxiline and eosin (H&E) staining (Merck).

Histological and immunohistological procedures

For histology, evaluation of cell viability and differentiation steps up to cell mineralization, the osteomicrospheres were harvested after three, seven, 14, and 28 days, respectively. The osteomicrospheres were fixed with 4% formalin, embedded in paraffin, sectioned by a microtome (Ultracut S, Reichert), and stained with H&E (Merck). Mouse monoclonal primary antibodies against collagen type I (Abcam) 1:100 diluted, osteocalcin (TaKaRa) 1:100 diluted, and a rabbit polyclonal antibody for osteonectin (Chemicon International) 1:50 diluted in 1 % BSA in PBS were used for immunohistochemistry. Secondary antibodies used were anti-mouse (Dako K-4001) and anti-rabbit antibodies (Dako K-4002). Sections were deparaffinized by immersing the samples twice in xylene for 5 min, followed by gradual dehydration of the sections through immersion in ethanol. Finally, sections were washed with distilled water.

Transmission electron microscopy (TEM)

Samples were fixed in 0.1 M phosphate buffer containing 2.5% glutaraldehyde (Sigma-Aldrich) dehydrated in an ascending series of ethanol and embedded in araldite resin (Plano). Ultrathin sections of 70 to 90 nm were cut (Leica RML 155) and stained for 30 min with 2% uranyl acetate (Plano) followed by incubation in Reynolds's lead citrate (Merck) for 3 min. Sections were examined with a Zeiss TEM 902 at an acceleration voltage of 80 kV.

Scanning electron microscopy (SEM)

Osteomicrospheres cultivated for three days on the fibroid biomaterial scaffold were fixed with glutaraldehyde (Sigma), followed by dehydration in graded series of ethanol, and chemical drying through the iterative transfer into hexamethyldisilazane (HMDS; Fluka). The samples were fixed on SEM stubs and sputtered with platinum.

RESULTS

After seven days of micromass formation, the H&E staining of paraffin sections (Figure 1a) revealed the integrity of highly aggregated osteoblast-like cells which formed round-shaped osteomicrospheres. Sphere diameters varied from 250 to 300 µm. Cell nuclei were visible in all regions of the microspheres. Where the inner core appeared less organized, the outer regions were even more oriented to form a dense layer of elongated cells on the surface of the osteomicrospheres (Figure 1a). The sphere's surface was built up from three to four cell layers superimposed by each other. Some regions of lower cell density appeared as bright spots distributed evenly throughout the sphere. Intracellular space occurred as thin layers containing extracellular matrix.

The *in vitro* immunoexpression of extracellular matrix proteins demonstrated an increased level of collagen type I during 28 days of sphere formation and cultivation (Figure 1b). After three days, first spots became visible and developed a heterogeneous pattern until day seven. After 14 days the distribution of collagen type I was homogeneous over the whole cross-section of the osteomicrospheres. The total expression of the non-collagenous matrix protein osteonectin varied slightly during cultivation time. Obviously, the distribution of the protein was enhanced in the core region of the microspheres compared to the outer regions. The expression of osteocalcin was evenly after 3 days. However, the development to a core region of high expression level and an outer shell of lower osteoclacin expression level lasted unit day seven. Microsphere maturation led to increased OC expression throughout the whole sphere. Cell differentiation and outgrowth over 28 days of cultivation varified activity and viability of the cells in the all regions of the spheres.

Proceeding outgrowth of single cells during cultivation caused decreasing sphere sizes, particularly starting from day seven. This process is also visualized in Figures 3b and c, showing cells outgrowing from a sphere and migrating radially to form a layer after seven days of cultivation on glass. Ultrastructural analysis carried out using transmission electron microscopy confirmed the sphere's shell to be composed of flattened, thin and elongated osteoblasts containing few organelles. Osteoblasts located in the core of the osteomicrospheres did not exhibit a directed orientation but exhibited large oval shaped nuclei and an active-state organelle system: extended rough endoplasmic reticulum, mitochondria,

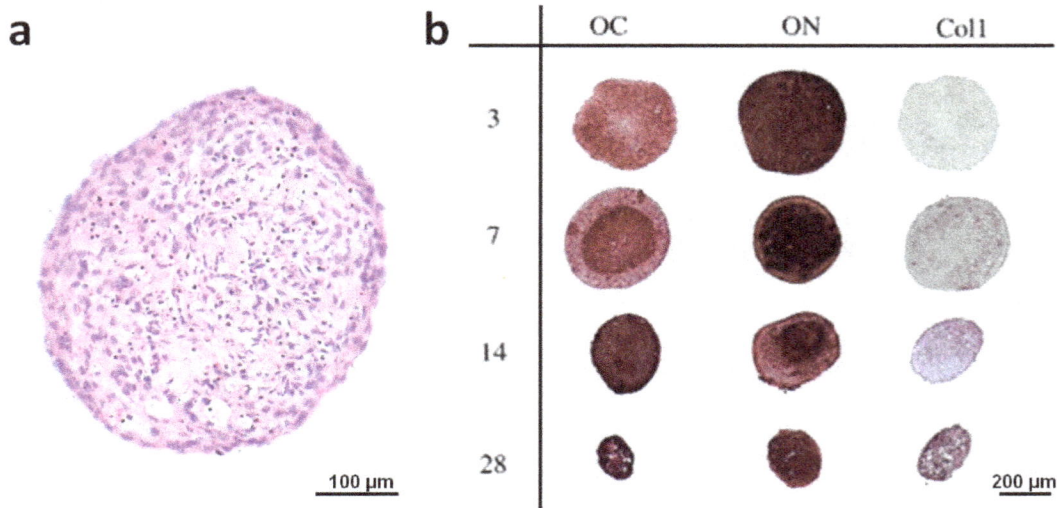

Figure 1. a) H&E staining of a paraffin section of a osteomicrosphere formed after seven days of cultivation with elongated cells forming the sphere's shell in the outer region. b) Micrographs of osteomicrospheres formed after 3, 7, 14, and 28 days of micromass cultivation stained for osteocalcin (OC), osteonectin (ON), and collagen I (Col I), respectively.

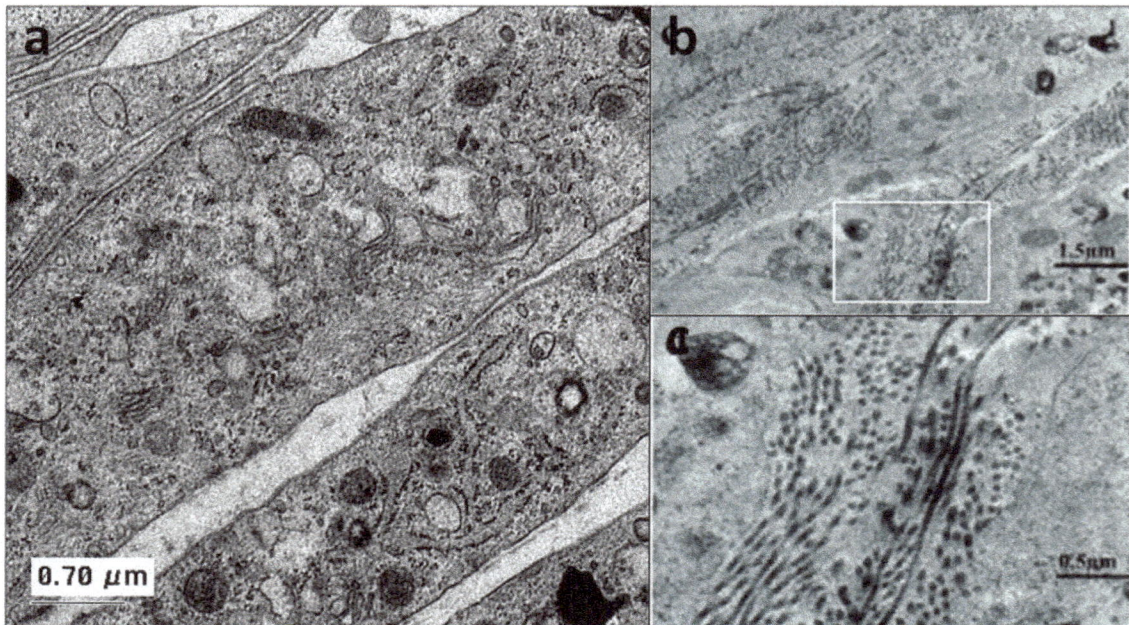

Figure 2. TEM image showing a) elongated cells b) developed, organized collagen bundles in a 21 day old osteomicrosphere and c) magnification of b).

secretory vesicles, polysomes, and gap junctions, all these features proving high metabolic activity (Figure 2).

The occurrence of extracellular matrix formation varied according to the location in the sphere, being concentrated to the core and less abundant in the outer layers. Additionally, extracellular matrix was less prominent in younger osteomicrospheres and more apparent in older microspheres. Continuous collagen fiber formation in the extracellular matrix of the microspheres was demonstrated throughout the entire cultivation time. Orientation of the fibrils was similar to the *in vivo* situation (Figures 2b and c). Directly after transfer

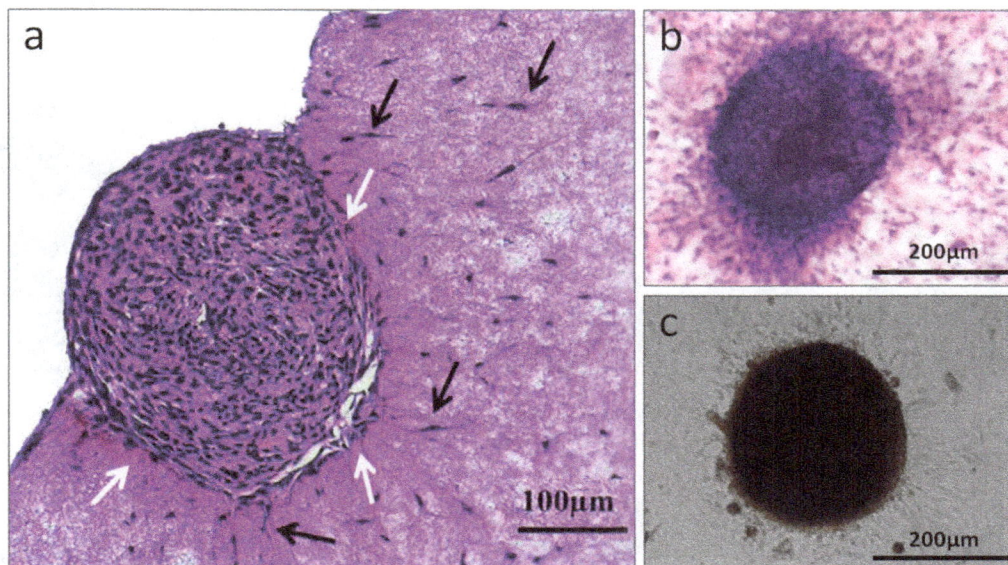

Figure 3. Imaging of cell migration of 7 days old osteomicrosphere after 3 days of incubation on fibrin membrane after H&E-staining (a). Several cells spread out of the sphere to the interface (white arrows) and infiltrated the artificial matrix (black arrows). Micromass culture after seven days of cultivation on glass H&E-stained (b) and without staining (c).

to the fibrin, the osteomicrospheres attached to the artificial matrix. After seven days, the cultivation of osteomicrospheres in the matrix led to an intimate contact of the sphere's cells with the fibrin surface. Migration of osteoblasts on the interface (white arrows in Figure 3a) as well as into the fibrin matrix (black arrows Figure 3a) was observed histologically.

Figures 3b and c show the behavior of the agglomerated cells on untreated glass slides. The tendency of the individual cells to grow out of the organized osteomicrospheres and to colonize the surrounding area or substrate is obvious. Subsequently, osteomicrospheres were assessed in contact with an artificial environment which facilitates the observation of three-dimensional sphere attachment as well as cell spreading. Therefore, seven days old microspheres were transferred to a fiber-like scaffold instead of a conventional foam-like scaffold and were evaluated morphologically using SEM (Figure 4). After three days of cultivation, osteomicrospheres attached to the silica fiber scaffold, showed the attempt to build up larger tissue-like formations by forming agglomerates of several osteomicrospheres (Figure 4a) and enveloped single fibers (Figure 4b). This resulted in covering of the fiber material by outgrown cells and corresponding extracellular matrix.

DISCUSSION

Edwards and Mason (2006) postulated that the use of passive three-dimensional scaffolds, inductive strategies

in which additional growth factors are incorporated into scaffold/matrix-systems to modify cell behavior, or strategies to form vital constructs of cells are important future strategies for tissue regeneration. The present study focuses on the third point.

The use of micromass cultures is a recently elaborated technique in bone biology since basic principles of cell and tissue development in three-dimensional space can be evaluated and basic tissue engineering techniques can be approached. Recent studies focus on two aspects: to initiate and investigate the basic principles of cellular self-assembly, and the scaffold-free formation of three-dimensional tissue (Battistelli et al., 2005, Tare et al., 2005). In the literature, different methods are reported which facilitate the formation of spherical-shaped micromass cultures (Handschel et al., 2007, Kale et al., 2000, Tortelli and Cancedda, 2009). These should mimic the targeted tissues; resemble their organization, their mechanical properties, and their physiological response to different stimuli (Tortelli and Cancedda, 2009).

In the present study osteomicrospheres were derived from conventional monolayers of primary osteoblast-like cells and showed bone tissue-like differentiation. The scaffold-free three-dimensional tissue was built up from differentiated osteoblast-like cells in high density and showed tissue-like behavior also during cultivation on artificial biomaterials.

Concerning the usage of cell spheroids, different studies constituted that three-dimensional cultures exert higher proliferation rates than cells cultured in two-dimensional systems. Furthermore their cell-specific differentiation and tissue-like migration is more distinct

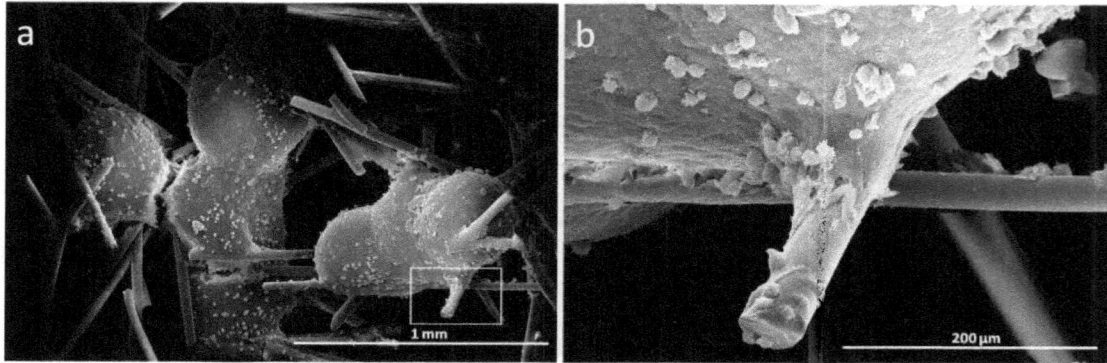

Figure 4. SEM overview a) of seven days old osteomicrospheres after three days of cultivation on a silica fiber scaffold. The osteomicrospheres showed strong attachment to the underlying fiber material. b) Outgrowing sphere cells covering the fiber surface.

(Abbott, 2003; Cukierman et al., 2001; Handschel et al., 2007). Moreover, the cells are flexible to change their shape and are able to react in a more physiological way when they are cultured in three-dimensional spheres; reasons are cell-matrix and cell-cell contacts supporting specific cell signals and migration of the cells (Handschel et al., 2011, Sivaraman et al., 2005; Weaver et al., 1997; Yamada and Geiger, 1997).

The micromass culture technique represents an alternative for substituting artificial scaffolds and a potent method for biomaterials testing. In contrast to two-dimensional monolayers, the three-dimensional culture technique shows more similarities to tissues under *in-vivo* conditions (Handschel et al., 2007). It shall be deemed to be proven that a lot of cellular functions, e.g. differentiation, migration, and proliferation, rely on intact cell-matrix and cell-cell interactions and an intimate attachment to components of the extracellular matrix. In the three-dimensional spheres, the cells can interact with each other and maintain these essential interactions (Boudreau and Jones, 1999; Cukierman et al., 2001; Yamada and Geiger, 1997).

These results correspond to findings concerning the spheroid behavior presented by Gerber et al. (2005). They also highlight faster differentiation and mineralization of micromass-cultures compared to conventional two-dimensional culture methods of the same cell types. Whereas the formation of osteomicrospheres is seen regularly in the non-attachment chambers, we observed a limited control over shape or size of the produced tissue. This may be based on insufficient non-attachment behavior of agarose, being not resistant to the inevitable precipitation of medium proteins forming a new interface layer. An increased cell detachment from the sphere and attachment onto the agarose-protein layer is observed over time. During the cultivation of about 28 days (Figure 1b), the size of the spheres decreased, especially between day seven and 14. In addition to the reason already discussed in the manuscript, it could be explained by the declining

prevention of cell adhesion to the agarose. During the repeated culture medium changes during the sphere cultivation, the initial function of the agarose declined, cells grew out of the spheres and the diameter of the agglomerates decreased. Cell apoptosis cannot be ruled out to be another reason for the decreasing size of the spheres and appears as a physiological process after 14 to 28 days.

The outgrowth of cells out of the microspheres was also recognized by the osteomicrospheres cultivated on glass chamber slides as visualized, showing tissue like behavior by colonizing the adjacent material. In the inner space of the osteomicrospheres, predominantly cubic cells were recognized, similar to the morphology of mature osteoblasts as described in literature (Heinemann et al., 2011). These are cuboidal, 20 to 30 µm wide cells with an oval nucleus, which are organized palisade-like around the matrix they synthesize. Furthermore, these cells are active, expressing extracellular matrix proteins namely collagen type I and osteocalcin as revealed in the present study by immunohistochemistry. The expression of osteonectin confirms the results of other studies demonstrating protein expression at earlier stages of osteoblastic cell lineage differentiation, and being a stimulating factor in the mineralization process. Osteocalcin, one of the latest osteoblast markers, indicates differentiation gradients and the transition to mineral formation (Owen et al., 1990). Although these macromolecules can be found in other tissues, their expression by osteoblasts and deposition into the matrix reflects the biosynthetic repertoire of the osteomicrospheres. In this study, an early and strong osteoblastic differentiation and expression of osteocalcin was detected starting in the core region of the microspheres. The same process was also described for embryonic stem cell micromasses, where the mineralization also started in the core of the spheres after osteogenic differentiation (Handschel et al., 2011). In the present study, osteocalcin expression in the core of the spheres represents beginning mineralization and emphasizes the tissue-like behavior of these three-dimensional constructs.

The osteomicrosphere's surrounding outer layer of flat elongated cells is similar to that described by Aubin et al. (1998) as elongated and thin bone-lining cells. This cell type is a rather inactive form, contains less cell organelles (Owen et al., 1990) and is typically present on the surface of the bone. In the present study, the cells of the outer layers of osteomicrospheres seem to have an "epithelial" function and they were able to grow out of the cell sphere and to migrate on biomaterial surfaces. The activity of the surface cells and also the core cells was shown by the ability to spread out and grow around biocompatible materials by maintaining their spheroid shape. The fibrin scaffold was chosen for this investigation, because fibrin as a material produced naturally in the body and furthermore is an adequate scaffold material in tissue engineering. The morphological adaption of the microspheres by spreading of single cells out of the cell complex highlights the biological activity of this tissue-like cell constructs.

Conclusion

The ability to study osteoblasts in an environment close to native tissue is one of the major advantages of the presented micromass culture system. The possibility to assure the reproducible formation of viable osteomicrospheres within a defined period of time that withstand handling procedures may open new paths in extra-corporal bone tissue engineering. Furthermore, the results of the present study suggest the use of the concept as a new test method classified between the conventional monolayer techniques and small animal studies. The cell osteomicrospheres show tissue-like behavior and differentiation performances, the physiological ability to interact with other osteomicrospheres to larger tissue-like complexes and to invade, envelope or encapsulate artificial biomaterials.

ACKNOWLEDGEMENT

We thank the members of the Department of Tissue Engineering and Biomineralisation of the University of Muenster, for excellent technical assistance.

REFERENCES

Abbott A (2003). Cell culture: biology's new dimension. Nature 424:870-872.

Aubin JE (1998). Advances in the osteoblast lineage. Biochem. Cell Biol. 76:899-910.

Bates RC, Edwards NS, Yates JD (2000). Spheroids and cell survival. Crit. Rev. Oncol. Hematol. 36:61-74.

Battistelli M, Borzi RM, Olivotto E, Vitellozzi R, Burattini S, Facchini A, Falcieri E (2005). Cell and matrix morpho-functional analysis in chondrocyte micromasses. Microsc. Res. Tech. 67:286-295.

Boudreau NJ, Jones PL (1999). Extracellular matrix and integrin signalling: the shape of things to come. Biochem. J. 339 (Pt 3):481-488.

Chen CS, Mrksich M, Huang S, Whitesides GM, Ingber DE (1997). Geometric control of cell life and death. Science 276:1425-1428.

Cukierman E, Pankov R, Stevens DR, Yamada KM (2001). Taking cell-matrix adhesions to the third dimension. Science 294:1708-1712.

Edwards PC, Mason JM (2006). Gene-enhanced tissue engineering for dental hard tissue regeneration: (1) overview and practical considerations. Head Face Med. 2:12.

Gerber I, ap Gwynn I, Alini M, Wallimann T (2005). Stimulatory effects of creatine on metabolic activity, differentiation and mineralization of primary osteoblast-like cells in monolayer and micromass cell cultures. Eur. Cell Mater. 10:8-22.

Handschel J, Naujoks C, Depprich R, Lammers L, Kubler N, Meyer U, Wiesmann HP (2011). Embryonic stem cells in scaffold-free three-dimensional cell culture: osteogenic differentiation and bone generation. Head Face Med. 7:12.

Handschel JG, Depprich RA, Kubler NR, Wiesmann HP, Ommerborn M, Meyer U (2007). Prospects of micromass culture technology in tissue engineering. Head Face Med. 3:4.

Heinemann C, Heinemann S, Worch H, Hanke T (2011). Development of an osteoblast/osteoclast co-culture derived by human bone marrow stromal cells and human monocytes for biomaterials testing. Eur. Cell Mater. 21:80-93.

Iwasaki A, Matsumoto T, Tazaki G, Tsuruta H, Egusa H, Miyajima H, Sohmura T (2009). Mass fabrication of small cell spheroids by using micro-patterned tissue culture plate. Adv. Eng. Mater. 11:801-804.

Kale S, Biermann S, Edwards C, Tarnowski C, Morris M, Long MW (2000). Three-dimensional cellular development is essential for ex vivo formation of human bone. Nat. Biotechnol. 18:954-958.

Langenbach F, Berr K, Naujoks C, Hassel A, Hentschel M, Depprich R, Kubler NR, Meyer U, Wiesmann HP, Kogler G, Handschel J (2011). Generation and differentiation of microtissues from multipotent precursor cells for use in tissue engineering. Nat. Protoc. 6:1726-1735.

Meyer U, Joos U, Wiesmann HP (2004). Biological and biophysical principles in extracorporal bone tissue engineering. Part I. Int. J. Oral Maxillofac. Surg. 33:325-332.

Owen TA, Aronow M, Shalhoub V, Barone LM, Wilming L, Tassinari MS, Kennedy MB, Pockwinse S, Lian JB, Stein GS (1990). Progressive development of the rat osteoblast phenotype in vitro: reciprocal relationships in expression of genes associated with osteoblast proliferation and differentiation during formation of the bone extracellular matrix. J. Cell Physiol. 143:420-430.

Rossi MI, Barros AP, Baptista LS, Garzoni LR, Meirelles MN, Takiya CM, Pascarelli BM, Dutra HS, Borojevic R (2005). Multicellular spheroids of bone marrow stromal cells: a three-dimensional in vitro culture system for the study of hematopoietic cell migration. Braz. J. Med. Biol .Res. 38:1455-1462.

Sivaraman A, Leach JK, Townsend S, Iida T, Hogan BJ, Stolz DB, Fry R, Samson LD, Tannenbaum SR, Griffith LG (2005). A microscale in vitro physiological model of the liver: predictive screens for drug metabolism and enzyme induction. Curr. Drug Metab. 6: 569-591.

Tare RS, Howard D, Pound JC, Roach HI, Oreffo RO (2005). Tissue engineering strategies for cartilage generation--micromass and three dimensional cultures using human chondrocytes and a continuous cell line. Biochem. Biophys. Res. Commun. 333:609-621.

Tortelli F, Cancedda R (2009). Three-dimensional cultures of osteogenic and chondrogenic cells: a tissue engineering approach to mimic bone and cartilage in vitro. Eur. Cell Mater. 17:1-14.

Weaver VM, Petersen OW, Wang F, Larabell CA, Briand P, Damsky C, Bissell MJ (1997). Reversion of the malignant phenotype of human breast cells in three-dimensional culture and in vivo by integrin blocking antibodies. J. Cell Biol. 137:231-245.

Wiesmann HP, Joos U, Meyer U (2004). Biological and biophysical principles in extracorporal bone tissue engineering. Part II. Int. J. Oral Maxillofac. Surg. 33:523-530.

Yamada KM, Geiger B (1997). Molecular interactions in cell adhesion complexes. Curr. Opin. Cell Biol. 9:76-85.

Zhang L, Su P, Xu C, Yang J, Yu W, Huang D (2010). Chondrogenic differentiation of human mesenchymal stem cells: a comparison between micromass and pellet culture systems. Biotechnol. Lett. 32:1339-1346.

23

Future of dentistry, nanodentistry, ozone therapy and tissue engineering

Maryam Moezizadeh

Department of Operative Dentistry, School of Dentistry, Shahid Beheshti University of Medical Sciences, Tehran, Iran.
E-mail: mamoezizadeh@yahoo.com.

Future of dentistry has become one of the most unique resources and providers of this oral systemic approach. In determining the patient's unique needs, it is becoming increasingly important for the dentist to work with other cutting-edge, integrative practitioners to fully assess the individual requirements of the patient before, during, and after dental care. Witnessing the beginning of truly groundbreaking advances in technology is a rare opportunity. Skepticism is a natural reaction when we are presented with a radically new method and its potential uses. Skepticism helps us filter the valuable from the worthless, the permanent from the ephemeral, and the rational from the preposterous. The present article describes a brief review on some of the breakthrough advances in dentistry regarding tissue engineering, nanodentistry and ozone therapy.

Key words: Nanodentistry, ozone therapy, tissue engineering.

INTRODUCTION

Patients today are increasingly taking more responsibility for their own health care. They tend to be well-read, educated by previous health-care experiences, and much more discriminating in their own choice of practitioners and treatment methods. As health practitioners struggling with the complex task of helping these patients achieve optimum wellness, we should constantly search for the information that best answers their questions and offers solutions to their health concerns (Kumar et al., 2010).

Nanodentistry will make possible the maintenance of comprehensive oral health by employing nanomaterials, biotechnology including tissue engineering, and ultimately, dental nanorobotics (nanomedicine). When the first micron-size dental nanorobots can be constructed in 10 to 20 years, these devices will allow precisely controlled oral analgesia, dentition replacement therapy using biologically autologous whole replacement teeth manufactured during a single office visit, and rapid nanometer-scale precision restorative dentistry. New treatment opportunities may include dentition renaturalization, permanent hypersensitivity cure, complete orthodontic realignments during a single office visit, covalently-bonded diamondized enamel, and continuous oral health maintenance using mechanical dentifrobots (Sharma et al., 2010).

Most of people possess a fear towards dentistry. On account of this fear, they avoid the dental treatment. Infact, people fear injections and drills that are used in dental clinics. Although, in recent time, dentistry has been experiencing a period of dynamic changes and growth, perhaps like no other time before. The use of ozone in dental treatment is the result of this dynamics and growth. Incorporation of ozone in dental clinic set-ups would eradicate the feeling of pain during dental treatment and also cut off the treatment time, significantly. Ozone has been shown to stimulate remineralization of recent caries-affected teeth after a period of about six to eight weeks. Scientific support, as suggested by demonstrated studies, for ozone therapy presents a potential for an atraumatic, biologically-based treatment for conditions encountered in dental practice (Garg and Tandon, 2009).

A new direction in the field of vital pulp therapy is given by the introduction of tissue engineering as an emerging science. It aims to regenerate a functional tooth-tissue structure by the interplay of three basic key elements: stem cells, morphogens and scaffolds. It is a multidisciplinary approach that combines the principles of biology, medicine, and engineering to repair and/or regenerate a damaged tissue and/or organ (Malhotra et al., 2009).

The present review article describes the future of dentistry and possible different treatment modalities in the future including nanodentistry, ozone therapy and tissue engineering. We hope this mini review article be useful for all dentists and our colleagues and open a new era in field of dentistry.

NANODENTISTRY

The human characteristics of curiosity, wonder, and ingenuity are as old as mankind. People around the world have been harnessing their curiosity into inquiry and the process of scientific methodology. Recent years have witnessed an unprecedented growth in research in the area of nanoscience. There is increasing optimism that nanotechnology applied to medicine and dentistry will bring significant advances in the diagnosis, treatment and prevention of disease. Growing interest in the future medical applications of nanotechnology is leading to the emergence of a new field called nanomedicine. Nanomedicine needs to overcome the challenges for its application, to improve the understanding of pathophysiologic basis of disease, bring more sophisticated diagnostic opportunities, and yield more effective therapies and preventive properties. Molecular technology is destined to become the core technology underlying all of 21st century medicine and dentistry (Mallanagouda et al., 2008).

Nano is derived from the Greek word for dwarf, and usually is combined with a noun to form words such as nanometer, nanotechnology or nanorobot. A nanometer is 10–9 meter, or one-billionth of a meter. Since it is not easy to visualize the scale of a nanometer, a comparison with concepts and objects of appreciable dimensions is helpful. If the height of an average human being were scaled up to stretch from the earth to the moon, then each of the person's atoms would be about the size of a baseball (approximately 10 centimeters in diameter). A nanometer then would be about five baseballs in a row (Jhaver, 2005).

Nanodentistry will make possible the maintenance of comprehensive oral health by employing nanomaterials, biotechnology, including tissue engineering and ultimately, dental nanorobotics. New potential treatment opportunities in dentistry may include, local anesthesia, dentition renaturalization, permanent hypersensitivity cure, complete orthodontic realignments during a single office visit, covalently bonded diamondised enamel, and continuous oral health maintenance using mechanical dentifrobots (Freitas, 2005; Chen et al., 2005).

When the first micro-size dental nanorobots can be constructed, dental nanorobots might use specific motility mechanisms to crawl or swim through human tissue withnavigational precision, acquire energy, sense, and manipulate their surroundings, achieve safe cytopenetra-

tion and use any of the multitude techniques to monitor, interrupt, or alter nerve impulse traffic in individual nerve cells in real time. These nanorobot functions may be controlled by an onboard nanocomputer that executes preprogrammed instructions in response to local sensor stimuli (Mallanagouda et al., 2008; Jhaver, 2005).

Inducing anesthesia

One of the most common procedures in dental practice, to make oral anesthesia, dental professionals will instill a colloidal suspension containing millions of active analgesic micron-sized dental nanorobot 'particles' on the patient's gingiva. After contacting the surface of the crown or mucosa, the ambulating nanorobots reach the dentin by migrating into the gingival sulcus and passing painlessly through the lamina propria or the 1 to 3-micron thick layer of loose tissue at the cementodentinal junction. On reaching dentin, the nanorobots enter dentinal tubules holes that are 1 to 4 microns in diameter and proceed toward the pulp, guided by a combination of chemical gradients, temperature differentials, and even positional navigation, all under the control of the onboard nanocomputer as directed by the dentist. There are many pathways to choose from, near to cemento-enamel junction (CEJ), midway between junction and pulp, and near to pulp.

Tubules diameter increases as it nears the pulp, which may facilitate nanorobot movement, although circumpulpal tubule openings vary in numbers and size (tubules number density 22,000 mm dentin-enamel junction (DEJ), 37,000 mm square midway, and 48000 mm square near to pulp). Tubules branching patterns, between primary and irregular secondary dentin, regular secondary dentin in young and old teeth (sclerosing) may present a significant challenge to navigation.

The presence of natural cells that are constantly in motion around and inside the teeth including human gingival and pulpal fibroblasts, cementoblasts of the cementum-dentine junction (CDJ), bacteria inside dentinal tubules, odontoblasts near the pulp dentin border, and lymphocytes within the pulp or lamina propria suggested that such journey should be feasible by cell-sized nanorobots of similar mobility. Once installed in the pulp and having established control over nerve impulse traffic, the analgesic dental nanorobots may be commanded by the dentist to shut down all sensitivity in any particular tooth that requires treatment. When on the hand-held controller display, the selected tooth immediately becomes numb. After the oral procedures are completed, the dentist orders the nanorobots to restore all sensation, to relinguish control of nerve traffic and to engress, followed by aspiration. Nanorobotic analgesics offer greater patient comfort and reduced anxiety, no needles, greater selectivity, and controllability of the analgesic effect, fast and completely reversible switchable action and avoidance of most side effects and complications (Freitas, 2000).

Tooth repair

Nanorobotic manufacture and installation of a biologically autologous whole replacement tooth that includes both mineral and cellular components, that is, 'complete dentition replacement therapy' should become feasible within the time and economic constraints of a typical office visit through the use of an affordable desktop manufacturing facility, which would fabricate the new tooth in the dentist's office. Chen et al. (2005) took advantage of these latest developments in the area of nanotechnology to simulate the natural biomineralization process to create the hardest tissue in the human body, dental enamel, by using highly organized micro architectural units of nanorod-like calcium hydroxyapatite crystals arranged roughly parallel to each other.

Dentin hypersensitivity

Natural hypersensitive teeth have eight times higher surface density of dentinal tubules and diameter with twice as large as nonsensitive teeth. Reconstructive dental nanorobots, using native biological materials, could selectively and precisely occlude specific tubules within minutes, offering patients a quick and permanent cure (Mallanagouda et al., 2008; Jhaver, 2005; Freitas, 2005).

Tooth repositioning

Orthodontic nanorobots could directly manipulate the periodontal tissues, allowing rapid and painless tooth straightening, rotating and vertical repositioning within minutes to hours (Whitesides and Love, 2001).

Tooth renaturalization

This procedure may become popular, providing perfect treatment methods for esthetic dentistry. This trend may begin with patients who desire to have their; (1) old dental amalgams excavated and their teeth remanufactured with native biological materials, and (2) full coronal renaturalization procedures in which all fillings, crowns, and other 20 th century modifications to the visible dentition are removed with the affected teeth remanufactured to become indistinguishable from original teeth (Freitas, 2005; Chen et al., 2005).

Dental durability and cosmetics

Durability and appearance of tooth may be improved by replacing upper enamel layers with covalently bonded artificial materials such as sapphire or diamond, which have 20 to 100 times the hardness and failure strength of natural enamel or contemporary ceramic veneers and good biocompatibility. Pure sapphire and diamond which are brittle and prone to fracture, can be made more fracture resistant as part of a nanostructured composite material that possibly includes embedded carbon nanotubes (Jayraman et al., 2004).

Nanorobotic dentifrice (dentifrobots) delivered by mouthwash or toothpaste could patrol all supragingival and subgingival surfaces at least once a day metabolizing trapped organic matter into harmless and odorless vapors and performing continous calculus debridement. Properly configured dentifrobots could identify and destroy pathogenic bacteria residing in the plaque and elsewhere, while allowing the 500 species of harmless oral microflora to flourish in a healthy ecosystem. Dentifrobots also would provide continuous barriers to halitosis, since bacterial putrification is the central metabolic process involved in oral malodor. With this kind of daily dental care available from an early age, conventional tooth decay and gingival diseases will disappear into the annals of medical history.

Potential benefits of nanotechnology are its ability to exploit the atomic or molecular properties of materials and the development of newer materials with better properties. Nanoproducts can be made by building-up particles by combining atomic elements and using equipments to create mechanical nanoscale objects. Nanotechnology has improved the properties of various kinds of fibers. Polymer nanofibers with diameters in the nanometer range, possess a larger surface area per unit mass and permit an easier addition of surface functionalities compared to polymer microfibers (Freitas, 2000; Whitesides and Love, 2001; Jayraman et al., 2004).

Polymer nanofiber materials have been studied as drug delivery systems, scaffolds for tissue engineering and filters. Carbon fibers with nanometer diamensions showed a selective increase in osteoblast adhesion necessary for successful orthopedic/dental implant applications due to a high degree of nanometer surface roughness (Freitas, 2000).

Nonagglomerated discrete nanoparticles are homogenously manufactured in resins or coatings to produce nanocomposites. The nanofiller used include an aluminosilicate powder having a mean particles size of about 80 nm and 1:4 M ratio of alumina to silica. Advantages include; superior hardness, flexible strength, modulus of elasticity, translucency and esthetic appeal, excellent color density, high polish, polish retention, and excellent handling properties (Yunshin et al., 2005; Price et al., 2004). Nanosolutions produce unique and dispersible nanoparticles that can be added to various solvents, paints and polymers in which they are dispersed homogenously. Nanotechnology in bonding agents ensures homogeneity and so the operator can now be totally confident that the adhesive is perfectly mixed every time. Nanofillers are integrated in the vinylsiloxanes, producing a unique addition siloxane impression material. Better flow, improved hydrophilic properties, hence fewer voids at margin and better model pouring, enhanced detail precision (Saravanakumar and Vijayalakshmi, 2006).

Ozone therapy

Most people suffer anxieties about being treated for tooth decay or more precisely; they fear the injections and drills, but, now, with ozone treatment, this is all the thing of the past. Studies have shown that 99% of all the bacteria causing tooth decay have been eliminated after 10 s of ozone exposure and even 99.9% bacteria after 20 s exposure. Thus, treating patients with ozone cuts off the treatment time with a great deal of difference, it eliminates the bacterial count more precisely and moreover, it is completely painless, therefore, increasing the patients' acceptability and compliance (Edward et al., 2008; Holmes, 2003; Lynch et al., 2004). Ozone can now be incorporated in various other treatment modalities also, like bleaching of discoloured teeth, root canal treatment, desensitization and treatment of some soft tissue infections (Bogra and Nikhil, 2003; Celiberti et al., 2006). Ozone, definitely, seems to be a promising treatment modality for various dental problems in future.

Studies have shown that: 1) ozone quickly dissipates in water and kills micro organisms via a mechanism involving the rupture of their membranes in such lesions, 2) It is a strong oxidizer to cell walls and cytoplasmic membrane of bacteria, 3) ozone treatment leads to oxidative decarboxylation of plaque pyruvate generating acetate and carbon dioxide as bye product, 4) it oxides volatile sulphur compounds precursor methionine to its corresponding sulphoxide and thus prevents malodour associated with root caries, 5) It also oxidized poly unsaturated fatty acids, and 6) ozone has little influence on the oxidation of dental alloys (Bogra and Nikhil, 2003).

Ozone is a gas composed of three atoms of oxygen and present naturally in the upper layer of atmosphere in abundance. It has got the capacity to absorb the harmful ultra-violet rays present in the light spectrum from the sun. Thus, ozone filters the light spectrum high up in the atmosphere and protects the living creatures from the ultra-violet rays (Nagayoshi et al., 2004). Ozone is an unstable gas and it quickly gives up nascent oxygen molecule to form oxygen gas. Due to the property of releasing nascent oxygen, it has been used in human medicine since long back to kill bacteria, fungi, to inactivate viruses and to control hemorrhages. Medical grade ozone is made from pure medical oxygen because oxygen concentration in the atmospheric air is variable. Atmospheric air is made up of nitrogen (71%), oxygen

(28%), and other gasses (1%) including ozone which is altered by processes related to altitude, temperature and air pollution (Baysan and Beighton, 2007).

Very recently, in dentistry, ozone has got its role in various dental treatment modalities. Interest in ozone use in dentistry is due to the infectious diseases associated with the oral cavity. Ozone therapy presents great advantages when used as a support for conventional treatments, for example, to dental caries, periodontal procedures and endodontic treatment (Nogales et al., 2008). The ozone therapy for managing caries is considered a breakthrough that is expected to be a cornerstone of dental care in the near future.

TISSUE ENGINEERING

New advances in stem cell biology and tissue engineering are leading to the development of cutting edge approaches to dentistry both in the repair and replacement of teeth (Koussoulakou and Koussoulakos, 2009).

Caries, pulpitis, and apical periodontitis increase health care costs and attendant loss of economic productivity. They ultimately result in premature tooth loss and therefore diminishing the quality of life. Advances in vital pulp therapy with pulp stem/progenitor cells might give impetus to regenerate dentin-pulp complex without the removal of the whole pulp. Tissue engineering is the science of design and manufacture of new tissues to replace lost parts because of diseases including cancer and trauma. The three key ingredients for tissue engineering are signals for morphogenesis, stem cells for responding to morphogens and the scaffold of extracellular matrix. In preclinical studies, cell therapy and gene therapy have been developed for many tissues and organs such as bone, heart, liver, and kidney as a means of delivering growth factors, cytokines, or morphogens with stem/progenitor cells in a scaffold to the sites of tissue injury to accelerate and/or induce a natural biological regeneration (Nakashima and Akamine, 2005; Lacerda-Pinheiro Set al., 2008)

The pulp tissue contains stem/progenitor cells that potentially differentiate into odontoblasts in response to bone morphogenetic proteins (BMPs). There are two strategies to regenerate dentin. First, is in vivo therapy, where BMP proteins or BMP genes are directly applied to the exposed or amputated pulp. Secondly, is ex vivo therapy which consists of isolation of stem/progenitor cells from pulp tissue, differentiation into odontoblasts with recombinant BMPs or BMP genes and finally transplanted autogenously to regenerate dentin (Koussoulakou and Koussoulakos, 2009).

Many strategies have evolved to engineer new tissues and organs, but virtually all combine a material with either bioactive molecules that induce tissue formation or cells grown in the laboratory. The bioactive molecules are frequently growth factor proteins that are involved in natural tissue formation and remodeling. The basic hypothesis underlying this approach is that the local delivery of an appropriate factor at a correct dose for a defined period of time can lead to the recruitment, proliferation and differentiation of a patient's cells from adjacent sites. These cells can then participate in tissue repair and/or regeneration at the required anatomic locale (Baum and David, 2000).

The second general strategy uses cells grown in the laboratory and placed in a matrix at the site where new tissue or organ formation is desired. These transplanted cells usually are derived from a small tissue biopsy specimen and have been expanded in the laboratory to allow a large organ or tissue mass to be engineered. Typically, the new tissue will be formed in part from these transplanted cells (Kim and Mooney, 1998).

With both approaches, specific materials deliver the molecules or cells to the appropriate anatomic site and provide mechanical support to the forming tissue by acting as a scaffold to guide new tissue formation. Currently, most tissue engineering efforts use biomaterials already approved for medical indications by the U.S. Food and Drug Administration (FDA). The most widely used synthetic materials are polymers of lactide and glycolide, since these are commonly used for biodegradable sutures. Both polymers have a long track record for human use and are considered biocompatible and their physical properties (for example, degradation rate and mechanical strength) can be readily manipulated. A natural polymer—type 1 collagen is often used because of its relative biocompatibility and ability to be remodeled by cells. Other polymers familiar to dentistry, including alginate, are also being used (Kim and Nikolovski, 1999).

New technology continually has had a major impact on dental practice, from the development of high-speed handpieces to modern restorative materials. Tissue engineering in the broadest sense unquestionably will affect dental practice significantly within the next 25 years. As an interdisciplinary endeavor, tissue engineering brings the power of modern biological, chemical and physical science to real clinical problems. The impact of tissue engineering likely will be most significant with mineralized tissues, already the focus of substantial research efforts.

These efforts will yield numerous clinical dental benefits, including improved treatments for intraosseous periodontal defects, enhanced maxillary and mandibular grafting procedures, perhaps more biological methods to repair teeth after carious damage and possibly even regrowing lost teeth.

In addition, it is expect to see a range of other tissue-engineering applications that may promote more rapid healing of oral wounds and ulcers, as well as the use of gene-transfer methods to manipulate salivary proteins and oral microbial colonization patterns. Less common, but still a treatment consideration for the dental profession, will be devices such as the artificial salivary gland and muscle (tongue) or mucosal grafts to replace tissues lost through surgery or trauma. This is an exciting time for biomedical science and its application. Clinical dental practice in 2025 will certainly be different (Baum and David, 2000).

CONCLUSION

The visions described in this article may sound unlikely, implausible, or even heretic. Yet, the theoretical and applied research to turn them into reality is progressing rapidly. Genetic engineering, nanotechnology and ozone therapy will change dentistry, healthcare, and human life more profoundly than many developments of the past. As with all technologies, these technologies carry a significant potential for misuse and abuse on a scale and scope never seen before. However, they also have potential to bring about significant benefits, such as improved health, better use of natural resources, and reduced environmental pollution. These truly are the days of 'Miracle and Wonder'. It is a bright future that lies ahead in dental field, but we shall all have to work very long and very hard to make it come to pass.

REFERENCES

Baum BJ, David J (2000). The impact of tissue engineering on dentistry. J. Am. Dent. Assoc. 131:309-318.
Baysan A, Beighton D (2007). Assessment of the ozone-mediated killing of bacteria in infected dentine associated with non-cavitated occlusal carious lesions. Caries Res. 41:337-341.
Bogra P, Nikhil V(2003). Ozone therapy for dental caries – A revolutionary treatment for future. JIDA 74:41-45.
Celiberti P, Pazera P, Lussi A (2006). The impact of ozone treatment on enamel physical properties. Am. J. Dent. 19:67-72.
Chen HF, Clarkson BH, Sunk, Mansfield JF (2005). Self assembly of synthetic hydroxyaptite nanorods into enamel prism like structure. J. Colloid Interf. Sci. 188:97-103.
Edward Lynch MA, Edward J (2008). Swift Jr. Evidence-based efficacy of ozone for root canal irrigation. J. Esth. Rest Dent. 20:287-293.
Freitas R Jr. (2005). Nanotechnology, nanomedicine and nanosurgery. Int. J. Surg. 3:243-246.
Freitas RA (2000) . Nanodentistry Fact or fiction? J. Am. Dent. Assoc. 131:1559-1565.
Garg R, Tandon S (2009). Ozone: A new face of dentistry. Int. J. Dental Sci. 7:34-40.
Holmes J (2003). Clinical reversal of root caries using ozone, double-blind, randomised, controlled 18-month trial. Gerodontology 20:106-14.
Jayraman K, Kotaki M, Zhang Y, Ramakrishna S (2004). Recent advances in Polymer nanofibers. J. Nanosci. Nanotechnol. 4:52-65.
Jhaver HM (2005). Nanotechnology: The future of dentistry. J. Nanosci. Nanotechnol. 5:15-17.
Kim BS, Mooney DJ(1998). Development of biocompatible synthetic extracellular matrices for tissue engineering. Trends Biotechnol. 16:224-230.
Kim BS, Nikolovski J (1999). Engineered smooth muscle tissues: regulating cell phenotype with the scaffold. Exp. Cell Res. 251:318-328.
Koussoulakou DS, Koussoulakos SL (2009). A curriculum vitae of teeth: evolution, generation, regeneration. Int. J. Biol. Sci. 5:226-243.
Kumar Verma S, Prabhat KC, Goyal LA (2010). Critical review of the

implication of nanotechnology in modern dental practice. Natl. J. Maxillofac. Surg. 1:41-44.

Lacerda-Pinheiro S, Septier D, Benhamou L, Kellermann O, Goldberg M (2008). An *in vivo* model for short-term evaluation of the implantation effects of biomolecules or stem cells in the dental pulp. Open Dent. J. 2:67-72.

Lynch E, Holmes, Steier L, Megighian G (2004). Integration into general dental practice. Successful treatment of Caries using the HealOzone. Part 1. Dent. Horizons 2:23-27.

Malhotra N, Kundabala M, Acharya S (2009). Current strategies and applications of tissue engineering in dentistry--a review part 1. Dent. Update 36:577-579.

Mallanagouda P, Dhoom Singh M, Sowjanya G (2008). Future impact of nanotechnology on medicine and dentistry. J. Ind. Soc. Periodontol. 12:34-40.

Nagayoshi M, Kitamura C, Fukuzumi T, Nishihara T (2004). Antimicrobial effect of ozonated water on bacteria invading dentinal tubules. J. Endod. 30:778-781.

Nakashima M, Akamine A (2005). The application of tissue engineering to regeneration of pulp and dentin in endodontics. J. Endod. 31:711-718.

Nogales CG, Ferrari PA, Kantorovich EO, Lage-Marques JL (2008). Ozone Therapy in Medicine and Dentistry. J. Contemp. Dent. Pract. 4:75-84.

Price RL, Ellison K, Haberstroh KM, W ebster TJ (2004). Nano-meter surface roughness increases select osteoblasts adhesion on carbon nanofiber compacts. J. Biomed. Mater. Res. 70:129-138.

Saravanakumar R, Vijayalakshmi R (2006). Nanotechnology in Dentistry. Ind. J. Dent. Res. 17:62-65.

Sharma S, Cross SE, W alli RP(2010). Nanocharacterization in dentistry. Int. J. Mol. Sci. 11:2523-2545.

Whitesides GM, Love JC (2001). The Art of Building Small. Sci. Am. 285:38-47.

Yunshin S, Park HN, Kim KH (2005). Biologic evaluation of chitosan nanofiber membrane for guided bone regeneration. J. Periodontol. 76:1778-1784.

24

Markers are shared between adipogenic and osteogenic differentiated mesenchymal stem cells

Melanie Köllmer[1], Jason S. Buhrman[1], Yu Zhang[1] and Richard A. Gemeinhart[1,2,3]

[1]Department of Biopharmaceutical Sciences, University of Illinois, Chicago, IL 60612-7231, USA. [2]Department of Bioengineering, University of Illinois, Chicago, IL 60607-7052, USA.
[3]Department of Ophthalmology and Visual Sciences, University of Illinois, Chicago, IL 60612-4319, USA.

The stem cell differentiation paradigm is based on the progression of cells through generations of daughter cells that eventually become restricted and committed to one lineage resulting in fully differentiated cells. Herein, we report on the differentiation of adult human mesenchymal stem cells (hMSCs) towards adipogenic and osteogenic lineages using established protocols. Lineage specific genes were evaluated by quantitative real-time PCR relative to two reference genes. The expression of osteoblast-associated genes (alkaline phosphatase, osteopontin, and osteocalcin) was detected in hMSCs that underwent adipogenesis. The expression of adipocyte marker genes (adiponectin, fatty acid binding protein P4, and leptin) increased in a time-dependent manner during adipogenic induction. Adiponectin and leptin were also detected in osteoblast-induced cells. Lipid vacuoles that represent the adipocyte phenotype were only present in the adipogenic induction group. Conforming to the heterogeneous nature of hMSCs and the known plasticity between osteogenic and adipogenic lineages, these data indicate a marker overlap between MSC-derived adipocytes and osteoblasts. We propose a careful consideration of experimental conditions such as investigated time points, selected housekeeping genes and the evidence indicating lack of differentiation into other lineages when evaluating hMSC differentiation.

Key words: Mesenchymal stem cell, differentiation markers, cell plasticity, differentiation.

INTRODUCTION

The human bone marrow stroma contains multipotent mesenchymal cells that give rise to adipocytes and osteoblasts, as well as many other lineages (Caplan and Dennis, 2006). Cells isolated based on adherence to the tissue culture substrate do not represent a homogenous population of mesenchymal progenitors rather subpopulations of cells with variable differentiation potential (Muraglia et al., 2000; Pittenger et al., 1999). Most of the clones derived from bone marrow stromal cells possess osteogenic and adipogenic differentiation potential but some are only able to differentiate towards osteoblasts. The mechanisms of the differentiation process from

precursor to fully differentiated mature cells are still not fully understood (Discher et al., 2009; Hwang et al., 2008; Scadden, 2006).

Methods have been developed for differentiating cells into specific differentiated cell types expressing the markers and phenotypes of the desired tissues (Pittenger et al., 1999). New materials-based and soluble factor-based differentiation protocols are constantly being developed to control the differentiation potential of all stem cell types (Fekete et al., 2012; Hoshiba et al., 2012; Keskar et al., 2009; Kollmer et al., 2012; Vater et al., 2011). Many of these protocols are validated to confirm the presence of specific differentiation markers, but frequently alternative differentiation pathways are not excluded. In addition, many in the field of tissue engineering utilize a small subset of differen-tiation markers (Bakhshandeh et al., 2012; Choi et al., 2010; He et al., 2012; Henderson et al., 2008; Hess et al., 2012; Marion et al., 2006; Pountos et al., 2007; Wiren et al., 2011; Zhang et al., 2012) despite the potential for expression of these markers in other lineages.

There are several examples in the literature showing the expression of a differentiation marker by multiple cell types. Leptin, an adipokine produced by adipocytes was observed on the mRNA level in human osteoblasts during the mineralization period (Reseland et al., 2001) as well as in hMSCs that underwent osteogenesis (Noh, 2012). Leptin has pleiotropic effects on other bone marrow cells, including osteoblasts (Noh, 2012; Nuttall and Gimble, 2004) and was shown to promote osteogenesis and to inhibit adipogenesis in immortalized human marrow stro-mal cells (Thomas et al., 1999). This could be paracrine communication controlling the growth and differentiation of adipocytes and signaling osteogenesis when sufficient adipocytes are present.

Alkaline phosphatase, widely used as a biochemical marker of bone turnover, also plays a role in adipoge-nesis. Inhibition of tissue-nonspecific alkaline phospha-tase resulted in a decreased accumulation of lipid va-cuoles during adipogenic differentiation of a murine preadipocyte cell line (Ali et al., 2005). Similarly, osteopontin (OPN) is not solely a key regulator of bone development, rather a multifunctional extracellular matrix (ECM) associated protein involved in inflammatory processes, tumorigenesis, cardiac fibrosis and obesity (Sodek et al., 2000). Upregulated OPN mRNA levels have been detected in adipose tissue of obese patients (Chapman et al., 2010). Osteocalcin, a non-collagenous protein found in mineralized adult bone, is another widely used bone marker. However, constitutive osteocalcin mRNA and protein expression by adipose stromal cells implicates that nonosteogenic cells of the marrow stroma also secrete osteocalcin (Benayahu et al., 1997). Exposure to the glucocorticoid dexamethasone which is a constituent of both, osteogenic and adipogenic differentiation media, has been shown to increase osteocalcin expression in cultured stromal cells (Leboy et al., 1991).

We can infer from these reports that those markers need further validation as tissue-specific differentiation markers. In the present study, we assessed the suitability of fatty acid binding protein 4, adiponectin and leptin as adipogenic differentiation markers and alkaline phos-phatase, collagen type I, osteocalcin and osteopontin as osteogenic differentiation markers by evaluating the expression of these markers during adipogenic and osteogenic culture conditions. We show that these mar-kers are not selectively expressed when cells are differentiated using common differentiation protocols.

MATERIALS AND METHODS

Human mesenchymal stem cell isolation and differentiation

Human bone marrow aspirates were obtained from AllCells, LLC (Emeryville, CA) and isolated by density gradient centrifugation utilizing Ficoll-Paque^{TM} PLUS solution followed by cell-surface marker negative selection with RosetteSep® Human Mesenchymal Stem Cell Enrichment Cocktail (Stem Cell Technologies, Vancouver, BC, Canada) according to the manufacturer's protocol. For each experiment, hMSCs isolated from one of three donors (non-smoker males ranging in age from 20 to 31 years old) were used with no cells used beyond passage four.

Cells were harvested using 0.25% trypsin with 1.0 M EDTA, centrifuged, and expanded in basal medium which consists of high glucose Dulbecco's Modified Eagle's Medium (DMEM) supplemented with 10% fetal bovine serum (FBS), 100 unit/mL penicillin and 100 unit/mL streptomycin (basal medium). Medium was changed every third day. Adipogenic differentiation was initiated by culturing 2×10^5 hMSCs in a well of a 6 well plate in MesenCult® adipogenic induction medium (Stem Cell Technologies, Vancouver, BC, Canada). The composition of the adipogenic medium is proprietary.

Adipogenic differentiation protocols routinely involve combinations of dexamethasone, 3-isobutyl-1-methyl-xanthine (IBMX), insulin, and indomethacin (Vater et al., 2011). Osteogenic differentiation was initiated by culturing 3×10^4 hMSCs in a well of a 6 well plate in Poietics^{TM} osteogenic induction medium (Lonza, Walkersville, MD, USA) containing dexamethasone, ascorbic acid and β-glycerophosphate supplements (Pittenger et al., 1999; Vater et al., 2011). Controls were maintained in basal medium for the entire culture period of 4 weeks.

Quantitative real-time PCR

After 14 and 28 days, total RNA was extracted using the TRIzol® reagent (Life Technologies, Grand Island, NY, USA) in combination with the PureLink^{TM} RNA Mini Kit (Life Technologies, Grand Island, NY, USA) according to manufacturer's instructions. To diminish genomic DNA contamination, RNA was treated with Turbo^{TM} DNAse (Life Technologies, Grand Island, NY, USA) according to manufacturer's instructions. The purified RNA (10 ng/mL) was reverse transcribed with the High Capacity cDNA Reverse Transcription (RT) Kit (Life Technologies, Grand Island, NY, USA) under the following conditions: 25 °C for 10 min, 37 °C for 120 min followed by 85 °C for 5 min.

To identify potential genomic DNA contamination, controls with no enzyme were evaluated. The PCR reactions were performed on an Applied Biosystems StepOnePlus^{TM} PCR machine using 5 µL SYBR® Green PCR Master Mix (Life Technologies, Grand Island, NY, USA), 2 µL sequence specific primers (0.5 mM, GAPDH was used at 0.25 mM) (Table 1) and 3 µL cDNA (cDNA dilutions: ADIPOQ, ALPL, FABP4, OPN: 10 fold, BGLAP: five-fold, COL1A1, LEP: 31 fold) under the following conditions: 95 °C for 10 min followed by 40 cycles of 15 s of denaturation at 95 °C and 60 s of annealing and elongation at 60 °C.

A melting curve analysis was performed after each run to confirm product specificity. The delta-delta-Ct method (Livak and Schmittgen, 2001) was employed to determine the relative gene expression level of the gene of interest normalized to the endogenous controls glyceraldehyde-3-phosphate (GAPDH) and ribosomal protein L13A (RPL13A). Statistical significance was determined using first a two-way ANOVA comparing the treatments and time followed by Newman-Keuls' post-hoc comparison of groups.

Table 1. Genes and primers used for qPCR.

Gene	Full Name	Sequences 5´-> 3´	Accession number/ Reference
ADIPOQ	Adiponectin	For: AGG GTG AGA AAG GAG ATC C Rev: GGC ATG TTG GGG ATA GTA A	NM_004797
ALPL	Alkaline Phosphatase	For: ATT TCT CTT GGG CAG GCA GAG AGT Rev: ATC CAG AAT GTT CCA CGG AGG CTT	NM_000478.4
BGLAP	Bone gamma-carboxyglutamate (Osteocalcin)	For: CAG CGA GGT AGT GAA GAG AC Rev: TGA AAG CCG ATG TGG TCA G	NM_199173
COL1A1	Collagen type I	For: TGT GGC CCA GAA GAA CTG GTA CAT Rev: ACT GGA ATC CAT CGG TCA TGC TCT	NM_000088
FABP4	Fatty acid binding protein 4	For: TGG TTG ATT TTC CAT CCC AT Rev: TAC TGG GCC AGG AAT TTG AC	NM_001442
GAPDH	Glyceraldehyde-3-phosphate dehydrogenase	For: TTC GAC AGT CAG CCG CAT CTT CTT Rev: GCC CAA TAC GAC CAA ATC CGT TGA	NM_002046.4
LEP	Leptin	For: CTG ATG CTT TGC TTC AAA TCC A Rev: GCT TTC AGC CCT TTG CGT T	NM_000230
OPN	Osteopontin	For: AGA ATG CTG TGT CCT CTG AAG Rev: GTT CGA GTC AAT GGA GTC CTG	NM_001251830
RPL13α	Ribosomal protein L13 α	For: CAT AGG AAG CTG GGA GCA AG Rev: GCC CTC CAA TCA GTC TTC TG	NM_012423

Immunofluorescence

Briefly, the samples were fixed in 4% paraformaldehyde, rinsed with 1X PBS and incubated with blocking solution (1% BSA in 1X PBS) for 30 min on a shaker. Osteocalcin mouse monoclonal IgG antibody (25μL; 100 μg/mL; Santa Cruz Biotech, Santa Cruz, CA, USA) was added directly to 500 μL blocking solution and samples were further incubated for 3 h at room temperature (RT). Samples were washed with 1X PBS and incubated with 5 μL AlexaFluor® 488-labeled goat anti-mouse secondary antibody (2 mg/mL; Molecular Probes, Carlsbad, CA) in 500 μL 1X PBS at RT and protected from light. After 25 min, 0.5 μL H33258 (1 mg/mL; Life Technologies, Grand Island, NY, USA) was added to the solution and incubated for an additional 5 min at RT. The secondary antibody solution was discarded after 30 min and the samples were washed with 1X PBS. Images were taken with an Olympus IX70 inverted microscope and processed using QCapture Pro software.

Sudan III staining

To stain lipid vacuoles, the samples were rinsed with 1X PBS (pH 7.4) and incubated with Sudan III solution (0.3% w/v of Sudan III in 70% ethanol) for 3 min. After several washes with double deionized water (DDIW), Harris hematoxylin solution was added and incubated for 1 min. Samples were destained in fresh acid ethanol (0.5% 1 N HCL in 70% EtOH) for 1 min. Afterwards, the wells were rinsed with DDIW until the water ran clear. Images were taken under bright field with an Olympus IX70 inverted microscope.

RESULTS

Depending on the reference genes, sometimes referred to as housekeeping genes, used in the qPCR analysis,

different gene expression patterns have been observed during osteogenic differentiation (Quiroz et al., 2010). Instabilities in gene expression levels during cell differentiation must be taken into account. The genes of interest were normalized to the most commonly used reference genes, GAPDH and RPL13A. These reference genes were also chosen based upon their recent assessment as reference genes in hMSCs (Curtis et al., 2010).

When normalized against GAPDH, there is significant variation in the results and two of the three genes examined lacked statistical significant (Figure 1A, C and E). When normalized against RPL13A, adipocyte differentiation markers (ADIPOQ, FABP4; LEP) were significantly upregulated ($P < 0.5$) under adipogenic conditions at day 28 (Figure 1B, D and F). Consistent with the postulate of an inverse relationship between adipo- and osteogenesis (Nuttall and Gimble, 2004), adipocyte-associated genes were not significantly upregulated in osteogenic differentiation groups. ADIPOQ and LEP were still detectable in cells that underwent osteogenic differentiation (Figure 1B, C, D and F), but at a much lower level. Lipid vacuoles that represent the adipogenic phenotype were only present in the adipogenic induction group (Figure 2). No lipid vacuoles could be detected in the osteogenic induction groups at any time point.

In contrast, osteogenic genes were upregulated during adipogenesis. With both reference genes, alkaline phosphatase (ALPL) mRNA levels were upregulated under adipogenic and osteogenic differentiation conditions

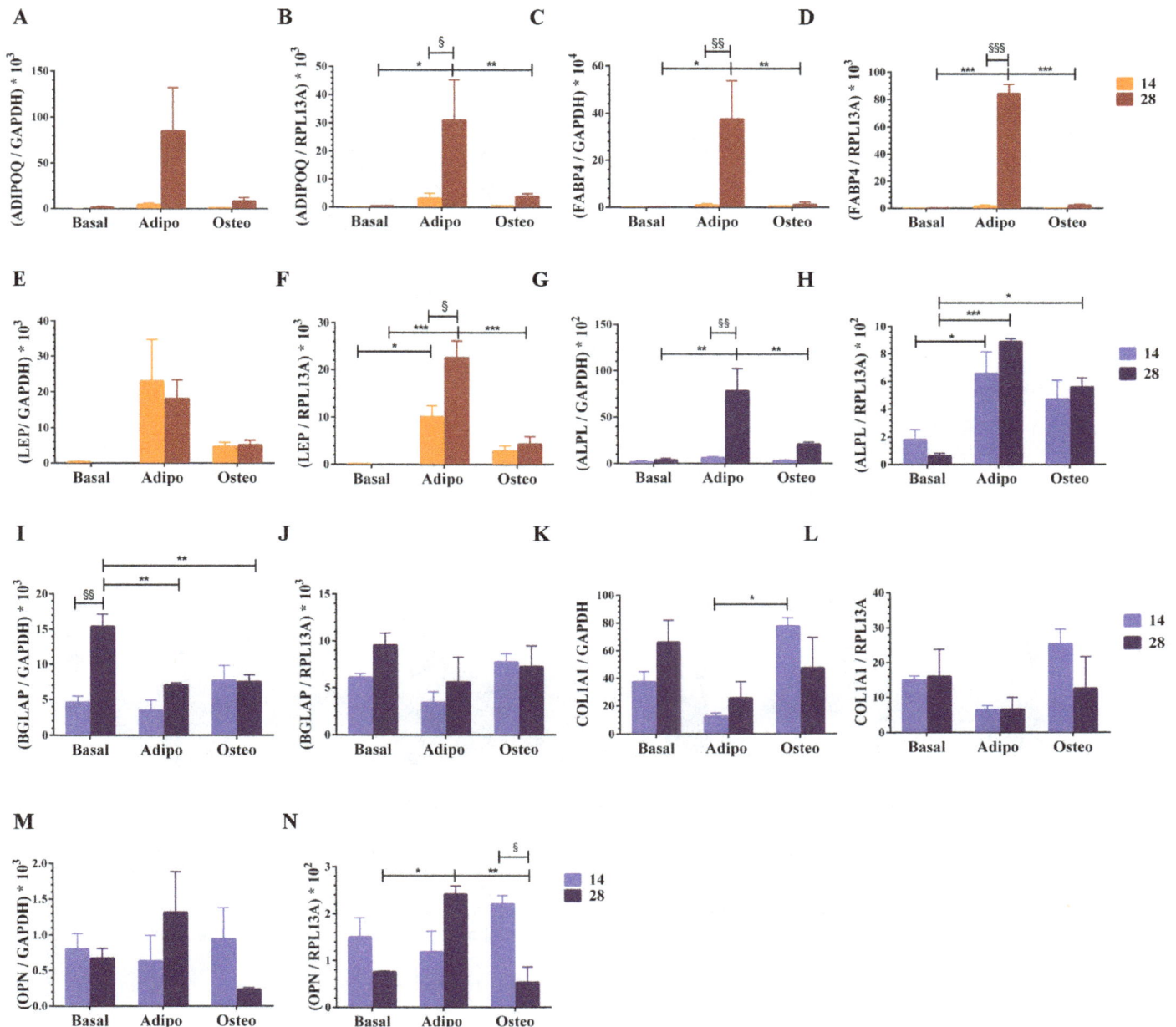

Figure 1. Expression profiles of genes encoding for adipocyte- and osteoblast-related genes. Expression of adipocyte marker genes (A-B) adiponectin (*ADIPOQ*), (B-C) fatty acid binding protein 4 (*FABP4*), (D-E) leptin (*LEP*), and osteoblast marker genes (G–H) alkaline phosphatase (*ALPL*), (I–J) osteocalcin (*BGLAP*), (K-L) collagen type I (*COL1A1*), (M-N) osteopontin (*OPN*) in hMSCs cultured in basal, adipogenic (adipo) and osteogenic (osteo) induction medium for 14 and 28 days; mRNA levels were normalized to the expression of endogenous control genes glyceraldehyde-3-phosphate (*GAPDH*) and ribosomal protein L13A (*RPL13A*). Values are presented as mean plus or minus (±) standard error of the mean (n=3 donors). Statistical significance is indicated as (*) for differences between treatment groups at day 14 and 28, respectively and (§) for the difference between day 14 and 28 for a given treatment and the number of symbols indicating level of significance with one, two, and three symbols indicating $p < 0.05$, $p < 0.01$, and $p < 0.001$, respectively.

(Figure 1G, H). Significantly elevated osteopontin (*OPN*) mRNA levels ($P < 0.5$) were observed in cells that were cultured in adipogenic medium for 28 days compared to the non-induced control and the osteogenic induction group when *OPN* was normalized against *RPL13A* (Figure 1N). At day 14, *OPN* levels were higher in the os-

teogenic induction group. Early bone marker, collagen type I (*COL1A1*) and mRNA levels were significantly elevated ($P < 0.5$) in the osteogenic induction group compared to the adipogenic group at day 14 when normalized to *GAPDH* (Figure 1K). Due to variations between donors, no statistically significant differences in the expres-

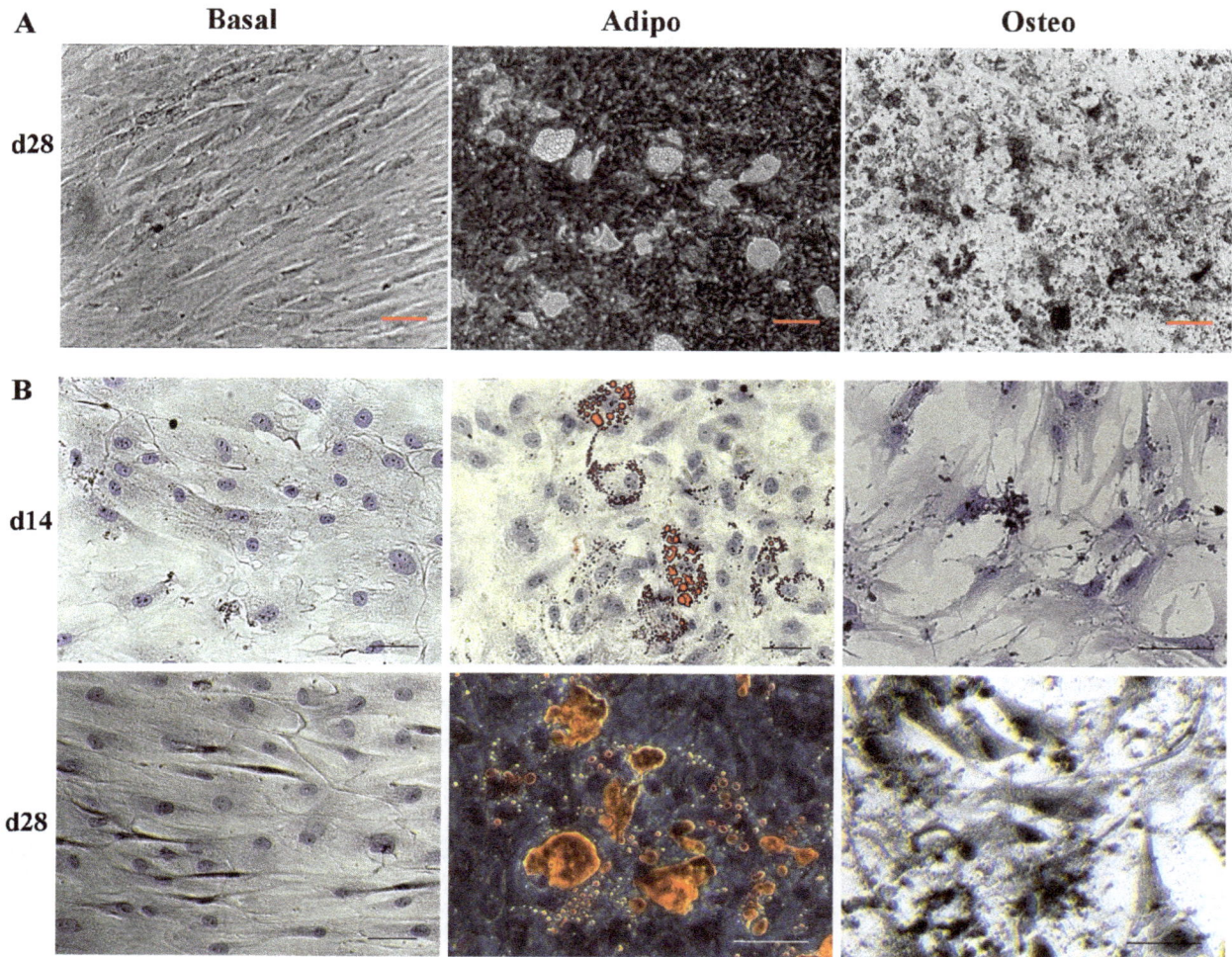

Figure 2. Morphology of differentiated cells. **(A)** Cells in the untreated control group (basal) maintained an undifferentiated phenotype with fibroblast-like cells, black regions within the cell monolayer indicate calcification in the osteogenic induction group (osteo) and lipid vacuoles were visible in the adipogenic induction group (adipo) at day 28 of differentiation. **(B)** To evaluate adipogenic differentiation, Sudan III staining was conducted. The presence of lipid vacuoles was observed in the adipo group at day 14 and 28. No lipid vacuoles could be seen in the basal and in the osteo group. The scale bar in each image is 50 µm.

sion of the late osteoblast marker osteocalcin (*BGLAP*) were observed between adipogenic and osteogenic treatment groups (Figure 1I, J). Similar to the *COL1A1* expression pattern, *BGLAP* appears to be lower in adipocyte cultures at day 14. Interestingly, *BGLAP* was significantly upregulated under basal conditions at day 28 when normalized to *GAPDH* (Figure 1I). Quiroz et al. (2010) obtained similar results due to the high variability of *GAPDH* under basal culture. The relative quantification with *RPL13A* seemed to produce lower standard deviations in the expression levels of adipogenic genes and thus gave more significance to the results. The overall expression patterns of adipogenic and osteogenic genes were similar with both housekeeping genes.

Immunostaining was carried out to detect and visualize the expression of osteocalcin (Figure 3). Similar to the mRNA expression data, osteocalcin was detected in the adipogenic induction group as well as in the osteogenic induction group at both time points (Figure 3). Under both conditions, osteocalcin expression levels decreased from day 14 to day 28. In contrast to the mRNA data, no osteocalcin protein expression was detected in the basal control groups.

DISCUSSION

Dexamethasone is a mutual component of adipogenic and osteogenic induction medium and has been shown to increase osteocalcin and leptin production in hMSCs (Leboy et al., 1991; Noh, 2012). Although additional compounds are added to the differentiation cocktails, heterogeneity in the cultures still exists. (Pittenger et al., 1999; Vater et al., 2011). Whether this phenomenon

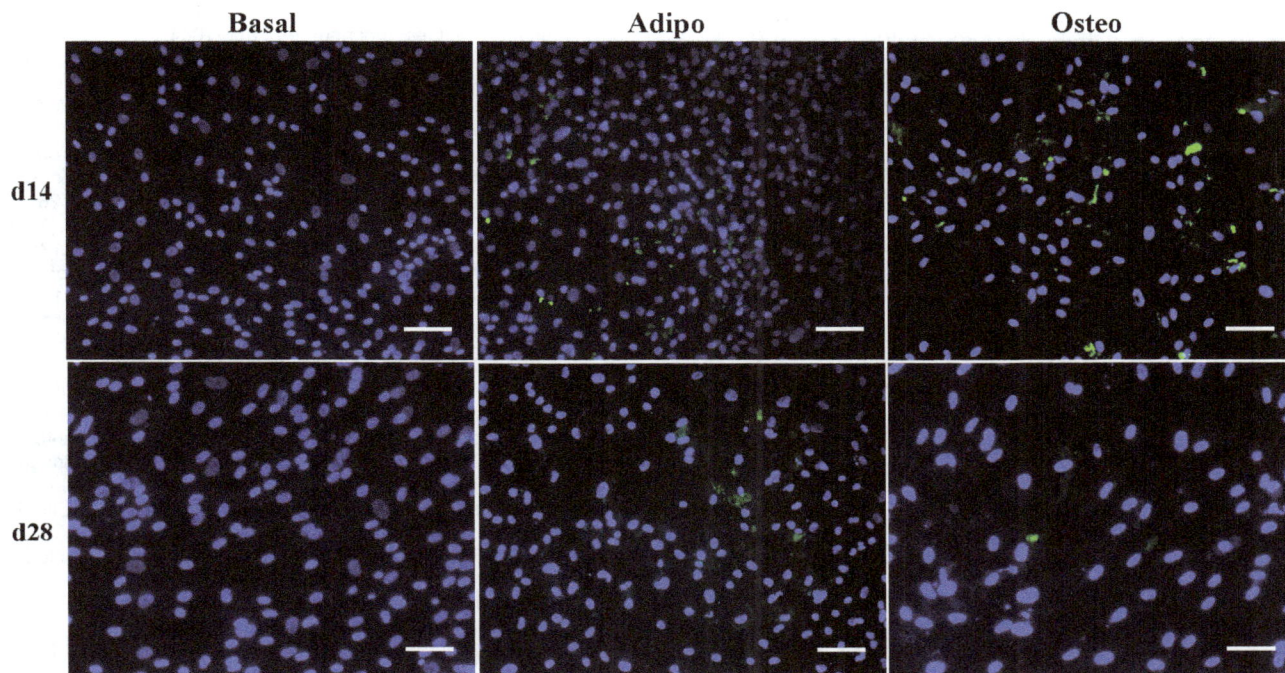

Figure 3. Representative immunofluorescence micrographs of osteocalcin expression in hMSCs, hMSC-derived adipocytes (adipo), and hMSC-derived osteoblasts (osteo). Osteocalcin expression (green) was detected to a comparable level in hMSC-derived adipocytes and osteoblasts. The expression in both groups decreased at day 28 compared to day 14. No osteocalcin was detected in the basal control. H33258 (blue) was used as a nuclear stain. Scale bar is 50 μm.

reprogress towards one lineage can transdifferentiate into another lineage, or whether the current set of markers is not definitive enough for one lineage, still needs further investigation. Similar to our observation, simultaneous expression of osteoblast markers (alkaline phosphatase and osteocalcin) and adipocyte markers (peroxisome proliferator-activated receptor γ 2 and lipoprotein lipase) was confirmed on a single cell level in hMSC-derived osteoblasts (Ponce et al., 2008).

When hMSCs differentiate into mature osteoblasts, they have to pass through several maturation stages that are characterized by a time-dependent expression of gene markers. Alkaline phosphatase and collagen type I mRNA levels are upregulated during early stages of bone formation (Jaiswal et al., 1997; Jikko et al., 1999). Osteopontin expression peaks twice during bone development, in the proliferation phase (~ day 4) and in the mineralization phase (~ day 14-21) (Aubin, 2001). Osteocalcin and osteopontin are highly expressed during the last stage of bone formation, the mineralization period. Although commonly used as an indicator for osteogenic differentiation, almost all of these markers are not bone specific. Alkaline phosphatase, an ubiquitous cellular protein, was upregulated during adipogenic and osteogenic differentiation (Figure 1B, H). Alkaline phosphatase does confirm initiation of differentiation but cannot be considered as lineage-specific. Collagen type I is the main component of bone extracellular matrix (ECM)

but has been identified in a number of unrelated cell types (Hing, 2004). In our study, COL1A1 expression was upregulated under osteogenic and downregulated under adipogenic conditions at day 14 (Figure 1K). At later time points, we observed no differences in the COL1A1 expression levels between adipocyte and osteocyte cultures. Osteopontin cannot be considered as a bone-specific marker either since it regulates cell adhesion, migration and survival in other tissues as well (Sodek et al., 2000). OPN was detected in adipogenic and osteogenic lineages but gene expression patterns were distinct from each other with OPN being upregulated at earlier time points in osteoblast cultures and at later stages in adipocyte cultures (Figure 1N). In contrast to the protein levels (Figure 3), osteocalcin mRNA levels were not significantly upregulated during osteogenic induction (Figure 1I, J). We obtained similar results when hMSCs underwent osteogenic differentiation within a 3D hydrogel scaffold (Kollmer et al., 2012). Although Ca^{2+}-levels were significantly upregulated, no increase in osteogenic gene expression was observed. A discrepancy between mineralization and gene expression data indicated that an up regulation in osteogenic genes in hMSCs does not correlate with their ability to differentiate towards the osteogenic lineage (Shafiee et al., 2011). Glucocorticoid-mediated down regulation of osteocalcin mRNA levels has also been reported (Viereck et al., 2002).

The presence of osteocalcin in hMSC-derived adipocytes is in concert with a recently published study where steocalcin was detected in human preadipocytes and to a lesser extent in fully differentiated adipocytes (Foresta et al., 2010). Osteocalcin's role in human physiology has been further expanded as a circulating hormone influencing beta-cell proliferation, glucose intolerance, and insulin resistance has recently arisen (Lee et al., 2007).

Conclusion

The current work and previous reports clearly indicate that many of the markers used for determining the end fate of adipocytic and osteoblastic differentiation are shared between adipogenic and osteogenic differentiated hMSCs. Adipocytes and osteoblasts share a common pool of precursor cells and their plasticity is regulated by activation or silencing of genes, signaling molecules and transcription factors (Garces et al., 1997; Gimble et al., 1996). Our data indicate the need for a better understanding of the conditions and molecular regulators involved in controlling the plasticity of hMSCs. This knowledge is a prerequisite to manipulate adult stem cells for engineering functional tissues in regenerative medicine and to shed light into the pathogenesis of metabolic and skeletal disorders, like atherogenesis, diabetes and osteoporosis. Furthermore, the present study suggests that differentiation towards one lineage should be accompanied by evidence indicating lack of differentiation into other lineages.

ACKNOWLEDGEMENTS

This investigation was conducted in a facility constructed with support from Research Facilities Improvement Program Grant C06 RR15482 from the National Centre for Research Resources, NIH. This research has been funded, in part, by the University of Illinois at Chicago Center for Clinical and Translational Science (CCTS) award RR029879 and the NIH NS055095. The authors also thank Dr. Debra A. Tonetti for use of instruments.

REFERENCES

Ali AT, Penny CB, Paiker JE, van Niekerk C, Smit A, Ferris WF, Crowther NJ (2005). Alkaline phosphatase is involved in the control of adipogenesis in the murine preadipocyte cell line, 3T3-L1. Clin. Chim. Acta 354:101-109.

Aubin JE (2001). Regulation of osteoblast formation and function. Rev Endocr. Metab. Disord. 2:81-94.

Bakhshandeh B, Soleimani M, Hafizi M, Paylakhi SH, Ghaemi N (2012). MicroRNA signature associated with osteogenic lineage commitment. Mol. Biol. Rep. 39:7569-7581.

Benayahu D, Shamay A, Wientroub S (1997). Osteocalcin (BGP), gene expression, and protein production by marrow stromal adipocytes. Biochem. Biophys. Res. Commun. 231:442-446.

Caplan AI, Dennis JE (2006). Mesenchymal stem cells as trophic mediators. J. Cell. Biochem. 98:1076-1084.

Chapman J, Miles PD, Ofrecio JM, Neels JG, Yu JG, Resnik JL, Wilkes J, Talukdar S, Thapar D, Johnson K, Sears DD (2010). Osteopontin is required for the early onset of high fat diet-induced insulin resistance in mice. PloS one 5:e13959.

Choi JH, Gimble JM, Lee K, Marra KG, Rubin JP, Yoo JJ, Vunjak-Novakovic G, Kaplan DL (2010). Adipose tissue engineering for soft tissue regeneration. Tissue Eng. Part B Rev. 16:413-426.

Curtis KM, Gomez LA, Rios C, Garbayo E, Raval AP, Perez-Pinzon, MA, Schiller PC (2010). EF1alpha and RPL13a represent normalization genes suitable for RT-qPCR analysis of bone marrow derived mesenchymal stem cells. BMC Mol. Biol. 11:61.

Discher DE, Mooney DJ, Zandstra PW (2009). Growth Factors, Matrices, and Forces Combine and Control Stem Cells. Science 324: 1673-1677.

Fekete N, Rojewski MT, Furst D, Kreja L, Ignatius A, Dausend J, Schrezenmeier H (2012). GMP-compliant isolation and large-scale expansion of bone marrow-derived MSC. PloS one 7:e43255.

Foresta C, Strapazzon G, De Toni L, Gianesello L, Calcagno A, Pilon, C, Plebani M, Vettor R (2010). Evidence for osteocalcin production by adipose tissue and its role in human metabolism. J. Clin. Endocrinol. Metab. 95:3502-3506.

Garces C, Ruiz-Hidalgo MJ, Font de Mora J, Park C, Miele L, Goldstein J, Bonvini E, Porras A, Laborda J (1997). Notch-1 controls the expression of fatty acid-activated transcription factors and is required for adipogenesis. J. Biol. Chem. 272:29729-29734.

Gimble JM, Robinson CE, Wu X, Kelly KA, Rodriguez BR, Kliewer SA, Lehmann JM, Morris DC (1996). Peroxisome proliferator-activated receptor-gamma activation by thiazolidinediones induces adipogenesis in bone marrow stromal cells. Mol. Pharmacol. 50:1087-1094.

He X, Yang X, Jabbari E (2012). Combined effect of osteopontin and BMP-2 derived peptides grafted to an adhesive hydrogel on osteogenic and vasculogenic differentiation of marrow stromal cells. Langmuir : the ACS Journal of Surfaces and Colloids 28:5387-5397.

Henderson JA, He X, Jabbari E (2008). Concurrent differentiation of marrow stromal cells to osteogenic and vasculogenic lineages. Macromol. Biosci. 8:499-507.

Hess R, Jaeschke A, Neubert H, Hintze V, Moeller S, Schnabelrauch M, Wiesmann HP, Hart DA, Scharnweber D (2012). Synergistic effect of defined artificial extracellular matrices and pulsed electric fields on osteogenic differentiation of human MSCs. Biomaterials 33:8975-8985.

Hing KA (2004). Bone repair in the twenty-first century: biology, chemistry or engineering? Philos. Trans. A Math. Phys. Eng. Sci. 362: 2821-2850.

Hoshiba T, Kawazoe N, Chen G (2012). The balance of osteogenic and adipogenic differentiation in human mesenchymal stem cells by matrices that mimic stepwise tissue development. Biomaterials 33:2025-2031.

Hwang NS, Varghese S, Elisseeff J (2008). Controlled differentiation of stem cells. Adv. Drug Deliv. Rev. 60:199-214.

Jaiswal N, Haynesworth SE, Caplan AI, Bruder SP (1997). Osteogenic differentiation of purified, culture-expanded human mesenchymal stem cells in vitro. J. Cell Biochem. 64:295-312.

Jikko A, Harris SE, Chen D, Mendrick DL, Damsky CH (1999). Collagen integrin receptors regulate early osteoblast differentiation induced by BMP-2. J Bone Miner. Res. 14:1075-1083.

Keskar V, Marion NW, Mao JJ, Gemeinhart RA (2009). In vitro evaluation of macroporous hydrogels to facilitate stem cell infiltration, growth, and mineralization. Tissue engineering. Part A. 15:1695-1707.

Kollmer M, Keskar V, Hauk TG, Collins JM, Russell B, Gemeinhart RA (2012). Stem Cell-Derived Extracellular Matrix Enables Survival and Multilineage Differentiation within Superporous Hydrogels. Biomacromolecules 13:963-973.

Leboy PS, Beresford JN, Devlin C, Owen ME (1991). Dexamethasone induction of osteoblast mRNAs in rat marrow stromal cell cultures. J. Cell. Physiol. 146:370-378.

Lee NK, Sowa H, Hinoi E, Ferron M, Ahn JD, Confavreux C, Dacquin R, Mee PJ, McKee MD, Jung DY, Zhang Z, Kim JK, Mauvais-Jarvis F, Ducy P, Karsenty G (2007). Endocrine regulation of energy metabolism by the skeleton. Cell 130:456-469.

Livak KJ, Schmittgen TD (2001). Analysis of relative gene expression data using real-time quantitative PCR and the 2(-Delta Delta C(T)) Method. Methods 25:402-408.

Marion NW, Mao JJ, Klimanskaya I, Lanza RL (2006). Mesenchymal Stem Cells and Tissue Engineering. Methods Enzymol. 420:339-361.

Muraglia A, Cancedda R, Quarto R (2000). Clonal mesenchymal progenitors from human bone marrow differentiate in vitro according to a hierarchical model. J. Cell Sci. 113 (Pt 7):1161-1166.

Noh M (2012). Interleukin-17A increases leptin production in human bone marrow mesenchymal stem cells. Biochem. Pharmacol. 83:661-670.

Nuttall ME, Gimble JM (2004). Controlling the balance between osteoblastogenesis and adipogenesis and the consequent therapeutic implications. Curr. Opin. Pharmacol. 4:290-294.

Pittenger MF, Mackay AM, Beck SC, Jaiswal RK, Douglas R, Mosca JD, Moorman MA, Simonetti DW, Craig S, Marshak DR (1999). Multilineage potential of adult human mesenchymal stem cells. Science 284:143-147.

Ponce ML, Koelling S, Kluever A, Heinemann DE, Miosge N, Wulf G, Frosch KH, Schutze N, Hufner M, Siggelkow H (2008). Coexpression of osteogenic and adipogenic differentiation markers in selected subpopulations of primary human mesenchymal progenitor cells. J. Cell. Biochem. 104:1342-1355.

Pountos I, Corscadden D, Emery P, Giannoudis PV (2007). Mesenchymal stem cell tissue engineering: Techniques for isolation, expansion and application. Injury 38:S23-S33.

Quiroz FG, Posada OM, Gallego-Perez D, Higuita-Castro N, Sarassa C, Hansford DJ, Agudelo-Florez P, Lopez LE (2010). Housekeeping gene stability influences the quantification of osteogenic markers during stem cell differentiation to the osteogenic lineage. Cytotechnology 62:109-120.

Scadden DT (2006). The stem-cell niche as an entity of action. Nature 441:1075-1079.

Shafiee A, Seyedjafari E, Soleimani M, Ahmadbeigi N, Dinarvand P, Ghaemi N (2011). A comparison between osteogenic differentiation of human unrestricted somatic stem cells and mesenchymal stem cells from bone marrow and adipose tissue. Biotechnol. Lett. 33: 1257-1264.

Sodek J, Ganss B, McKee MD (2000). Osteopontin. Crit. Rev. Oral Biol. Med. 11:279-303.

Thomas T, Gori F, Khosla S, Jensen MD, Burguera B, Riggs BL (1999). Leptin acts on human marrow stromal cells to enhance differentiation to osteoblasts and to inhibit differentiation to adipocytes. Endocrinology 140:1630-1638.

Vater C, Kasten P, Stiehler M (2011). Culture media for the differentiation of mesenchymal stromal cells. Acta Biomaterialia 7:463-477.

Viereck V, Siggelkow H, Tauber S, Raddatz D, Schutze N, Hufner M (2002). Differential regulation of Cbfa1/Runx2 and osteocalcin gene expression by vitamin-D3, dexamethasone, and local growth factors in primary human osteoblasts. J. Cell. Biochem. 86:348-356.

Wiren KM, Hashimoto JG, Semirale AA, Zhang XW (2011). Bone vs. fat: embryonic origin of progenitors determines response to androgen in adipocytes and osteoblasts. Bone 49:662-672.

Zhang Y, Khan D, Delling J, Tobiasch E (2012). Mechanisms underlying the osteo- and adipo-differentiation of human mesenchymal stem cells. Sci. World J. 2012:793823.

25

Mesenchymal stem cells and acquisition of a bone phenotype: An ion channel overview

Priscilla C. Aveline, Julie I. Bourseguin and Gaël Y. Rochefort*

Inserm Unit U658, Hospital Porte Madeleine, Orleans, France.

Bone remodeling is a physiological process determined by the sequential and coordinated interaction of osteocytes, osteoclasts, osteoblasts and angiogenesis. During bone repair, osteoblastic cells, originated from mesenchymal stem cells (MSCs), are highly regulated to proliferate and to produce an osteogenic matrix, thus forming a new bone. MSCs are multipotential and undifferentiated cells that are present in the adult bone marrow. They serve *in vitro* and *in vivo* as precursors for bone marrow stroma, bone, fat, cartilage, muscle (smooth, cardiac and skeletal) and neural cells. MSCs are usually isolated from adult bone marrow but can also be isolated from several other tissues, such as fetal liver, adult circulating blood, umbilical cord blood, placenta or adipose tissue. In the bone marrow, MSCs give rise to mesenchymal cells residing in the bone (osteogenic, chondrogenic and adipogenic cells) and also support hematopoiesis. Therefore, MSCs regulate both osteogenesis and hematopoiesis, and they are responsible in part for the regenerative capacity of bone tissue. Acquisition of such a phenotype may be also regarded through the modification of the ionic channel expression. This review highlights current status and progresses in the differentiation MSCs along the osteoblastic/osteocytic pathways and the ionic channel expression and evolution during this differentiation.

Key words: Mesenchymal stem cells, osteoblastic differentiation, ionic channels

INTRODUCTION

Stem cells are generally described as clonogenic and undifferentiated cells that are able to self-renew and to differentiate into one or more types of differentiated and committed cells (Reyes-Botella et al., 2000; Jiang et al., 2002). To date, stem cells have been isolated and characterized from tissues of all ages, including embryonic, fetal and adult tissues. Among the general term of adult (postnatal) stem cells, mesenchymal stem cells (MSCs) represent a population of multipotential cells which are currently defined by a combination of morphologic, phenotypic, and functional properties, and which are capable of giving rise to at least mesenchymal-derived tissues, including bone, cartilage, fat, tendon and muscle (Friedenstein et al., 1974; Dazzi et al., 2006). MSCs have been identified into a large number of adult tissues, including the bone marrow where they provide the cellular microenvironment supporting hematopoiesis.

As a part of the stromal fraction, MSCs also regulate osteogenesis and are responsible, in part, for the regenerative capacity of bone tissue (Friedenstein et al., 1974). Since MSCs differentiate into osteoblasts, the major bone-forming cells, they have been used to develop new clinical therapies to treat or attenuate a number of skeletal disorders. Furthermore, osteoblasts and osteocytes possess ion channels and channels have different functions. It has been reported that K^+ and Ca^{++} channels have roles in proliferation, differentiation and apoptosis (Tao et al., 2008), in the modulation of the progression of the cell cycle and in affecting the proliferation of MSCs (Wang et al., 2008)(Park et al., 2008). There is a variable expression of K^+ channel with cell cycle of MSCs and this way contributes to the cell cycle progression. On the other hand, Ca^{++} channels contribute to the bone renewal and they are different expression during the osteogenic differentiation. Therefore, the purpose of the article is to review the literature on the adult MSC biology and their ion channel profile variations and modifications during the bone phenotype acquisition.

---

*Corresponding author. E-mail: gael.rochefort@gmail.com.

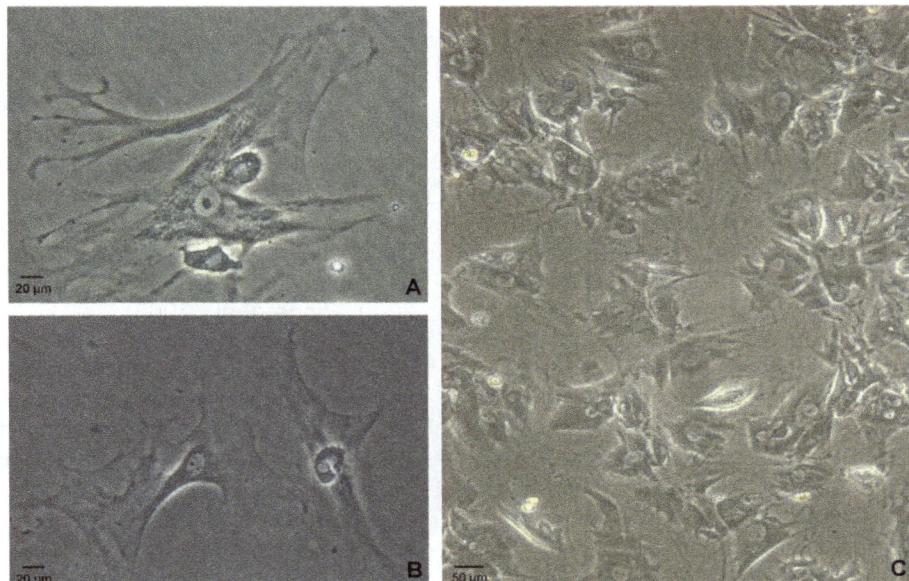

Figure 1. Mesenchymal stem cell shape in culture. Rat mesenchymal stem cells (A and B) were cultured in DMEM 20% fetal calf serum until sub-confluence (C), before being passaged. These plastic adherent bone marrow cells extract from rat bones are elongated cells that look like fibroblasts and that formed discrete colonies after few days.

What is a mesenchymal stem cell?

Definition and ontogenesis of a mesenchymal stem cell

MSCs, as the other types of stem cells, own the same characteristics and meet the classic criteria that define a stem cell. One of the defining criteria of stem cells is the self-renewal ability or the ability to generate identical copy of themselves with the same capacities through mitotic division over extended time periods (clonality). In culture, after they have been passed several times, MSCs have self-renewal activity. *In vivo*, cells collected from an initial transplant recipient could rise to a multiple cell type within a secondary recipient (Reyes-Botella et al., 2000; Jiang et al., 2002).

Then, the basic characteristics of stem cells are their multipotenciality where a single cell can differentiate into a variety of lineage cells. MSCs are multipotent stem cells and have the fundamental capacity to differentiate into a limited range of cell lineages (Gronthos et al., 1996; Prockop, 1997). The multilineage differentiation potentials of MSCs are study *in vitro* since their discovery, five decades ago. These studies, *in vitro* and *in vivo*, demonstrated that bone marrow MSCs, from many species, are able to differentiate at a single-cell level into limb-bud mesodermal cell type: osteoblasts, chondroblasts, adipocytes, fibroblasts and skeletal myoblasts. MSCs can also acquire characteristics of non-mesodermal lineages, such as endothelial cells, neural cells and endoderm, *in vitro* and *in vivo* (Pittenger et al., 1999; Schwartz et al., 2002; Verfaillie, 2002; Verfaillie et al., 2002).

Finally, these cells have also, *in vivo*, the capacities to generate a functional reconstitution of a given tissue when they are transplanted in a damage recipient or in the absence of tissue damage (Friedenstein et al., 1974; Latsinik et al., 1986; Haynesworth et al., 1992). Tissue regeneration is a stem cell property that has been partially confirmed for MSCs, especially for bone repair (osteogenesis imperfecta, bone fracture consolidation…). MSCs may also be used in several other diseases because of their large differentiation potential (vascular, cardiac…) (Orlic et al., 2002; Toma et al., 2002; Togel et al., 2005).

Such capacities of differentiation into a variety of connective tissue cells type make them an excellent candidate source for clinical tissue regeneration (Brazelton et al., 2000; Sanchez-Ramos et al., 2000).

The historical identification of mesenchymal stem cells

Alexander Friedenstein and colleagues, in 1974, were the first to identify mesenchymal stem cells (MSC) (Friedenstein et al., 1974; Friedenstein et al., 1974). They noticed, upon the other plastic adherent bone marrow cells extract from rabbit and rodent bones, non phagocytic, elongated cells that look like fibroblasts that formed discrete colonies after few days (Figure 1). Such colony was generated from a single cell and thus was called colony-forming unit-fibroblasts (CFU-F). Thereafter, *in vitro*, clonal cultures derived from individual

CFU-F were introduced into diffusion chambers in experimental models where the formation of bone, cartilage and stromal elements were observed. *In vivo*, when a colony is seeded under the renal capsule of semi-syngeneic animals, cells generated, after few weeks, fibrous tissue, bone and bone-containing bone marrow. Among these chimerical animals, Friedenstein observed that fibrous tissue and bone cells were of donor origin, while marrow hematopoietic cells within the bony spaces were provided by the host. Thus, Friedenstein conjectured that the bone marrow contains a population of progenitor cells able to generate *in vivo* fibrous tissue and bone and that these cells proving the adequate microenvironnement for HSC homing and growth. Friedenstein was the first to consider the existence of stem cell niches within the bone marrow. Few years later, in 1980, Castro-Malaspina showed that the colonies described are of fibroblastic nature by immunologic studies (Castro-Malaspina et al., 1980; Dazzi et al., 2006).

Arnold Caplan, during the 1990s, defined MSCs as cells that could give rise to bone and marrow stroma, but also to cartilage, tendon and muscle (Dennis et al., 1999; Caplan et al., 2006). The development of novel approaches to isolate and purify populations of MSCs has furthered our understanding of MSC biology but has also created several designation and abbreviations for describing these cells (Baksh et al., 2004). In the end of the 1888, Maureen Owen used the term of "Stromal Stem Cells" to show that MSCs take part of the stromal layer of the bone marrow and did not belong to the haematopoietic population (Owen et al., 1988). Teen years later, Darwin Prockop proposed the abbreviation of MSC for "Mesenchymal Stem Cell" or "Marrow stromal cells" because of their ability to differentiate into mesenchymal tissue and to serve also as a niche for other type of stem cells such as HSCs (Prockop, 1997; Kopen et al., 1999). In 1999, James Dennis indicated that this cell type may not represent an authentic category of stem cell but in closer to progenitor cell situated downstream of stem cell compartment and call these cells "Mesenchymal Progenitor Cell" (MPCs) (Dennis et al., 1999). Pamela Gehron Robey and Paolo Bianco used the term of Skeletal Stem Cells, in the early 2000s to show that MSCs were able to give rise to the components of skeletal tissues (Bianco et al., 2000). Catherine's Verfaillie and colleagues described culture-derived bone marrow-derived progenitor cells that may represent a more primitive cell type with different differentiation potential larger than that of MSCs, and they designated them as MAPCs for "Multipotent Adult Progenitor Cells" and "Mesodermal Progenitor Cells" (Jiang et al., 2002).

Diversity of sources of mesenchymal stem cells

MSCs can be isolated from many different common species. Among all theses species, the human (Pittenger et al., 1999; Zvaifler et al., 2000; Kuznetsov et al., 2001; Covas et al., 2005), the murine (Phinney et al., 1999; Baddoo et al., 2003) and the rat (Santa Maria et al., 2004; Rochefort et al., 2005; Rochefort et al., 2006) MSCs are the best characterized. Next to these three species, MSCs can also be isolated from guinea pigs, cats (Martin et al., 2002), baboons (Devine et al., 2001), sheep (Airey et al., 2004), dogs (Silva et al., 2005), pigs (Moscoso et al., 2005; Bosch et al., 2006), cows (Bosnakovski et al., 2005) and horses (Worster et al., 2000; Ringe et al., 2003).

MSC might be found with a variable proportion in different fetal and adult tissues but these cells often represent a small portion of these tissues.

Usually, they are isolated from the stromal fraction of adult bone marrow. Indeed, the bone marrow source is the most well studied and accessible but MSCs form a rare population of the bone marrow microenvironnement and may represent only 0.01 - 0.0001% of the adult human bone marrow nucleated cells. This is considerably lower than the proportion of hematopoietic stem cells that represent about 1% of the marrow nucleated cells. These cells are obtained by flushing the marrow out of animal bones with culture medium or from human bone marrow aspirates and transferred into a culture dish (Phinney et al., 1999; Pittenger et al., 1999; Santa Maria et al., 2004; Tropel et al., 2004; Zhang et al., 2004; Rochefort et al., 2005; Miao et al., 2006; Rochefort et al., 2006).

Cells with mesenchymal stem cells characteristics were isolated from several adult tissues including spleen, pancreas (da Silva Meirelles et al., 2006; Seeberger et al., 2006), liver (Campagnoli et al., 2001; Dan et al., 2006), kidney (da Silva Meirelles et al., 2006), lung (da Silva Meirelles et al., 2006), smooth muscle (da Silva Meirelles et al., 2006), skeletal muscle (Howell et al., 2003; Barry et al., 2004; Yoshimura et al., 2007), aorta (da Silva Meirelles et al., 2006), vena cava (da Silva Meirelles et al., 2006), brain (da Silva Meirelles et al., 2006), thymus (da Silva Meirelles et al., 2006), dental pulp (Pierdomenico et al., 2005), deciduous teeth (Barry et al., 2004), scalp tissue and hair follicle (Shih et al., 2005), periosteum (Barry et al., 2004; Yoshimura et al., 2007), trabecular bone (Barry et al., 2004), adipose tissue (Barry et al., 2004; Yoshimura et al., 2007) and synovium (Barry et al., 2004; Yoshimura et al., 2007). MSCs have also been isolated from fetal tissues similar to the adult tissues but also, with a variable portion, from several part of the placenta (Igura et al., 2004; Miao et al., 2006) including chorionic vili, amniotic fluid (Tsai et al., 2004), fresh or cryopreserved umbilical cord blood (Erices et al., 2000; Erices et al., 2003) and umbilical cord vein (Covas et al., 2005).

MSCs were also described in the peripheral blood of normal adult and women during and after the pregnancy, from a fetal origin and may persist for at least 60 years (Zvaifler et al., 2000; O'Donoghue et al., 2004; Villaron et al., 2004; Dazzi et al., 2006). Blood samples represent a particular important source of MSC, more accessible in Human than the bone marrow source. MSCs can be mobilized into the bloodstream after treatment (chemotherapy, cytokine injection) or physiopathological events inducing a physiological release of stem cells from their reservoirs in responses of stress signals such as cytokines like granulocyte colony-stimulating factor (G-CSF) or granulocyte-macrophage colony-stimulating factor (GM-CSF). Recently, scientists show that several physiopathological circumstances, including hypoxia, myocardial infarction or encephalopathy can increase the release of MSCs in blood (Erices et al., 2000; Erices et al., 2003; Romanov et al., 2003; Lee et al., 2004).

The usual phenotype of mesenchymal stem cells

In the past, MSCs were selected by adherence to plastic surface and were considered as a homogeneous population. However, scientists observed differences between the morphology and the differentiation potential of the cultured cells. MSC markers have been identified that are suitable to show heterogeneity of these cells not only with regard to phenotype but also with regard to the differentiation capacity. Scientists bring to the fore distinct primary CSM subsets based on surface markers. Considerable progress has been made about characterizing the cell surface antigenic pattern of MSCs using fluorescence activated cell sorting and magnetic bead sorting techniques. Yet, phenotypic isolation of MSC is not possible because no specific MSC immuno-phenotyping markers have been identified. However, MSC express a wide variety of antigens presents by many cell types. Thus, their identification is based on an extensive combination of markers building up a pattern of typical surface molecules of MSC, such as differentiation and lineage specific markers, adhesion molecules, extracellular matrix and growth factor factors receptors, and a panel of monoclonal antibodies (Simmons et al., 1991; Simmons et al., 1992; Gronthos et al., 1995, 1996; Dazzi et al., 2006).

An important number of makers have been described that can be suitable for the isolation of MSCs from primary tissues. Currently, the most accepted profile for phenotype MSC is the co-expression of CD105 (SH2) and CD73 (SH3 and SH4). MSCs are positive for CD13, CD 29, CD 44, CD49a and CD49e, CD90, CD106, CD 166, CD349. MSC are negative for the hematopoietic and/or endothelial markers CD14, CD31, CD34, CD45, and CD133. MSCs constitutively express low surface density of MHC class I molecules (HLA-ABC) and are negative for MHC class II (HLA-DR). Therefore, MSCs can be considered as poorly immunologenic (Haynesworth et al., 1992; Pittenger et al., 1999; De Ugarte et al., 2003; Vogel 2003; Barry et al., 2004; Dazzi et al., 2006).

Nevertheless, some molecules, used as markers to identified MSC, are variably expressed during time in culture and depending on the extent of MSC multipotency. Moreover, some antigens appear to be present on cell subsets only. Last markers identified are CD140b, CD271 (bone marrow), SSEA-4 (placenta), the ganglioside GD2, CD146, CD200 and the integrin complex. These markers are very selective for the recognition of MSCs but they lack the ability to discriminate MSC subsets. Very recently, MSCA-1 and CD56 were identified as combination markers suitable to purify and characterize MSC populations from primary bone marrow that show distinct morphological features and differentiation capacities.

Others study that the gene expression pattern of MSC by serial analysis gene expression, restriction fragment differential display and DNA microarray and show the presence of multiple cell lineages as part of their transcription (Silva et al., 2003; Dazzi et al., 2006; Battula et al., 2009; Buhring et al., 2009).

The lineage differentiation abilities of mesenchymal stem cells

An embryo is composed to three layers (ectodermal, endodermal and mesodermal layers) and each layers gives several tissues. The ectodermal tissue gives the peripheral and central nervous systems and also the epithelium that become epidermal tissue, muscles (iris' muscles, the myoepithelial cells of the lacrimal salivary...) (Fuchs et al., 1994; Lo et al., 1997). To concern the mesoderm, it gives also muscles such as skeletal, cardiac, visceral and vascular smooth muscles, bone, cartilage, blood and connective tissue (Baron, 2001). Finally, the endoderm layer gives all the intestinal tissue (Soria, 2001; Gupta, 2002; Wobus et al., 2005).

Mechanism of the lineage differentiation

Stem cell can differentiate into different cells of the tissue in which they reside. They have a multi-lineage differentiation potential of several stem cells types including MSCs (Herzog et al., 2003). MSCs are able to differentiate into limb-bud in different cells like osteoblasts, chondroblasts, fibroblasts, adipocytes and skeletal myoblasts (Friedenstein et al., 1974; Haynesworth et al., 1992; Gronthos et al., 1996; Prockop, 1997; Pittenger et al., 1999). Also, they are able to acquire characteristics of cell lineage outside the limb-bud, such as endothelial cells and neural cells (Brazelton et al., 2000; Reyes-Botella et al., 2000; Sanchez-Ramos et al., 2000; Jiang et al., 2002; Zhao et al., 2004). This differentiation potential has been show in the early 2000s by Markus Loeffler and his team. They described the MSCs differentiation pathway that was reversible and flexible. Therefore, adipocytes became turn into osteoblasts, chondrocytes into adipocytes, adipocytes into chondrocytes, osteoblasts into chondrocytes, and so on (Loeffler et al., 2002; Baksh et al., 2004; Loeffler et al., 2004; Song et al., 2004). The MSCs express simultaneously the mRNA or protein level of adipocytes, osteoblasts or chondrocytes markers (Baksh et al., 2004). This plasticity is due to the de-differentiation and reprogramming of MSCs lineage by stimulation of different specifics transcription factors.

Plasticity is an important MSCs characteristic because whatever the differentiation level, MSCs are able to give all the cells of his lineage.

The ectodermal differentiation of mesenchymal stem cells

The ectodermal tissue is the covers of the body surface. It forms many tissues, including the central nervous system, cranial bones, ganglia and nerves, epidermis, hair, mammary glands.

The mesenchymal stem cells differentiation into central nervous system cells: The brain development is very crucial, neurons migrate and forms the neural tubes and they differentiate into several cells and satellite cells. In 1992, Reynolds and Weiss isolated for the first time multipotential stem cells from the striatum of adult human brain. Then they induced multipotential stem cells into neural cells in vitro (Reynolds et al., 1992). A novel site of mesenchymal stem cell was discovered: the neural stem cells (NSCs). NSCs are present in the brain during life span mammalian and humans (Brazelton et al., 2000; Sanchez-Ramos et al., 2000). These cells are one of the newest and most promising for treating several diseases like neurodegenerative diseases or injuries of central nervous system.

MSCs express neural markers and they could have an in vitro neural differentiation into neuron and glial-like cells. It is possible with a NSCs culture with a medium completed by neural factors. Many studies showed different MSCs-NSCs differentiation and MSCs had morphological characteristics, neural markers, electrophysiological properties as a neuron (Deng et al., 2001; Wehner et al., 2003; Kanemura, 2010).

The mesenchymal stem cells differentiation into skin-related cells: Skin is the largest organ of the body. It is composed by three layers: the epidermis, derma and hypoderm and they are several and different functions. The skin has many roles like temperature regulation, vitamin D and melanin synthesis. Also it has an immunity function, it is a barrier against pathogens and it is a sensitive organ.

In vitro, skin MSCs were studied many times. In fact it is a big challenge to understand all the MSCs mechanisms. Today, MSCs are used like a cellular

therapy when a burn injury. Bone marrow MSCs ware transplanted and dermis can be regenerate with a scar reduction (Deng et al., 2001; Gharzi et al., 2003; Chunmeng et al., 2004; Bey et al., 2010).

The endodermal differentiation of mesenchymal stem cells

During embryo development, endoderm layer forms a layer inside the gastrula. Endoderm gives many tissues such as digestive tube epithelial lining (except a part of mouth, pharynx and the rectum terminal), all digestive tube glands lining cells (liver and pancreas), tympanic cavity, trachea, bronchi, lung and so on.

The mesenchymal stem cells differentiation into gastrointestinal tract related cells: It is the organ system that takes in food, makes digestion to extract nutrients and energy, and expels waste. The gastro-intestinal tract important functions are digestion and excretion. It is connected with different organs that come from endoderm tissues too including liver which secretes bile, pancreas which secretes bicarbonate and many enzymes (trypsin, lipase, pancreatic amylase, and chymotrypsin).

Gastrointestinal tract measure nearly 6.5 m long (20 feet) in a normal adult male. It is divided into three parts: foregut, midgut and hindgut. The upper gastrointestinal tract including the mouth, pharynx, esophagus and stomach, correspond to foregut except a part of duodenum. The lower gastrointestinal tract are consisting of the intestines and anus that derives from the midgut for the lower duodenum, and are consisting of the transverse colon and upper part of the anal canal from the hindgut.

The mesenchymal stem cells differentiation into Liver cells: It plays an important role in metabolism and many functions in the body like plasma protein synthesis, glycogen storage and drug detoxification. Liver comes from endoderm part of the foregut (hepatic diverticulum). It produces bile and has several functions like carbohydrate metabolism regulation, lipid and cholesterol metabolism and insulin and coagulators factors production. At cellular level, liver is composed to hepatocytes and these cells are able to regenerate themselves, only 25% liver tissue can regenerate all liver.

In fetal and adult liver, many different cells are present: several types of bone marrow-derived stem cells that are able to differentiate into hepatocytes cells with some in vitro conditions. A MSCs culture with hepatocytes growth factor (HGF) in medium obtains hepatocytes cells (Oh et al., 2000; Avital et al., 2001; Okumoto et al., 2003; Lange et al., 2005).

In vivo, bone marrow and liver transplantation were used to treat several diseases or in order to reconstitute a liver. Different studies demonstrate that bone marrow-derived MSCs with some conditions give several types of liver cells such as hepatocytes, oval cells and cholangiocytes (Petersen et al., 1999; Herrera et al., 2006). Recently, MSCs differentiation into hepatocytes lineage was confirmed (Avital et al., 2001; Seo et al., 2005).

The mesenchymal stem cells differentiation into pancreas cells: It is an organ in the digestive and endocrine system which it is both exocrine (secrets pancreatic juice containing digestive enzymes) and endocrine (hormones secretions including insulin, glucagon and somatostatin that are produce by islet of Langerhans). Several diseases could alter pancreas functions including tumors, cancer, cystic fibrosis, diabetes, exocrine pancreatic insufficiency, and so on. Diabetes is responsible to a high morbidity and mortality in many countries caused by a destruction of insulin secreting pancreatic β cells (type 1) or by a relative deficiency due to decreased insulin sensitivity (type 2).

In vitro, MSCs are able to differentiate into insulin, glucagon or somatostatin-secreting cells with some conditions and can differentiate into cells that have the islet pancreatic transcription factors (Nkx-2.2, Nkx-6.1, Pax-4, Pax-6, Isl-1 and Ipf-1) (Hess et al., 2003; Moriscot et al., 2005; Eberhardt et al., 2006; Timper et al., 2006).

The mesenchymal stem cells differentiation into Lung-related cells: It represents the largest surface area, a human adult lung measure nearly 70 m² whereas skin surface area represents only 2 m². Lung has an essential function that is respiration; it is able to transport oxygen from the atmosphere into the bloodstream, to excrete carbon dioxide from the bloodstream into the atmosphere.

Lung comes from endoderm epithelium and its development takes place into four chronological stages: the pseudoglandular stage, the canalicular stage, the terminal sac stage and the alveolar stage (Gomi et al., 1994; Yamada et al., 2003).

The mesodermal differentiation of mesenchymal stem cells

The mesoderm layer is forming just in triphoblastic animals during gastrulation. Some cells migrate and contribute a supplementary layer between ectodermal and endodermal layers. Mesoderm formed a coelom that become several tissues including skeleton, dermis of skin, skeletal muscles, connective tissue, heart, blood, and so on.

The mesenchymal stem cells differentiation into adipocytes: Adipocytes are a reserve of energy, they stock lipids and when energy is required, they break down lipids into free fatty acids. They also have other

function that is important in control of metabolism by a paracrine secretion end endocrine hormones, thus regulating insulin sensitivity and secretion (Kershaw et al., 2004).

During embryogenesis, a population of MSCs of mesodermal layer migrates in the vascular stroma of adipose tissue. These cells have a multi-step of differentiation process to acquire an adipocytes phenotype (Otto et al., 2005; Rosen et al., 2006). MSCs express also transcription factors such as CCAAT-enhancer-binding protein α (C/EBPα) and peroxisome proliferator-activated receptor γ (PPARγ), these factors promote adipogenesis. Insulin is also a factor of adipocytes differentiation when using low concentrations, more, serums, dexamethasone, indomethacin and 3-isobutyl-1-methylxanthine (IBMX) induce MSCs differentiation into adipocytes.

The mesenchymal stem cells differentiation into osteogenic cells: Bones are rigid organ that form vertebrates skeleton. The most important functions of bone are the organs protection, hold up the body and support muscles. Bones protect the brain, spinal cord, lungs, heart, bone marrow, and so on. Bone permits skeletal muscle movements with the rigid attachment. Red bone marrow in bone is the major producer of blood cells and most of these cells are responsible of immunity. The skeleton has a mineral storage role, it stores calcium and phosphate and release them when the body's needs.

During early development, cells of neural crest give cranio-facial skeleton whereas the axial skeleton comes from sclerotome cells (somites) and the lateral plate of mesoderm gives the appendicular skeletal components (Olsen et al., 2000). But potential MSCs persist in bone marrow and have a role in bone growth, remodeling and bone repair. These MSCs could be differentiate into osteoblasts, *in vitro* osteoblasts derived MSCs can be obtain *in vitro* by medium with different transcription factors in order to induce osteogenic lineage such as transforming growth factor-β (TGF- β), interleukin-6 (IL-6), growth hormone, sortilin, leptin and transglutaminase (Taguchi et al., 1998; Weinreb et al., 1999; Ramoshebi et al., 2002; Canalis et al., 2003; Rawadi et al., 2003; Sykaras et al., 2003). Theirs products can be used to induce osteogenic lineage including prostaglandin E2, 1, 25-dihydroxyvitamin D3 (active form of vitamin D3) L-ascorbic acid (vitamin C), dexamethasone, β-glycerol phosphate, and so on (Raisz et al., 1993; Rogers et al., 1995; Weinreb et al., 1999; Rosa et al., 2003; Sottile et al., 2003).

The mesenchymal stem cells differentiation into chondrogenic cells: Cartilage is a connective tissue that is composed of collagenous fibers and/or elastic fibers and chondrocytes, all embedded in matrix of gel-like ground substance. Cartilage is not vascular and it is localized in many places in body like joints, rib cage, nose, ear, bronchial tubes and between inter-vascular discs. During embryonic development, cartilage cells precursors (chondroblasts) are enclosed in perichondrium. Perichondrium was replaced by bone or joint cartilage.

Functional differentiation and adaptation of mesenchymal stem cells

MSCs are present into different tissues and during cell differentiation; MSCs are modulating by different conditions (according to the place where MSCs are) like cell stress, bone resorption and bone formation, wound healing, and so on. During differentiation, MSCs membrane has changing in receptors and channels on their surface.

Identification of ion channels into mesenchymal stem cells

Two distinct outward currents, present either alone or in combination, are recorded in MSCs. Most MSCs are demonstrated outward current by K^+ channels blockers but they were not blocked by blockers of Cl^- currents. Therefore, MSCs currents may depend on K^+ channels and not on Cl^- channels.

The most important current present in MSCs is from K+ channels. These channels can be blocked by tetrathylammonium for example: Kv1.1 channels, Kv3 channel family, Kv7.2 channels and BKCa (MaxiK) that are channels voltage Ca^{++} activated K^+ channels of large conductance. There are three criteria to a large conductance Ca^{++} activated K^+ current: 1/ typical electrophysiological properties, 2/ noisy current traces due to large conductance single channel opening, and 3/ BKCa is sensitive to iberiotoxin (a specific BKCa blocker). BKCa channels are sensors of intracellular Ca^{++} that regulate membrane potential in a Ca^{++} dependant manner, and they are modulated by phosphate.

Functional differentiation and adaptation of mesenchymal stem cells into bone cells

MSCs give rise to an osteoblastic lineage: pre-osteoblasts, osteoblasts and osteocytes. Osteoblasts are the cells that produce mineralize matrix to become an osteocyte immured in this mineralize matrix.

During MSCs differentiation, these cells have many changes including forms, functions, phenotypes, and so on. To have osteoblastic lineage in cell culture, medium is constituted to dexamethasone, ascorbic acid and β-glycerophosphate.

Surface marker modifications into mesenchymal stem cells during the bone lineage differentiation: Osteoblasts lineage has different propriety because they

have many functions. Osteoblasts lineage express peroxisome proliferator-activated receptor γ (PPAR γ), bone sialoprotein (BSP), osteocalcin, osteopontin, phosphatase alkaline, collagen 1αA and 1α2. But they do not express osterix and lipoprotein lipase (LPL). Also there are modifications about surface markers. MSCs have many markers and during osteoblastic lineage, markers disappear or appear (Table 1).

MSCs, osteoblasts and osteocytes have common markers such as CD29, CD44, CD73 and CD105 and do not expressed CD3, CD4 and CD45. Others markers have a variable expression during MSCs differentiation. CD10, CD13 and HLA-DR are expressed by MSCs and osteoblasts but osteocytes they do not. CD40 is not expressed by MSCs but during their differentiation to osteoblastic lineage, osteoblasts and osteocytes express CD40. Others markers have a very variable expression, MSCs express CDx, osteoblasts do not express this CD and osteocytes express this, and vice versa.

Ion channel evolution into mesenchymal stem cells during the bone lineage differentiation: Like surface markers, ion channels have an evolution during MSCs differentiation to lineage osteoblastic (Table 2 and Figure 2). MSCs, osteoblasts and osteocytes have some common ions channels such as BKca, Kv1.3, Kv2.1, SK1 or SK2. However, a lot of ion channels have a variable expression during MSCs differentiation into osteoblasts and osteocytes. For example, Cav1.1, Cav1.2, Cav1.3, Cav3.1, Kv1.2, Kv3.1 or Kv3.2 are expressed by MSCs but there expression are decreasing in osteoblasts and/or osteocytes. At the opposite, TREK2, TRAAK, Kv1.6 and Kv2.2 are not expressed by MSCs but during their differentiation to osteoblastic lineage, osteoblasts and/or osteocytes express these ion channels. Others markers have a very variable expression, MSCs express some ion channels, osteoblasts do not express these ion channels and osteocytes express this, and vice versa.

Very few specific studies have demonstrated presence or absence during cell differentiation and there are very studies on osteocytes channels, especially only about calcium and sodium channels. Cav are channels that have to activate by a calcium currents in cells. There is one channel in common that is not present in MSCs, osteoblasts and osteocytes and this channel is SK3. Furthermore, osteoblasts and osteocytes possess ion channels and channels have different functions.

K^+ channel has roles in proliferation, differentiation and apoptosis: it has been reported that at least the intermediate-conductance Ca^{++}-activated potassium (IKCa) channel regulates the cell cycle progression and proliferation of mouse MSCs (Tao et al., 2008), whereas Kv channel activity (especially Kv1.2 and Kv2.1 associated with IKDR; Slo and KCNN4 associated with IKCa; and Kv1.4 and Kv4.3 associated with Ito) modulates the progression of the cell cycle and affects the proliferation of MSCs (Wang et al., 2008). A study has also showed that the functional potassium channel

Table 1. Relevant antigens that are strongly, weakly or not expressed in MSCs, osteoblasts and osteocytes.

	MSCs +/- ?	MSCs References	OSTEOBLASTS +/- ?	OSTEOBLASTS References	OSTEOCYTES +/- ?	OSTEOCYTES References
CD3	--	(Pittenger et al., 1999; Reyes-Botella et al., 2000; Ahuja et al., 2003; De Ugarte et al., 2003a)	--	(Reyes-Botella et al., 2000)	--	(Ahuja et al., 2003)
CD4	--	(Pittenger et al., 1999; De Ugarte et al., 2003; Katz et al., 2005)	--		?	(Ahuja et al., 2003)
CD10	--	(Battula et al., 2007)	++	(Reyes-Botella et al., 2000; Seshi et al., 2003; Garcia Ruiz et al., 2006)	++	(Tsai et al., 2004)
CD11b	--	(Katz et al., 2005; Aurich et al., 2007; Parekkadan et al., 2007; Yoshimura et al., 2007)	--	(Reyes-Botella et al., 2000; Marom et al., 2005)	±	(Ahuja et al., 2003; Tsai et al., 2004)
CD11c	--	(De Ugarte et al., 2003; Katz et al., 2005)	--		?	(Ahuja et al., 2003)
CD13	--	(Pittenger et al., 1999; De Ugarte et al., 2003b; Aurich et al., 2007; Parekkadan et al., 2007)	++	(Reyes-Botella et al., 2000)	±	(Ahuja et al., 2003)
CD14	--	(Pittenger et al., 1999; Zuk et al., 2001; De Ugarte et al., 2003; Aurich et al., 2007; Parekkadan et al., 2007)	--	(Reyes-Botella et al., 2000)	--	(Ahuja et al., 2003)
CD15	?	(Pittenger et al., 1999; De Ugarte et al., 2003)	--	(Reyes-Botella et al., 2000)	--	
CD19	?	(Mansilla et al., 2006)	++	(Seshi et al., 2003)	±	
CD29	++	(Pittenger et al., 1999; De Ugarte et al., 2003)	++	(Jiang et al., 2002)	++	(Tsai et al., 2004)
CD31	?	(Zuk et al., 2001; De Ugarte et al., 2003; Katz et al., 2005)	--	(Marom et al., 2005)	+	
CD34	--	(Aurich et al., 2007)	--	(Reyes-Botella et al., 2000)	++	(Jiang et al., 2002; Tsai et al., 2004)
CD38	?	(Pittenger et al., 1999; De Ugarte et al., 2003)	--	(Reyes-Botella et al., 2000)	++	
CD40	±	(Weiss et al., 2008)	--	(Ahuja et al., 2003)	++	(Ahuja et al., 2003)
CD40L	--	(Weiss et al., 2008)	--	(Reyes-Botella et al., 2000; Bonewald 2004)	++	(Ahuja et al., 2003)
CD44	++	(Pittenger et al., 1999; Zuk et al., 2001; Silva et al., 2003; Talens-Visconti et al., 2006; Jager et al., 2007)	++	(Noonan et al., 1996; Reyes-Botella et al., 2000; Garcia Ruiz et al., 2006)	++	(Noonan et al., 1996; Tsai et al., 2004; Yamasaki et al., 2005)
CD45	--	(Pittenger et al., 1999; Zuk et al., 2001; Jiang et al., 2002; Silva et al., 2003; Talens-Visconti et al., 2006; Jager et al., 2007)	--	(Brown et al., 1990; Reyes-Botella et al., 2000; Seshi et al., 2003)	--	(Ahuja et al., 2003)
CD54	?	(Pittenger et al., 1999; De Ugarte et al., 2003)	++	(Noonan et al., 1996; Reyes-Botella et al., 2000; Garcia Ruiz et al., 2006)	++	

Table 1. Contd.

Marker						
CD56	(Brooke et al., 2008)	++	(Reyes-Botella et al., 2000)	++		?
CD62e	(Pittenger et al., 1999; Zuk et al., 2001; Katz et al., 2005)	--	(Marom et al., 2005)	±		?
CD68	(Parekkadan et al., 2007; Brooke et al., 2008)	--	(Reyes-Botella et al., 2000)	--		?
CD73	(Pittenger et al., 1999)	++	(Brown et al., 1990)	++		++
CD79a	(Seshi et al., 2003)	?	(Seshi et al., 2003)	±		?
CD80	(Pittenger et al., 1999; Katz et al., 2005)	±	(Reyes-Botella et al., 2000; Garcia Ruiz et al., 2006)	++		?
CD86	(Weiss et al., 2008; Wang et al., 2009)	--	(Reyes-Botella et al., 2000)	++	(Jiang et al., 2002; Ahuja et al., 2003)	--
CD90	(Covas et al., 2005; Flores-Figueroa et al., 2005; Mansilla et al., 2006; Kang et al., 2008; Liu et al., 2008; Rose et al., 2008)	++	(Lee et al., 2004)	++		?
CD105	(Flores-Figueroa et al., 2005; Mansilla et al., 2006; Battula et al., 2007; Liu et al., 2008; Rose et al., 2008)	++	(Lee et al., 2004)	++	(Jiang et al., 2002; Tsai et al., 2004)	++
CD117	(Pittenger et al., 1999; Katz et al., 2005)	±	(Tsai et al., 2004)	?	(Tsai et al., 2004)	--
CD265	(Tobon-Arroyave et al., 2005)	++	(McCormick et al., 1988)	++		?
HLA-DR	(Panepucci et al., 2004; Ishii et al., 2005; Katz et al., 2005; Aurich et al., 2007)	±	(Reyes-Botella et al., 2000; Garcia Ruiz et al., 2006)	++	(Jiang et al., 2002; Tsai et al., 2004)	--
CMH II	(Wang et al., 2009)	--		?	(Ahuja et al., 2003; Tsai et al., 2004)	?
CMH I	(Kang et al., 2008)	++		++	(Jiang et al., 2002; Tsai et al., 2004)	++

The attribution concerning the degree of expression of some markers varies between different authors, because of different stem cell cultures and different antibodies used.

++: strong expression; ±: weak expression; --: absence of expression; ?: unknown expression.

profiling may depend on the cellular passage of MSCs (Park et al., 2008). There is a variable expression of K^+ channel with cell cycle of MSCs and this way contributes to the cell cycle progression.

Ca^{++} channels contribute to the bone renewal and they are different expression during the osteogenic differentiation. Ca^{++} entry across the plasma membrane is a main pathway for Ca^{++} signal, although it was reported that Ca^{++} release from intracellular stores plays important role in Ca^{++} oscillations in human mesenchymal stem cells (Kawano et al., 2002). One pathway of Ca^{++} entry across the plasma membrane is voltage-operated Ca^{++} channel which is well known to play

Table 2. Relevant channels that are strongly, weakly or not expressed in MSCs, osteoblasts and osteocytes.

Family	Group	Subtype	MSCs +/- ?	MSCs References	OSTEOBLASTS +/- ?	OSTEOBLASTS References	OSTEOCYTES +/- ?	OSTEOCYTES References
Calcic Channels	IcaL	Cav1.1	±	(Zahanich et al., 2005)	±	(Zahanich et al., 2005; Bergh et al., 2006)	--	(Bergh et al., 2006)
	IcaL	Cav1.2	+++	(Zahanich et al., 2005; Balana et al., 2006; Zeng et al., 2011)	++	(Gu et al., 2001; Zahanich et al., 2005; Bergh et al., 2006)	--	(Gu et al., 2001; Bergh et al., 2006)
	IcaL	Cav1.3	±	(Zahanich et al., 2005)	++	(Gu et al., 2001; Bergh et al., 2006)	--	(Gu et al., 2001)
	IcaP/Q	Cav2.1	?		++	(Zahanich et al., 2005)	?	
	IcaN	Cav2.2	?		++	(Selim et al., 2006)	?	
	IcaR	Cav2.3	?		--	(Selim et al., 2006)	?	
	IcaT	Cav3.1	+	(Zahanich et al., 2005)	+	(Zahanich et al., 2005)	--	(Gu et al., 2001)
	IcaT	Cav3.2	?		++	(Bergh et al., 2006)	+	(Bergh et al., 2006)
	IcaT	Cav3.3	?		-	(Gu et al., 2001)	?	
Potassic Channels	2TM	Kir1.1 (Kir1.x)	+	(Li et al., 2006)	?		?	
	2TM	Kir2.1 (Kir2.x)	+	(Li et al., 2006; Park et al., 2007; Park et al., 2008)	?		?	
	2TM	Kir2.2	-	(Li et al., 2006; Park et al., 2007)	?		?	
	2TM	Kir6.1 (Kir6.x)	?		++	(Gu et al., 2001)	?	
	2TM	Kir6.2	?		++	(Gu et al., 2001)	?	
	4TM	TWIK1 (TWIK)	++	(Park et al., 2007)	?	(Hughes et al., 2006)	?	
	4TM	TREK1 (TREK)	++	(Magra et al., 2007)	++	(Chen et al., 2005; Hughes et al., 2006)	?	
	4TM	TREK2	--	(Magra et al., 2007)	++	(Wu et al., 2001; Hughes et al., 2006)	?	
	4TM	TRAAK	--	(Magra et al., 2007)	++		?	
	6TM	Kv1.1 (Kv1.x)	++	(Heubach et al., 2004; Balana et al., 2006; Park et al., 2007; Bonnet et al., 2008; Park et al., 2008)	++	(Yellowley et al., 1998)	?	
	6TM	Kv1.2	++	(Bonnet et al., 2008; Wang et al., 2008; Wang et al., 2009; Zeng et al., 2011)	--	(Yellowley et al., 1998)	?	
	6TM	Kv1.3	+	(Li et al., 2005; Zeng et al., 2011)	++	(Yellowley et al., 1998)	?	
	6TM	Kv1.4	+	(Wang et al., 2008)	?		?	
	6TM	Kv1.5	+	(Balana et al., 2006; Bonnet et al., 2008)	?		?	
	6TM	Kv1.6	--	(Li et al., 2005)	++	(Yellowley et al., 1998)	?	
	6TM	Kv2.1 (Kv2.x)	++	(Heubach et al., 2004; Bonnet et al., 2008; Wang et al., 2008)	++	(Yellowley et al., 1998)	?	
	6TM	Kv2.2	--	(Li et al., 2005)	++	(Yellowley et al., 1998)	?	

Table 2. Contd.

Group	Channel	Expression	References	Expression	References	Expression
Kv3.x	Kv3.1	++	(Heubach et al., 2004)	--	(Yellowley et al., 1998)	?
	Kv3.2	++	(Li et al., 2005)	--	(Yellowley et al., 1998)	?
	Kv3.3	?		++	(Yellowley et al., 1998)	?
	Kv3.4	?		++	(Yellowley et al., 1998)	?
Kv4.x	Kv4.2	+	(Heubach et al., 2004; Park et al., 2007; Park et al., 2008; Benzhi et al., 2009)			?
	Kv4.3	++	(Heubach et al., 2004; Balana et al., 2006; Wang et al., 2008)			?
Kv7.x	Kv7.1	--	(Heubach et al., 2004)			?
	Kv7.2	+	(Heubach et al., 2004)			?
	Kv7.3	+	(Heubach et al., 2004)			?
KCNQ	KCNQ2	?		++	(Yellowley et al., 1998)	?
	KCNQ4	?		++	(Yellowley et al., 1998)	?
	KCNQ5	?		++	(Yellowley et al., 1998)	?
EAG	Eag1	++	(Li et al., 2006; Park et al., 2008)			?
	Eag2	--	(Li et al., 2006)			?
Slo	BKCa	++	(Balana et al., 2006; Li et al., 2006; Park et al., 2007; Bonnet et al., 2008; Park et al., 2008; Wang et al., 2008)	++	(Ypey et al., 1992; Yellowley et al., 1998)	±
SK	SK1	?		++	(Gu et al., 2001)	++
	SK2	?		++	(Gu et al., 2001)	++
	SK3	--	(Li et al., 2006)	?	(Gu et al., 2001)	--
	SK4	++	(Li et al., 2006)	+	(Yellowley et al., 1998)	?
Sodium Channels	Nav1.1	++	(Li et al., 2006; Mareschi et al., 2009)			?
	Nav1.2	++	(Li et al., 2006; Mareschi et al., 2009)			?
	Nav1.3	++	(Mareschi et al., 2009)			
	Nav1.5	--	(Park et al., 2007)			?
	Nav1.6	++	(Zeng et al., 2011)			?
	Nav1.7	++	(Park et al., 2007)			?
	hNE-Na	+	(Park et al., 2008)			?
	EAAT1	?	(Marom et al., 2005)	++	(Marom et al., 2005)	++

The attribution concerning the degree of expression of some channels varies between different authors, because of different cultivation media, various ages of the donors or prolonged stem cell cultures and different antibodies used.

++: very strong expression; +: positive expression; ±: weak expression; --: absence of expression; ?: unknown expression.

an important role for Ca^{++} entry across the plasma membrane in excitable cells and it is deemed to contribute to cellular differentiation or proliferation. However, it is still not clear which functions voltage-operated Ca^{++} channels perform in non-excitable cells, especially in the undifferentiated stem cells.

Bone formation depends on proper channels functions: L type Ca^{2+} channels blockers could stimulate alkaline phospholipase activity and mineralization. The channels might come into play at a later stage of differentiation.

Conclusions and future expectation

Mesenchymal Stem Cells differentiation concern the entire organism and during all the organism life. They are defined by different properties like the capacity to differentiate *in vitro* and *in vivo* into many cells include osteoblasts, adipocytes, chondroblasts, like proliferation adherent cells and a defined immunophenotype (markers and channels). MSCs are the most promising cells for the future for therapeutic applications with their capacity to modulate the immune system. It is regarded as a highly interesting topic how different patterns of ion channels diversity may differentially regulate intracellular pathways leading to different physiological responses during mesenchymal stem cell differentiation for its high potential in medical application. It is still unclear how ion channel profiling are differentially expressed during MSC differentiation and how they are decoded, and most importantly, what is the actual relationship between ion channel signals and mesenchymal stem cells differentiation along to the osteogenic differentiation. Further studies are needed to fully clarify this issue.

REFERENCES

Ahuja SS, Zhao S, Bellido T, Plotkin LI, Jimenez F, Bonewald LF (2003). CD40 ligand blocks apoptosis induced by tumor necrosis factor alpha, glucocorticoids, and etoposide in osteoblasts and the osteocyte-like cell line murine long bone osteocyte-Y4. Endocrinol., 144(5): 1761-1769.

Airey JA, Almeida-Porada G, Colletti EJ, Porada CD, Chamberlain J, Movsesian M, Sutko JL, Zanjani ED (2004). Human mesenchymal stem cells form Purkinje fibers in fetal sheep heart. Circ., 109(11): 1401-1407.

Aurich I, Mueller LP, Aurich H, Luetzkendorf J, Tisljar K, Dollinger MM, Schormann W, Walldorf J, Hengstler JG, Fleig WE, Christ B (2007). Functional integration of hepatocytes derived from human mesenchymal stem cells into mouse livers. Gut, 56(3): 405-415.

Avital I, Inderbitzin D, Aoki T, Tyan DB, Cohen AH, Ferraresso C, Rozga J, Arnaout WS, Demetriou AA (2001). Isolation, characterization, and transplantation of bone marrow-derived hepatocyte stem cells. Biochem. Biophy. Res. Commun., 288(1): 156-164.

Baddoo M, Hill K, Wilkinson R, Gaupp D, Hughes C, Kopen GC, Phinney DG (2003). Characterization of mesenchymal stem cells isolated from murine bone marrow by negative selection. J. Cell Biochem., 89(6): 1235-1249.

Baksh D, Song L, Tuan RS (2004). Adult mesenchymal stem cells: characterization, differentiation, and application in cell and gene therapy. J. Cell Mol. Med., 8(3): 301-316.

Balana B, Nicoletti C, Zahanich I, Graf EM, Christ T, Boxberger S, Ravens U (2006). 5-Azacytidine induces changes in electrophysiological properties of human mesenchymal stem cells. Cell Res., 16(12): 949-960.

Baron M (2001). Induction of embryonic hematopoietic and endothelial stem/progenitor cells by hedgehog-mediated signals. Differ., 68(4-5): 175-185.

Barry FP, Murphy JM (2004). Mesenchymal stem cells: clinical applications and biological characterization. Int. J. Biochem. Cell Biol., 36(4): 568-584.

Battula VL, Bareiss PM, Treml S, Conrad S, Albert I, Hojak S, Abele H, Schewe B, Just L, Skutella T, Buhring HJ (2007). Human placenta and bone marrow derived MSC cultured in serum-free, b-FGF-containing medium express cell surface frizzled-9 and SSEA-4 and give rise to multilineage differentiation. Differ., 75(4): 279-291.

Battula VL, Treml S, Bareiss PM, Gieseke F, Roelofs H, de Zwart P, Muller I, Schewe B, Skutella T, Fibbe WE, Kanz L, Buhring HJ (2009). Isolation of functionally distinct mesenchymal stem cell subsets using antibodies against CD56, CD271, and mesenchymal stem cell antigen-1. Haematol., 94(2): 173-184.

Benzhi C, Limei Z, Ning W, Jiaqi L, Songling Z, Fanyu M, Hongyu Z, Yanjie L, Jing A, Baofeng Y (2009). Bone marrow mesenchymal stem cells upregulate transient outward potassium currents in postnatal rat ventricular myocytes. J. Mol. Cell Cardiol., 47(1): 41-48.

Bergh JJ, Shao Y, Puente E, Duncan RL, Farach-Carson MC (2006). Osteoblast Ca(2+) permeability and voltage-sensitive Ca(2+) channel expression is temporally regulated by 1,25-dihydroxyvitamin D(3). Am. J. Physiol. Cell Physiol., 290(3): C822-831.

Bey E, Prat M, Duhamel P, Benderitter M, Brachet M, Trompier F, Battaglini P, Ernou I, Boutin L, Gourven M, Tissedre F, Crea S, Mansour CA, de Revel T, Carsin H, Gourmelon P, Lataillade JJ (2010). Emerging therapy for improving wound repair of severe radiation burns using local bone marrow-derived stem cell administrations. Wound Repair. Regen., 18(1): 50-58.

Bianco P, Gehron Robey P (2000). Marrow stromal stem cells. J. Clin. Invest., 105(12): 1663-1668.

Bonewald LF (2004). Osteocyte biology: its implications for osteoporosis. J. Musculoskelet Neuronal Interact 4(1): 101-104.

Bonnet P, Awede B, Rochefort GY, Mirza A, Lermusiaux P, Domenech J, Eder V (2008). Electrophysiological maturation of rat mesenchymal stem cells after induction of vascular smooth muscle cell differentiation in vitro. Stem Cells Dev., 17(6): 1131-1140.

Bosch P, Pratt SL, Stice SL (2006). Isolation, characterization, gene modification, and nuclear reprogramming of porcine mesenchymal stem cells. Biol. Reprod., 74(1): 46-57.

Bosnakovski D, Mizuno M, Kim G, Takagi S, Okumura M, Fujinaga T (2005). Isolation and multilineage differentiation of bovine bone marrow mesenchymal stem cells. Cell Tissue Res., 319(2): 243-253.

Brazelton TR, Rossi FM, Keshet GI, Blau HM (2000). From marrow to brain: expression of neuronal phenotypes in adult mice. Sci., 290(5497): 1775-1779.

Brooke G, Tong H, Levesque JP, Atkinson K (2008). Molecular trafficking mechanisms of multipotent mesenchymal stem cells derived from human bone marrow and placenta. Stem Cells Dev., 17(5): 929-940.

Brown SD, Piantadosi CA (1990). In vivo binding of carbon monoxide to cytochrome c oxidase in rat brain. J. Appl. Physiol., 68(2): 604-610.

Buhring HJ, Treml S, Cerabona F, de Zwart P, Kanz L, Sobiesiak M (2009). Phenotypic characterization of distinct human bone marrow-derived MSC subsets. Ann. N. Y. Acad. Sci., 1176: 124-134.

Campagnoli C, Roberts IA, Kumar S, Bennett PR, Bellantuono I, Fisk NM (2001). Identification of mesenchymal stem/progenitor cells in human first-trimester fetal blood, liver, and bone marrow. Blood 98(8): 2396-2402.

Canalis E, Economides AN, Gazzerro E (2003). Bone morphogenetic proteins, their antagonists, and the skeleton. Endocr. Rev., 24(2): 218-235.

Caplan AI, Dennis JE (2006). Mesenchymal stem cells as trophic mediators. J. Cell Biochem., 98(5): 1076-1084.

Castro-Malaspina H, Gay RE, Resnick G, Kapoor N, Meyers P, Chiarieri D, McKenzie S, Broxmeyer HE, Moore MA (1980). Characterization of human bone marrow fibroblast colony-forming cells (CFU-F) and their progeny. Blood 56(2): 289-301.

Chen X, Macica CM, Ng KW, Broadus AE (2005). Stretch-induced PTH-related protein gene expression in osteoblasts. J. Bone Miner Res., 20(8): 1454-1461.

Chunmeng S, Tianmin C (2004). Effects of plastic-adherent dermal multipotent cells on peripheral blood leukocytes and CFU-GM in rats. Transplant Proc., 36(5): 1578-1581.

Covas DT, Piccinato CE, Orellana MD, Siufi JL, Silva WAJr, Proto-Siqueira R, Rizzatti EG, Neder L, Silva AR, Rocha V, Zago MA (2005). Mesenchymal stem cells can be obtained from the human saphena vein. Exp. Cell Res., 309(2): 340-344.

da Silva Meirelles L, Chagastelles PC, Nardi NB (2006). Mesenchymal stem cells reside in virtually all post-natal organs and tissues. J. Cell Sci., 119(Pt 11): 2204-2213.

Dan YY, RiehleKJ, Lazaro C, Teoh N, Haque J, Campbell JS, Fausto N (2006). Isolation of multipotent progenitor cells from human fetal liver capable of differentiating into liver and mesenchymal lineages. Proc. Natl. Acad. Sci USA., 103(26): 9912-9917.

Dazzi F, Ramasamy R, Glennie S, Jones SP, Roberts I (2006). The role of mesenchymal stem cells in haemopoiesis. Blood Rev., 20(3): 161-171.

De Ugarte DA, Alfonso Z, Zuk PA, Elbarbary A, Zhu M, Ashjian P, Benhaim P, Hedrick MH, Fraser JK (2003a). Differential expression of stem cell mobilization-associated molecules on multi-lineage cells from adipose tissue and bone marrow. Immunol. Lett., 89(2-3): 267-270.

De Ugarte DA, Morizono K, Elbarbary A, Alfonso Z, Zuk PA, Zhu M, Dragoo JL, Ashjian P, Thomas B, Benhaim P, Chen I, Fraser J, Hedrick MH (2003b). Comparison of multi-lineage cells from human adipose tissue and bone marrow. Cells Tissues Organs., 174(3): 101-109.

Deng W, Obrocka M, Fischer I, Prockop DJ (2001). In vitro differentiation of human marrow stromal cells into early progenitors of neural cells by conditions that increase intracellular cyclic AMP. Biochem. Biophy. Res. Commun., 282(1): 148-152.

Dennis JE, Merriam A, Awadallah A, Yoo JU, Johnstone B, Caplan AI (1999). A quadripotential mesenchymal progenitor cell isolated from the marrow of an adult mouse. J. Bone Miner. Res., 14(5): 700-709.

Devine SM, Bartholomew AM, Mahmud N, Nelson M, Patil S, Hardy W, Sturgeon C, Hewett T, Chung T, Stock W, Sher D, Weissman S, Ferrer K, Mosca J, Deans R, Moseley A, Hoffman R (2001). Mesenchymal stem cells are capable of homing to the bone marrow of non-human primates following systemic infusion. Exp. Hematol., 29(2): 244-255.

Eberhardt M, Salmon P, von Mach MA, Hengstler JG, Brulport M, Linscheid P, Seboek D, Oberholzer J, Barbero A, Martin I, Muller B, Trono D, Zulewski H (2006). Multipotential nestin and Isl-1 positive mesenchymal stem cells isolated from human pancreatic islets. Biochem. Biophys. Res. Commun., 345(3): 1167-1176.

Erices A, Conget P, Minguell JJ (2000). Mesenchymal progenitor cells in human umbilical cord blood. Br. J. Haematol., 109(1): 235-242.

Erices AA, Allers CI, Conget PA, Rojas CV, Minguell JJ (2003). Human cord blood-derived mesenchymal stem cells home and survive in the marrow of immunodeficient mice after systemic infusion. Cell Transplant 12(6): 555-561.

Flores-Figueroa E, Arana-Trejo RM, Gutierrez-Espindola G, Perez-Cabrera A, Mayani H (2005). Mesenchymal stem cells in myelodysplastic syndromes: phenotypic and cytogenetic characterization. Leuk. Res., 29(2): 215-224.

Garcia Ruiz PJ, Ruiz Ezquerro JJ, Garcia Torres A, Fanjul S (2006). Ancient descriptions of movement disorders: Cathedral el Burgo de Osma (Soria, Spain). J. Neurol., 253(6): 731-734.

Gharzi A, Reynolds AJ, Jahoda CA (2003). Plasticity of hair follicle dermal cells in wound healing and induction. Exp. Dermatol., 12(2): 126-136.

Gomi T, Kimura A, Adriaensen D, Timmermans JP, Scheuermann DW, De Groodt-Lasseel MH, Kitazawa Y, Kikuchi Y, Naruse H, Kishi K (1994). Stages in the development of the rat lung: morphometric, light and electron microscopic studies. Kaibogaku Zasshi 69(4): 392-405.

Gronthos S, Simmons PJ (1995). The growth factor requirements of STRO-1-positive human bone marrow stromal precursors under serum-deprived conditions in vitro. Blood 85(4): 929-940.

Gronthos S, Simmons PJ (1996). The biology and application of human bone marrow stromal cell precursors. J. Hematother., 5(1): 15-23.

Gu Y, Preston MR, El Haj AJ, Howl JD, Publicover SJ (2001). Three types of K(+) currents in murine osteocyte-like cells (MLO-Y4). Bone 28(1): 29-37.

Gupta S (2002). Hepatocyte transplantation. J Gastroenterol Hepatol 17 Suppl 3: S287-293.

Haynesworth SE, Baber MA, Caplan AI (1992). Cell surface antigens on human marrow-derived mesenchymal cells are detected by monoclonal antibodies. Bone 13(1): 69-80.

Herrera MB, Bruno S, Buttiglieri S, Tetta C, Gatti S, Deregibus MC, Bussolati B, Camussi G (2006). Isolation and characterization of a stem cell population from adult human liver. Stem Cells 24(12): 2840-2850.

Herzog EL, Chai L, Krause DS (2003). Plasticity of marrow-derived stem cells. Blood 102(10): 3483-3493.

Hess D, Li L, Martin M, Sakano S, Hill D, Strutt B, Thyssen S, Gray DA, Bhatia M (2003). Bone marrow-derived stem cells initiate pancreatic regeneration. Nat. Biotechnol., 21(7): 763-770.

Heubach JF, Graf EM, Leutheuser J, Bock M, Balana B, Zahanich I, Christ T, Boxberger S, Wettwer E, Ravens U (2004). Electrophysiological properties of human mesenchymal stem cells. J. Physiol., 554(Pt 3): 659-672.

Howell JC, Lee WH, Morrison P, Zhong J, Yoder MC, Srour EF (2003). Pluripotent stem cells identified in multiple murine tissues. Ann. N. Y. Acad. Sci., 996: 158-173.

Hughes S, Magnay J, Foreman M, Publicover SJ, Dobson JP, El Haj AJ (2006). Expression of the mechanosensitive 2PK+ channel TREK-1 in human osteoblasts. J. Cell Physiol., 206(3): 738-748.

Igura K, Zhang X, Takahashi K, Mitsuru A, Yamaguchi S, Takashi TA (2004). Isolation and characterization of mesenchymal progenitor cells from chorionic villi of human placenta. Cytotherapy., 6(6): 543-553.

Ishii M, Koike C, Igarashi A, Yamanaka K, Pan H, Higashi Y, Kawaguchi H, Sugiyama M, Kamata N, Iwata T, Matsubara T, Nakamura K, Kurihara H, Tsuji K, Kato Y (2005). Molecular markers distinguish bone marrow mesenchymal stem cells from fibroblasts. Biochem. Biophys. Res. Commun., 332(1): 297-303.

Jager M, Krauspe R (2007). Antigen expression of cord blood derived stem cells under osteogenic stimulation in vitro. Cell Biol. Int., 31(9): 950-957.

Jiang Y, Jahagirdar BN, Reinhardt RL, Schwartz RE, Keene CD, Ortiz-Gonzalez XR, Reyes M, Lenvik T, Lund T, Blackstad M, Du J, Aldrich S, Lisberg A, Low WC, Largaespada DA, Verfaillie CM (2002). Pluripotency of mesenchymal stem cells derived from adult marrow. Nat., 418(6893): 41-49.

Kanemura Y (2010). Development of cell-processing systems for human stem cells (neural stem cells, mesenchymal stem cells, and iPS cells) for regenerative medicine. Keio J. Med., 59(2): 35-45.

Kang JW, Kang KS, Koo HC, Park JR, Choi EW, Park YH (2008). Soluble factors-mediated immunomodulatory effects of canine adipose tissue-derived mesenchymal stem cells. Stem Cells Dev., 17(4): 681-693.

Katz AJ, Tholpady A, Tholpady SS, Shang H, Ogle RC (2005). Cell surface and transcriptional characterization of human adipose-derived adherent stromal (hADAS) cells. Stem Cells 23(3): 412-423.

Kawano S, Shoji S, Ichinose S, Yamagata K, Tagami M, Hiraoka M (2002). Characterization of Ca(2+) signaling pathways in human mesenchymal stem cells. Cell Calcium., 32(4): 165-174.

Kershaw EE, Flier JS (2004). Adipose tissue as an endocrine organ. J. Clin. Endocrinol. Metab., 89(6): 2548-2556.

Kopen GC, Prockop DJ, Phinney DG (1999). Marrow stromal cells migrate throughout forebrain and cerebellum, and they differentiate into astrocytes after injection into neonatal mouse brains. Proc. Natl. Acad. Sci. USA., 96(19): 10711-10716.

Kuznetsov SA, Mankani MH, Gronthos S, Satomura K, Bianco P, Robey PG (2001). Circulating skeletal stem cells. J. Cell Biol., 153(5): 1133-1140.

Lange C, Bassler P, Lioznov MV, Bruns H, Kluth D, Zander AR, Fiegel HC (2005). Hepatocytic gene expression in cultured rat mesenchymal stem cells. Transplant. Proc., 37(1): 276-279.

Latsinik NV, Gorskaia Iu F, Grosheva AG, Domogatskii SP, Kuznetsov SA (1986). [The stromal colony-forming cell (CFUf) count in the bone marrow of mice and the clonal nature of the fibroblast colonies they form]. Ontogenez 17(1): 27-36.

Lee OK, Kuo TK, Chen WM, Lee KD, Hsieh SL, Chen TH (2004). Isolation of multipotent mesenchymal stem cells from umbilical cord blood. Blood 103(5): 1669-1675.

Li J, Liu D, Ke HZ, Duncan RL, Turner CH (2005). The P2X7 nucleotide receptor mediates skeletal mechanotransduction. J. Biol. Chem., 280(52): 42952-42959.

Li WL, Su, Yao YC, Tao XR, Yan YB, Yu HY, Wang XM, Li JX, Yang YJ, Lau JT, JHu YP (2006). Isolation and characterization of bipotent liver progenitor cells from adult mouse. Stem Cells 24(2): 322-332.

Liu F, Akiyama Y, Tai S, Maruyama K, Kawaguchi Y, Muramatsu K, Yamaguchi K (2008). Changes in the expression of CD106, osteogenic genes, and transcription factors involved in the osteogenic differentiation of human bone marrow mesenchymal stem cells. J. Bone Miner. Metab., 26(4): 312-320.

Lo L, Sommer L, Anderson DJ (1997). MASH1 maintains competence for BMP2-induced neuronal differentiation in post-migratory neural crest cells. Curr. Biol., 7(6): 440-450.

Loeffler M, Roeder I (2002). Tissue stem cells: definition, plasticity, heterogeneity, self-organization and models--a conceptual approach. Cells Tissues Organs., 171(1): 8-26.

Loeffler M, Roeder I (2004). Conceptual models to understand tissue stem cell organization. Curr. Opin. Hematol., 11(2): 81-87.

Magra M, Hughes S, El Haj AJ, Maffulli N (2007). VOCCs and TREK 1 ion channel expression in human tenocytes. Am. J. Physiol. Cell Physiol., 292(3): C1053-1060.

Mansilla E, Marin GH, Drago H, Sturla F, Salas E, Gardiner C, Bossi S, Lamonega R, Guzman A, Nunez A, Gil MA, Piccinelli G, Ibar R, Soratti C (2006). Bloodstream cells phenotypically identical to human mesenchymal bone marrow stem cells circulate in large amounts under the influence of acute large skin damage: new evidence for their use in regenerative medicine. Transplant. Proc., 38(3): 967-969.

Mareschi K, Rustichelli D, Comunanza V, De Fazio R, Cravero C, Morterra G, Martinoglio B, Medico E, Carbone E, Benedetto C, Fagioli F (2009). Multipotent mesenchymal stem cells from amniotic fluid originate neural precursors with functional voltage-gated sodium channels. Cytotherapy., 11(5): 534-547.

Marom R, Shur I, Solomon R, Benayahu D (2005). Characterization of adhesion and differentiation markers of osteogenic marrow stromal cells. J. Cell Physiol., 202(1): 41-48.

Martin DR, Cox NR, Hathcock TL, Niemeyer GP, Baker HJ (2002). Isolation and characterization of multipotential mesenchymal stem cells from feline bone marrow. Exp. Hematol., 30(8): 879-886.

McCormick DP, Ivey FMJr, Gold DM, Zimmerman DM, Gemma S, Owen MJ (1988). The preparticipation sports examination in Special Olympics athletes. Tex. Med., 84(4): 39-43.

Miao Z, Jin J, Chen L, Zhu J, Huang W, Zhao J, Qian H, Zhang X (2006). Isolation of mesenchymal stem cells from human placenta: comparison with human bone marrow mesenchymal stem cells. Cell Biol. Int., 30(9): 681-687.

Moriscot C, de Fraipont F, Richard MJ, Marchand M, Savatier P, Bosco D, Favrot M, Benhamou PY (2005). Human bone marrow mesenchymal stem cells can express insulin and key transcription factors of the endocrine pancreas developmental pathway upon genetic and/or microenvironmental manipulation in vitro. Stem Cells 23(4): 594-603.

Moscoso I, Centeno A, Lopez E, Rodriguez-Barbosa JI, Santamarina I, Filgueira P, Sanchez MJ, Dominguez-Perles R, Penuelas-Rivas G, Domenech N (2005). Differentiation "in vitro" of primary and immortalized porcine mesenchymal stem cells into cardiomyocytes for cell transplantation. Transplant. Proc., 37(1): 481-482.

Noonan KJ, Stevens JW, Tammi R, Tammi M, Hernandez JA, Midura RJ (1996). Spatial distribution of CD44 and hyaluronan in the proximal tibia of the growing rat. J. Orthop. Res., 14(4): 573-581.

O'Donoghue K, Fisk NM (2004). Fetal stem cells. Best Pract. Res. Clin. Obstet. Gynaecol., 18(6): 853-875.

Oh SH, Miyazaki M, Kouchi H, Inoue Y, Sakaguchi M, Tsuji T, Shima N, Higashio K, Namba M (2000). Hepatocyte growth factor induces differentiation of adult rat bone marrow cells into a hepatocyte lineage in vitro. Biochem. Biophys. Res. Commun., 279(2): 500-504.

Okumoto K, Saito T, Hattori E, Ito JI, Adachi T, Takeda T, Sugahara K, Watanabe H, Saito K, Togashi H, Kawata S (2003). Differentiation of bone marrow cells into cells that express liver-specific genes in vitro: implication of the Notch signals in differentiation. Biochem. Biophys. Res. Commun., 304(4): 691-695.

Olsen BR, Reginato AM, Wang W (2000). Bone development. Annu. Rev. Cell Dev. Biol., 16: 191-220.

Orlic D, Hill JM, Arai AE (2002). Stem cells for myocardial regeneration. Circ. Res., 91(12): 1092-1102.

Otto TC, Lane MD (2005). Adipose development: from stem cell to adipocyte. Crit. Rev. Biochem. Mol. Biol., 40(4): 229-242.

Owen M, Friedenstein AJ (1988). Stromal stem cells: marrow-derived osteogenic precursors. Ciba. Found Symp., 136: 42-60.

Panepucci RA, Siufi JL, Silva WAJr, Proto-Siquiera R, Neder L, Orellana M, Rocha V, Covas DT, Zago MA (2004). Comparison of gene expression of umbilical cord vein and bone marrow-derived mesenchymal stem cells. Stem Cells 22(7): 1263-1278.

Parekkadan B, Sethu P, van Poll D, Yarmush ML, Toner M (2007). Osmotic selection of human mesenchymal stem/progenitor cells from umbilical cord blood. Tissue Eng., 13(10): 2465-2473.

Park KS, Choi MR, Jung KH, Kim S, Kim HY, Kim KS, Cha EJ, Kim Y, Chai YG (2008). Diversity of ion channels in human bone marrow mesenchymal stem cells from amyotrophic lateral sclerosis patients. Korean J. Physiol. Pharmacol., 12(6): 337-342.

Park KS, Jung KH, Kim SH, Kim KS, Choi MR, Kim Y, Chai YG (2007). Functional expression of ion channels in mesenchymal stem cells derived from umbilical cord vein. Stem Cells 25(8): 2044-2052.

Petersen BE, Bowen WC, Patrene KD, Mars WM, Sullivan AK, Murase N, Boggs SS, Greenberger JS, Goff JP (1999). Bone marrow as a potential source of hepatic oval cells. Sci., 284(5417): 1168-1170.

Phinney DG, Kopen G, Isaacson RL, Prockop DJ (1999). Plastic adherent stromal cells from the bone marrow of commonly used strains of inbred mice: variations in yield, growth, and differentiation. J. Cell Biochem., 72(4): 570-585.

Pierdomenico L, Bonsi L, Calvitti M, Rondelli D, Arpinati M, Chirumbolo G, Becchetti E, Marchionni C, Alviano F, Fossati V, Staffolani N, Franchina M, Grossi A, Bagnara GP (2005). Multipotent mesenchymal stem cells with immunosuppressive activity can be easily isolated from dental pulp. Transplant., 80(6): 836-842.

Pittenger MF, Mackay AM, Beck SC, Jaiswal RK, Douglas R, Mosca JD, Moorman MA, Simonetti DW, Craig S, Marshak DR (1999). Multilineage potential of adult human mesenchymal stem cells. Sci., 284(5411): 143-147.

Prockop DJ (1997). Marrow stromal cells as stem cells for nonhematopoietic tissues. Sci., 276(5309): 71-74.

Raisz LG, Pilbeam CC, Fall PM (1993). Prostaglandins: mechanisms of action and regulation of production in bone. Osteoporos Int. 3 Suppl., 1: 136-140.

Ramoshebi LN, Matsaba TN, Teare J, Renton L, Patton J, Ripamonti U (2002). Tissue engineering: TGF-beta superfamily members and delivery systems in bone regeneration. Expert Rev. Mol. Med., 4(20): 1-11.

Rawadi G, Vayssiere B, Dunn F, Baron R, Roman-Roman S (2003). BMP-2 controls alkaline phosphatase expression and osteoblast mineralization by a Wnt autocrine loop. J. Bone Miner. Res., 18(10): 1842-1853.

Reyes-Botella C, Montes MJ, Vallecillo-Capilla MF, Olivares EG, Ruiz C (2000). Expression of molecules involved in antigen presentation and T cell activation (HLA-DR, CD80, CD86, CD44 and CD54) by cultured human osteoblasts. J. Periodontol., 71(4): 614-617.

Reynolds BA, Weiss S (1992). Generation of neurons and astrocytes from isolated cells of the adult mammalian central nervous system. Sci. 255(5052): 1707-1710.

Ringe J, Haupl T, Sittinger M (2003). [Mesenchymal stem cells for tissue engineering of bone and cartilage]. Med. Klin Munich 98 Suppl., 2: 35-40.

Rochefort GY, Delorme B, Lopez A, Herault O, Bonnet P, Charbord P, Eder V, Domenech J (2006). Multipotential mesenchymal stem cells are mobilized into peripheral blood by hypoxia. Stem Cells 24(10): 2202-2208.

Rochefort GY, Vaudin P, Bonnet N, Pages JC, Domenech J, Charbord P, Eder V (2005). Influence of hypoxia on the domiciliation of mesenchymal stem cells after infusion into rats: possibilities of targeting pulmonary artery remodeling via cells therapies? Respir. Res., 6: 125.

Rogers JJ, Young HE, Adkison LR, Lucas PA, Black ACJr (1995). Differentiation factors induce expression of muscle, fat, cartilage, and bone in a clone of mouse pluripotent mesenchymal stem cells. Am. Surg., 61(3): 231-236.

Romanov YA, Svintsitskaya VA, Smirnov VN (2003). Searching for alternative sources of postnatal human mesenchymal stem cells: candidate MSC-like cells from umbilical cord. Stem Cells 21(1): 105-110.

Rosa AL, Beloti MM (2003). TAK-778 enhances osteoblast differentiation of human bone marrow cells. J. Cell Biochem., 89(6): 1148-1153.

Rose RA, Jiang H, Wang X, Helke S, Tsoporis JN, Gong N, Keating SC, Parker TG, Backx PH, Keating A (2008). Bone marrow-derived mesenchymal stromal cells express cardiac-specific markers, retain the stromal phenotype, and do not become functional cardiomyocytes in vitro. Stem Cells 26(11): 2884-2892.

Rosen ED, MacDougald OA (2006). Adipocyte differentiation from the inside out. Nat. Rev. Mol. Cell Biol., 7(12): 885-896.

Sanchez-Ramos J, Song S, Cardozo-Pelaez F, Hazzi C, Stedeford T, Willing A, Freeman TB, Saporta S, Janssen W, Patel N, Cooper DR, Sanberg PR (2000). Adult bone marrow stromal cells differentiate into neural cells in vitro. Exp. Neurol., 164(2): 247-256.

Santa Maria L, Rojas CV, Minguell JJ (2004). Signals from damaged but not undamaged skeletal muscle induce myogenic differentiation of rat bone-marrow-derived mesenchymal stem cells. Exp. Cell Res., 300(2): 418-426.

Schwartz RE, Reyes M, Koodie L, Jiang Y, Blackstad M, Lund T, Lenvik T, Johnson S, Hu WS, Verfaillie CM (2002). Multipotent adult progenitor cells from bone marrow differentiate into functional hepatocyte-like cells. J. Clin. Invest., 109(10): 1291-1302.

Seeberger KL, Dufour JM, Shapiro AM, Lakey JR, Rajotte RV, Korbutt GS (2006). Expansion of mesenchymal stem cells from human pancreatic ductal epithelium. Lab. Invest 86(2): 141-153.

Selim AA, Mahon M, Juppner H, Bringhurst FR, Divieti P (2006). Role of calcium channels in carboxyl-terminal parathyroid hormone receptor signaling. Am. J. Physiol. Cell Physiol., 291(1): C114-121.

Seo MJ, Suh SY, Bae YC, Jung JS (2005). Differentiation of human adipose stromal cells into hepatic lineage in vitro and in vivo. Biochem. Biophys. Res. Commun., 328(1): 258-264.

Seshi B, Kumar S, King D (2003). Multilineage gene expression in human bone marrow stromal cells as evidenced by single-cell microarray analysis. Blood Cells Mol. Dis., 31(2): 268-285.

Shih DT, Lee DC, Chen SC, Tsai RY, Huang CT, Tsai CC, Shen EY, Chiu WT (2005). Isolation and characterization of neurogenic mesenchymal stem cells in human scalp tissue. Stem Cells 23(7): 1012-1020.

Silva GV, Litovsky S, Assad JA, Sousa AL, Martin BJ, Vela D, Coulter SC, Lin J, Ober J, Vaughn WK, Branco RV, Oliveira EM, He R, Geng YJ, Willerson JT, Perin EC (2005). Mesenchymal stem cells differentiate into an endothelial phenotype, enhance vascular density, and improve heart function in a canine chronic ischemia model. Circul., 111(2): 150-156.

Silva WAJr, Covas DT, Panepucci RA, Proto-Siqueira R, Siufi JL, Zanette DL, Santos AR, Zago MA (2003). The profile of gene expression of human marrow mesenchymal stem cells. Stem Cells 21(6): 661-669.

Simmons PJ, Masinovsky B, Longenecker BM, Berenson R, Torok-Storb B, Gallatin WM (1992). Vascular cell adhesion molecule-1 expressed by bone marrow stromal cells mediates the binding of hematopoietic progenitor cells. Blood 80(2): 388-395.

Simmons PJ, Torok-Storb B (1991). Identification of stromal cell precursors in human bone marrow by a novel monoclonal antibody, STRO-1. Blood 78(1): 55-62.

Song L, Baksh D, Tuan RS (2004). Mesenchymal stem cell-based cartilage tissue engineering: cells, scaffold and biology. Cytotherapy., 6(6): 596-601.

Soria B (2001). In-vitro differentiation of pancreatic beta-cells. Differentiation 68(4-5): 205-219.

Sottile, V., A. Thomson and J. McWhir (2003). In vitro osteogenic differentiation of human ES cells. Cloning Stem Cells 5(2): 149-155.

Sykaras N, Opperman LA (2003). Bone morphogenetic proteins (BMPs): how do they function and what can they offer the clinician? J. Oral Sci., 45(2): 57-73.

Taguchi Y, Yamamoto M, Yamate T, Lin SC, Mocharla H, DeTogni P,

Nakayama N, Boyce BF, Abe E, Manolagas SC (1998). Interleukin-6-type cytokines stimulate mesenchymal progenitor differentiation toward the osteoblastic lineage. Proc. Assoc. Am. Phys., 110(6): 559-574.

Talens-Visconti R, Bonora A, Jover R, Mirabet V, Carbonell F, Castell JV, Gomez-Lechon MJ (2006). Hepatogenic differentiation of human mesenchymal stem cells from adipose tissue in comparison with bone marrow mesenchymal stem cells. World J. Gastroenterol., 12(36): 5834-5845.

Tao R, Lau CP, Tse HF, Li GR (2008). Regulation of cell proliferation by intermediate-conductance Ca2+-activated potassium and volume-sensitive chloride channels in mouse mesenchymal stem cells. Am. J. Physiol. Cell Physiol., 295(5): C1409-1416.

Timper K, Seboek D, Eberhardt M, Linscheid P, Christ-Crain M, Keller U, Muller B, Zulewski H (2006). Human adipose tissue-derived mesenchymal stem cells differentiate into insulin, somatostatin, and glucagon expressing cells. Biochem. Biophys. Res. Commun., 341(4): 1135-1140.

Tobon-Arroyave SI, Franco-Gonzalez LM, Isaza-Guzman DM, Florez-Moreno GA, Bravo-Vasquez T, Castaneda-Pelaez DA, Vieco-Duran B (2005). Immunohistochemical expression of RANK, GRalpha and CTR in central giant cell granuloma of the jaws. Oral Oncol., 41(5): 480-488.

Togel F, Hu Z, Weiss K, Isaac J, Lange C, Westenfelder C (2005). Administered mesenchymal stem cells protect against ischemic acute renal failure through differentiation-independent mechanisms. Am. J. Physiol. Renal. Physiol., 289(1): F31-42.

Toma C, Pittenger MF, Cahill KS, Byrne BJ, Kessler PD (2002). Human mesenchymal stem cells differentiate to a cardiomyocyte phenotype in the adult murine heart. Circ., 105(1): 93-98.

Tropel P, Noel D, Platet N, Legrand P, Benabid AL, Berger F (2004). Isolation and characterisation of mesenchymal stem cells from adult mouse bone marrow. Exp. Cell Res., 295(2): 395-406.

Tsai MS, Lee JL, Chang YJ, Hwang SM (2004). Isolation of human multipotent mesenchymal stem cells from second-trimester amniotic fluid using a novel two-stage culture protocol. Hum. Reprod., 19(6): 1450-1456.

Verfaillie CM (2002). Adult stem cells: assessing the case for pluripotency. Trends. Cell Biol., 12(11): 502-508.

Verfaillie CM, Pera MF, Lansdorp PM (2002). Stem cells: hype and reality. Hematol. Am. Soc. Hematol. Edu. Program.: 369-391.

Villaron EM, Almeida J, Lopez-Holgado N, Alcoceba M, Sanchez-Abarca LI, Sanchez-Guijo FM, Alberca M, Perez-Simon JA, San Miguel JF, Del Canizo MC (2004). Mesenchymal stem cells are present in peripheral blood and can engraft after allogeneic hematopoietic stem cell transplantation. Haematol., 89(12): 1421-1427.

Vogel G (2003). Biotechnology. Stem cells lose market luster. Sci., 299(5614): 1830-1831.

Wang M, Yang Y, Yang D, Luo F, Liang W, Guo S, Xu J (2009). The immunomodulatory activity of human umbilical cord blood-derived mesenchymal stem cells in vitro. Immunol., 126(2): 220-232.

Wang SP, Wang JA, Luo RH, Cui WY, Wang H (2008). Potassium channel currents in rat mesenchymal stem cells and their possible roles in cell proliferation. Clin. Exp. Pharmacol. Physiol., 35(9): 1077-1084.

Wehner T, Bontert M, Eyupoglu I, Prass K, Prinz M, Klett FF, Heinze M, Bechmann I, Nitsch R, Kirchhoff F, Kettenmann H, Dirnagl U, Priller J (2003). Bone marrow-derived cells expressing green fluorescent protein under the control of the glial fibrillary acidic protein promoter do not differentiate into astrocytes in vitro and in vivo. J. Neurosci., 23(12): 5004-5011.

Weinreb M, Grosskopf A, Shir N (1999). The anabolic effect of PGE2 in rat bone marrow cultures is mediated via the EP4 receptor subtype. Am. J. Physiol., 276(2 Pt 1): E376-383.

Weiss ML, Anderson C, Medicetty S, Seshareddy KB, Weiss RJ, VanderWerff I, Troyer D, McIntosh KR (2008). Immune properties of human umbilical cord Wharton's jelly-derived cells. Stem Cells 26(11): 2865-2874.

Wobus AM, Boheler KR (2005). Embryonic stem cells: prospects for developmental biology and cell therapy. Physiol. Rev., 85(2): 635-678.

Worster AA, Nixon AJ, Brower-Toland BD, Williams J (2000). Effect of transforming growth factor beta1 on chondrogenic differentiation of cultured equine mesenchymal stem cells. Am. J. Vet. Res., 61(9): 1003-1010.

Wu SN, Jan CR, Chiang HT (2001). Fenamates stimulate BKCa channel osteoblast-like MG-63 cells activity in the human. J. Investig. Med., 49(6): 522-533.

Yamada Y, Boo JS, Ozawa R, Nagasaka T, Okazaki Y, Hata K, Ueda M (2003). Bone regeneration following injection of mesenchymal stem cells and fibrin glue with a biodegradable scaffold. J. Craniomaxillofac. Surg., 31(1): 27-33.

Yamasaki T, Deie M, Shinomiya R, Izuta Y, Yasunaga Y, Yanada S, Sharman P, Ochi M (2005). Meniscal regeneration using tissue engineering with a scaffold derived from a rat meniscus and mesenchymal stromal cells derived from rat bone marrow. J. Biomed. Mater. Res. A., 75(1): 23-30.

Yellowley CE, Hancox JC, Skerry TM, Levi AJ (1998). Whole-cell membrane currents from human osteoblast-like cells. Calcif. Tissue Int., 62(2): 122-132.

Yoshimura H, Muneta T, Nimura A, Yokoyama A, Koga H, Sekiya I (2007). Comparison of rat mesenchymal stem cells derived from
bone marrow, synovium, periosteum, adipose tissue, and muscle. Cell Tissue Res., 327(3): 449-462.

Ypey DL, Weidema AF, Hold KM, Van der Laarse A, Ravesloot JH, Van Der Plas A, Nijweide PJ (1992). Voltage, calcium, and stretch activated ionic channels and intracellular calcium in bone cells. J. Bone Miner. Res. 7 Suppl., 2: S377-387.

Zahanich I, Graf EM, Heubach JF, Hempel U, Boxberger S, Ravens U (2005). Molecular and functional expression of voltage-operated calcium channels during osteogenic differentiation of human mesenchymal stem cells. J. Bone Miner. Res., 20(9): 1637-1646.

Zeng R, Wang LW, Hu ZB, Guo WT, Wei JS, Lin H, Sun X, Chen LX, Yang LJ (2011). Differentiation of human bone marrow mesenchymal stem cells into neuron-like cells in vitro. Spine (Phila Pa 1976) 36(13): 997-1005.

Zhang Y, Li CD, Jiang XX, Li HL, Tang PH, Mao N (2004). Comparison of mesenchymal stem cells from human placenta and bone marrow. Chin. Med. J., (Engl) 117(6): 882-887.

Zhao LX, Zhang J, Cao F, Meng L, Wang DM, Li YH, Nan X, Jiao WC, Zheng M, Xu XH, Pei XT (2004). Modification of the brain-derived neurotrophic factor gene: a portal to transform mesenchymal stem cells into advantageous engineering cells for neuroregeneration and neuroprotection. Exp. Neurol., 190(2): 396-406.

Zuk PA, Zhu M, Mizuno H, Huang J, Futrell JW, Katz AJ, Benhaim P, Lorenz HP, Hedrick MH (2001). Multilineage cells from human adipose tissue: implications for cell-based therapies. Tissue Eng., 7(2): 211-228.

Zvaifler NJ, Marinova-Mutafchieva L, Adams G, Edwards CJ, Moss J, Burger JA, Maini RN (2000). Mesenchymal precursor cells in the blood of normal individuals. Arthritis, Res., 2(6): 477-488.

26

Round-bottomed Honeycomb Microwells: Embryoid body shape correlates with stem cell fate

Silin Sa[1], Diep T. Nguyen[3], Jonathan D. Pegan[2], Michelle Khine[3] and Kara E. McCloskey[1,2]*

[1]Graduate Group in Biological Engineering and Small-scale Technologies, University of California, Merced, USA.
[2]School of Engineering, University of California, Merced, USA.
[3]Department of Biomedical Engineering, University of California, Irvine, USA.

The differentiation of embryonic stem cells (ESC) into tissue-specific cells utilizes either monolayer cultures or three-dimensional cell aggregates called embryoid bodies (EB). However, the generation of a large number of EB of controlled sizes can be challenging and labor intensive. Our laboratories have developed a simple, robust, ultra-rapid, and inexpensive design of Honeycomb Microwells for generation of EB. Here, we compare EB generated using (1) Honeycomb Microwells, (2) the commercially available AggreWell™400, and (3) the more traditional Hanging Drop method. We compared the efficiency, viability, quality, and control of EB sizes. Results indicate that the Honeycomb Microwell and AggreWell™400 efficiently generate small EB at approximately 500 cells per EB. However, the cone-bottomed AggreWell plate generates cone-shaped EB at 1000-2000 cells per EB. Moreover, the cone-shape correlates with a reduction in the formation of the primitive endoderm GATA-4+ cells (1% compared with 6-8% in spherical EB), but does not significantly affect mesoderm or ectoderm development. We conclude that the non-spherical EB shape correlates with a reduction in the development of primitive endoderm, and that use of these AggreWell plates should be avoided in deriving endoderm tissue products.

Key words: Embryonic stem cells, embryoid bodies, Microwell, AggreWell, hanging drop, primitive endoderm, GATA-4.

INTRODUCTION

Embryonic stem cells (ESC) and induced-pluripotent stem cells (iPSC) are both excellent *in vitro* cell culture systems for studying stem cell fate. These cells are especially attractive due to their unlimited *in vitro* expansion potential (Amit et al., 2000) while also maintaining the capacity to differentiate into a variety of cell types (Odorico et al., 2001). Moreover, once differentiation methodologies are developed and fully optimized, these stem cell sources could provide the specialized tissue-specific cells needed for a variety of applications in regenerative medicine. The most common methods of ESC differentiation require either the induction on monolayer cultures or the generation of 3-D structures, called embryoid bodies (EBs) (Itskovitz-Eldor et al., 2000), which recapitulate various aspects of embryogenesis. Subsequently, many methodologies for initial induction of ESC towards different tissue-specific cell fates require EB formation, reviewed in (Desbaillets et al., 2000).

Unfortunately, the generation of high numbers of reproducible EB is not trivial. EB can vary in size when formed using the traditional liquid suspension or methylcellulose culture (Kurosawa, 2007) methods for EB generation, and thus, affect the downstream stem cell differentiation products (Park et al., 2007). For the production of larger numbers of EB, stirred-suspension cultures using spinner flasks and bioreactors are usually employed (Dang et al., 2004; Dang and Zandstra, 2005; Hwang et al., 2009a; Lock and Tzanakakis, 2009;

*Corresponding author. E-mail: kmccloskey@ucmerced.edu.

Schroeder et al., 2005) even though these suspension cultures often gives rise to EB populations which are also heterogeneous in morphology and thus, differentiation capacity (Choi et al., 2010; Hwang et al., 2009b; Valamehr et al., 2008). The labor intensive hanging drop can yield relatively uniform EB, but this method requires culturing EB in droplets of medium that are less than 20 µL in size, making medium exchange and hence, long-term culture extremely difficult (Dang et al., 2004). The 96-well plate has also been used successfully to generate EB (Koike et al., 2005). This well-plate remains labor intensive, but does allow medium exchange for longer-term cultures. Lastly, it should be noted that the generation of uniform EB from human ESC is additionally challenging compared with mouse ESC due to the apoptotic nature of human ESC when dispersed into single cell suspensions.

Recently, STEMCELL Technologies introduced the AggreWell™400, the first commercially available plate for the generation of standardized EB from ESC and iPS cells. This product was followed by the AggreWell™800, which is a plate containing larger sized-wells for generating larger EB. Together, these plates allow the generation of large quantities of uniform-sized EB. Although the wells of the AggreWell are cone-shaped, and not spherical, the system allows very nice control over the size of EB by altering the number of cells added to each well.

In differentiation studies using the AggreWell™400, it was observed the larger EB would retain the cone-shaped morphology after transfer from the well plate into suspension culture. We suspected that this shape might affect the stem cell fate of the EB, and therefore, set out to study the germ layer development in the cone-shaped EB compared with the traditional hanging drop. We also include our laboratory's round-bottomed Honeycomb Microwells (Chen et al., 2008; Nguyen et al., 2009) and compared the efficiency, viability, aspect ratio, circularity, and control over EB sizes between the three methods of EB generation.

Results will show that the Honeycomb Microwell and the AggreWell plates are both able to efficiency generate EB of well-controlled sizes, but the round-bottomed Honeycomb Microwell generates more circular-shaped EB. Additionally, the cone-shaped EB from the AggreWell plate exhibit a significant reduction in the formation of the primitive endoderm in the EB.

METHODS

Murine embryonic stem cell culture

E14 mouse ESC transfected with green fluorescence protein (GFP) expression linked to the myosin heavy chain promoter (courtesy of Conklin Lab, UCSF) were maintained in Knockout Dulbecco's modification of Eagle Medium (DMEM: Gibco) supplemented with 15% Knockout Serum Replacement (KSR; Gibco), 100 mg/mL of penicillin-streptomycin (Invitrogen), 1 mM L-glutamine (Gibco), 0.1

mM nonessential amino acids (NEAA; Invitrogen), 0.1 mM betamercaptoethanol (Calbiochem) and 1000 U/mL leukemia inhibitory factor (LIF; Chemicon) and plated on tissue cultured plates coated with 0.1% Gelatin (Sigma-Aldrich). When the mouse ESC on the tissue culture plates were 80% confluent, they were detached using trypsin/EDTA (0.1%/1 mM), and dissociated into single cells using a pipette to assure uniform distribution of cells during the loading process. Mouse ESC were then spun down and re-suspended in differentiation medium, which has the same composition as ESC medium with the exclusion of LIF and supplemented with 20% KSR and 10 ng/ml bone morphogenic protein4 (BMP4; R&D System).

Human embryonic stem cell culture

H9 human ESC were maintained in mouse embryonic fibroblast (MEF)-conditioned medium supplemented with 5 ng/ml basic fibroblast growth factor (bFGF, Sigma) and plated on tissue cultured plates coated with Matrigel (Becton Dickenson). When the human ESC on the tissue culture plates were 80% confluent, they were detached using Accutase (Invitrogen), and dissociated into single cells using a pipette to assure uniform distribution of cells during the loading process. The human ESC were then spun down and re-suspended in high glucose Dulbecco's Modification of Eagle Medium (DMEM; Gibco) supplemented with 20% Knockout Serum Replacement (KSR; Gibco), and 25 ng/ml bone morphogenic protein-4 (BMP-4; R&D System). The Y-27632 ROCK inhibitor, final concentration of 10 µM, was added during EB formation to enhance cell survival during EB formation.

Fabrication of Honeycomb Microwells

Our ultra-rapid fabrication method for generating Honeycomb Microwells has been previously published (Chen et al., 2008; Nguyen et al., 2009). Briefly, a laser-jet printer was used to print desired patterns onto pre-stressed polystyrene (PS) sheets. These sheets were then heated to 155°C for approximately 5 m to induce shrinkage, forming closely arrayed Microwells (Honeycomb Microwells) of tunable sizes, and PDMS molded onto the PS master forming Microwells. After removing PDMS from micromolds, the Microwells were bonded to glass slides (to prevent floating) and inserted into standard 24-well culture plates.

Generation of embryoid bodies using Honeycomb Microwell

To load cells, the bottom of the Microwell was bonded to a piece of cover glass (Fisherbrand) using an O_2 plasma machine (SPI Supplies) and placed into each well of a standard 24-well plate containing 500 µL of differentiation medium. The initial 500 µL assisted in preventing air bubbles within the well and enhanced the adherence of the cover glass to the plate. Next an additional 500 µL of differentiation medium was placed into the well and was pipetted gently to remove any remaining air bubbles on the PDMS surface. ESC was added at concentrations of 1.75×10^5 cells/ml concentration to the 200, 300 and 400-µm Honeycomb Microwells for EB approximately 150, 250, and 500 cells each. For larger EB with 1000 and 2000 each, we used 3×10^5 cells per ml and 6×10^5 cells per ml concentration respectively in the 400-µm Honeycomb Microwell. The 1.75×10^5 cells/ml concentration was also used for the 200 and 300 µm Honeycomb Microwells. To achieve uniform EB size, 1 mL of the ESCs were then gently pipetted and dispensed drop-wise into each well of a 24-well plate (Chen et al., 2008; Nguyen et al., 2009). To prevent convective effects within each well of the 24-well plate, which may disrupt the uniform distribution, ESC were allowed to settle into the Honeycomb Microwells at room temperature for 15–30 m before being transferred into the incubator.

Generation of embryoid bodies using AggreWell™

The manufacturer's instructions from the AggreWell manual version 1.0.0 were initially followed for generating EB. Accordingly, 1 mL of differentiation medium was placed into each well of an AggreWell™400 plate (Stem Cell Technologies) and then centrifuged the AggreWell™400 plate at 3000g for 10 m in a swinging bucket rotor that was fitted with a plate holder to remove any small bubbles from the AggreWell. ESC were then added at concentrations of 3×10^5 cells/ml, 6×10^5 cells/ml, and 1.2×10^6 cells/ml to each well to make EBs with 500, 1000, and 2000 cells respectively. The AggreWell™400 plate was centrifuged at 200 g for 3 m to capture the cells in the wells. According to instructions, aggregates should be harvested 24 h after adding the ESC to the AggreWell™400 plate. Because some aggregates tended to break up if transferred to suspension culture after only 24 h, we increased this time to 48 h in the AggreWell plate before transfer to suspension culture.

Generation of embryoid bodies using hanging drops

For the generation of EB with 500, 1000, and 2000 cells using hanging drops, approximately 100,000, 200,000, and 400,000 ESC were suspended in 3 mL of differentiation medium without LIF respectively. Using a micro-multi-channel pipette, micro-drops were pipetted onto the lid of a culture dish. The lid was inverted and placed on the culture dish. Within 24 h, the ESC in the drops aggregated into the initial stage of EB formation.

Verification of the number of cells per EB

On day 2, 100 EB were collected from each Honeycomb Microwell, AggreWell, and Hanging Drop plates. The 100 EB in each group were pooled together and then dissociated into single cells. The total number of cells from each of the 100 EB-pooled samples was counted using a hemacytometer, and then this number was divided by 100 to calculate the average number of cells per EB.

Characterization of EB size and shape

EB were formed via Microwell, AggreWell, and Hanging Drop methods and imaged at 24 and 48 h after seeding. The EB were imaged in live mode with a digital camera attached to an inverted microscope (Fisher Scientific) and operated by imaging software (Micron, Westover Scientific). From these images, the diameters and perimeters were measured from 100 randomly selected EB formed from each method. From this data, the shape factor, aspect ratio, and circularity of the EB were also calculated.

Cell viability

EB formed by Microwell, AggreWell, and Hanging Drop methods were harvested after 48 h. Cell viability was determined by using a live/dead kit for mammalian cells (Invitrogen). Both whole EB and single cells dissociated from the EB were incubated in 2 M calcein-acetoxymethyl ester (AM) and 4 M ethidium homodimer in phosphate buffer saline (PBS) for 20 min at 37°C. After the incubation period, the stained cells were analyzed by Flow Cytometer LSR II (Becton Dickenson) and whole EBs imaged by laser scanning confocal microscopy. Live and dead cells were indicated by calcein AM (green) and ethidium homodimer (red) respectively.

Confocal microscopy

Laser scanning confocal microscopy was performed on a Nikon digital eclipse C1 confocal microscope equipped with a Nikon eclipse TE2000U inverted microscope using a 20× air objective for imaging. Laser beams with 488 nm excitation wavelengths and 515/30 nm band pass (BP) emission filters were used for Calcein AM and GATA-4 labeled cells. The EBs were also stained with mouse anti-GATA 4-Alexa Fluor® 488 following permeablization with 0.7% Triton X100 for 15 m. Laser beams with 543nm excitation wavelengths and 590/50 nm BP emission filters were used for ethidium homodimer-labeled EBs. Laser beams with 633 nm excitation wavelengths and 650 nm long pass emission filters were used for Draq5 labeled EBs. For EBs approximately 180 μm in diameter, typically 45 images were acquired at 4 μm☐ slice intervals, each slice being the average of three laser scans. Z-stacked images were processed by using the Java-based image analysis program ImageJ (http://rsb.info.nih.gov/ij/). Z-projected images were assembled for each time point to produce a single image based on the sum of pixel brightness values through the image stack (ImageJ: z-project).

Flow cytometry

Individual cells were isolated from EB with cell dissociation buffer and stained for calcein-AM (green) and ethidium homodimer (red) for analyzing live and dead cells within the EB. Cells were also stained with mouse anti-nestin-PE (R&D System), mouse anti-brachyury-PE (R&D System) and mouse anti-GATA 4-Alexa Fluor® 488 (Becton Dickenson) for 60 m after fixation and permeablization with 0.7% Triton X100 for 15 m.

Statistics

For statistical significance, data was acquired and pooled from 100 EB for each treatment group and all treatment groups for generating EB were repeated at least 3 times each. Statistical significance was measure using a Student's T-test individually between all groups.

RESULTS

According to the technical manual version 1.0.0 for the AggreWell plate, EB may be harvested from the AggreWell plate after 24 h. However, after only 24 h, the cells had not yet aggregated sufficiently to be removed as EB (Figures 1A and B) and often broke apart into smaller clusters once transferred to suspension cultures (not shown). Therefore, we also examined the ESC aggregates after 48 h in both AggreWell and Honeycomb Microwell and found that, after 48 h, the cells had aggregated enough to be removed and transferred into suspension cultures without breaking apart (Figures 1C and D). Additionally, we calculated aspect ratio of the EB in the wells using the diameters of the long- and short-axes for both the AggreWell and Honeycomb Microwell (Figure 1E). Both the AggreWell and Honeycomb

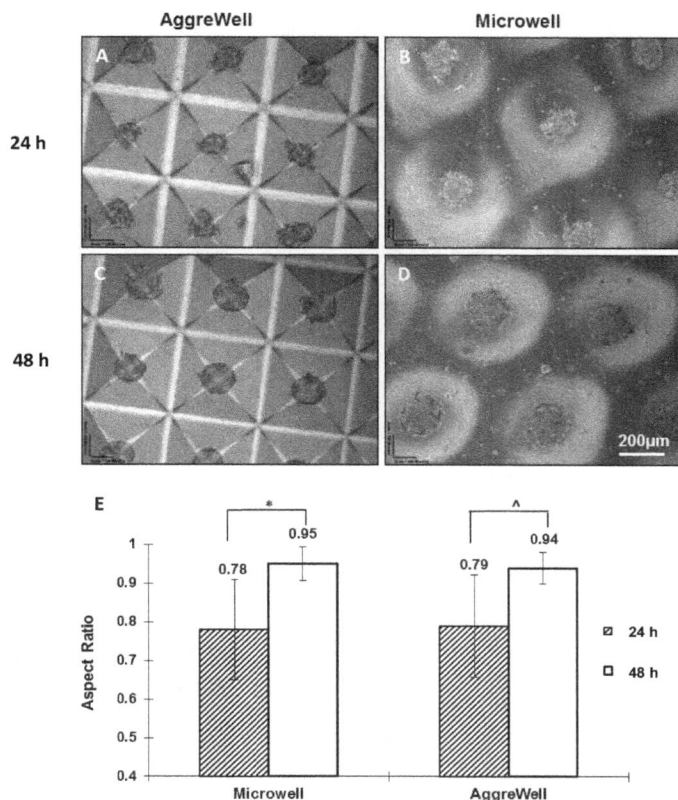

Figure 1. The aspect ratios of EB formed in the Honeycomb Microwell and AggreWell plates increase after 48 h compared with after only 24 h. Images of the EB formed in the AggreWell plates and the 400-μm sized Honeycomb Microwells. (**A**) Cell aggregates in AggreWell 24 h after seeding. (**B**) Cell aggregates in Honeycomb Microwell 24 h after seeding. (**C**) Cell aggregates in AggreWell 48 h after seeding. (**D**) Cell aggregates in Honeycomb Microwell 48 h after seeding. Note that the cells seeded 48 h prior formed larger aggregates that appear to be more circular compared with after only 24 h. (**E**) Aspect ratios of the EBs were also calculated. The long- and short-axis diameters were measured from EBs formed via AggreWell and Honeycomb Microwell. No significant differences were observed between the calculated aspect ratios for AggreWell and Honeycomb Microwell. Note that after 48 h, the aspect ratios of the EBs formed using both Honeycomb Microwell and AggreWell are both approaching 1, indicating a more circular formation compared with after only 24 h (* and ^ indicate comparisons with P < 0.0001).

Microwell are able to generate equivalent EB, with the EB generated from both methods exhibiting a more circular shape after 48 h compared with 24 h (P < 0.0001).

Circularity of EB

After transferring EB into suspension culture, we noticed morphological differences for larger sized EB ~ 1000 cells per EB (Figure 2). Specifically, the larger EB formed using the AggreWell plate appeared to retain the cone-shape from the bottom of the well plate (Figure 2E).

Moreover, the X-large EB in the AggreWell (~2000 cells per EB) were even more densely packed with increased cone-shapes (Figure 2H).

We then calculated the circularity, C, a more sensitive measurement than aspect ratio, for the 2 day old EB following transfer into suspension culture. The cross-sectional area and perimeter of each cell aggregate was acquired using Image J. From each image (Figure 3A), a high contrast black and white image (Figure 3B) was generated. Image J then created a high resolution outline of the cell aggregates (Figure 3C), and running this outline through a mean filter then smoothed the outlines of the cell aggregates (Figure 3 D and E). A radius of 20 pixels was used in order to remove excess surface aberrations in the aggregate outlines. This allows one to measure the general shape of the cell aggregate without incorporating the natural surface topology of the EB. Cross-sectional area, A, and perimeter, P, values were finally measured from these smoothed outlines and then used to calculate the circularity of individual cell aggregates (Figure 3 F and G).

$$C = 4\pi * [A/(P*r^2)] \tag{1}$$

The circularity calculations indicate that the Honeycomb Microwell and AggreWell plates generate more circular-shaped EB compared with the Hanging Drop for the smaller sized-EB, but that the Honeycomb Microwell generated more circular EB of the larger sized EB.

Controlling EB sizes

The diameters of the EB were measured after 2 and 7 days for the three different sized Honeycomb Microwells (Figure 4A), and using the Honeycomb Microwell, AggreWell, and Hanging Drop to generate both the smaller EB (500 cells; Figure 4B) and the larger EB (1000 cells; Figure 4C). All three methods of EB formation generated EB with well-controlled sizes (Figures 4A and B) as evident from the small standard deviations in these EB. The EB were then moved to suspension cultures and allowed to grow and differentiate for another 5 days. The diameters of the EB were measured again after 7 total days in culture. After 5 days with unconstrained growth, the variation in sizes of the EB had increased, but has remained proportional to the original EB sizes.

Cell Viability in EB

Cell viabilities in EB after 2 days of growth in the Honeycomb Microwell, AggreWell, and Hanging Drop were measured using a live/dead (red/green) staining kit. Z-stacked confocal microscopy images revealed a high number of live cells (green) compared with dead (red) cells (Figures 5A, B, and C). The EB were then

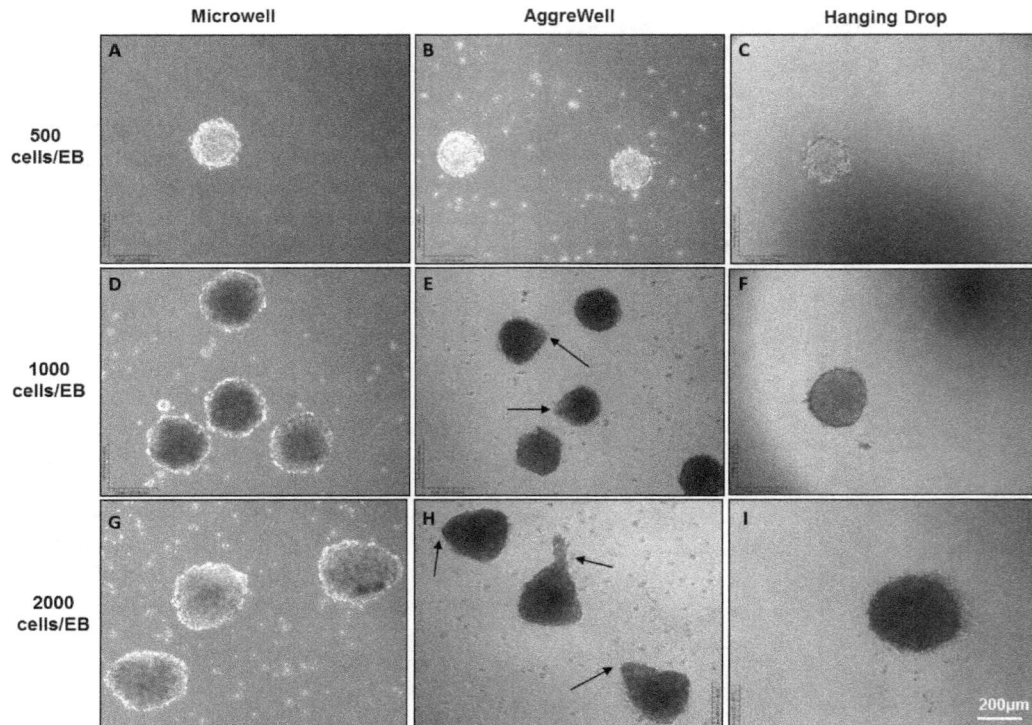

Figure 2. Representative images depicting the morphology of EB flipped into suspension after 48 h. The methods for generating EB with 500- and 1000-cell EB were optimized for the Honeycomb Microwell, AggreWell and Hanging Drop and imaged after transferring cells into suspension culture after 48 h. Images depict EB at 500 cells per EB generated using the A) Honeycomb Microwell, B) AggreWell and C) Hanging Drop, and the larger EB at 1000 cells per EB generated using the D) Honeycomb Microwell, E) AggreWell and F) Hanging Drop. The smaller EB with 500-cell EB appear to be approximately equivalent in size, shape, and cell density for all three generation methods, whereas the EB containing 1000 cells per EB generated using the AggreWell appear to be more densely packed, smaller in size, and exhibit cone-shaped peaks (indicated by arrows) that reflect the cone-shaped morphology of the wells in the AggreWell plate. EB were also generated at 2000 cells per EB in G) Honeycomb Microwells, H) AggreWell plates, and I) Hanging Drops in order to test the limits of the well plates. Note that at 2000 cells per EB, the AggreWell plate generated smaller, denser, and more distinctly cone-shaped (indicated by arrows) EB compared with the other methods.

disaggregated, stained and analyzed by flow cytometry, plotted as histograms expressing the calcein fluorescent dye indicating live cells (not shown). The data indicates that after 48 h, the EB formed in the AggreWell and Honeycomb Microwell still contained greater than 95% live cells (Figure 5D).

Primitive mesoderm, ectoderm, and endoderm in EB

While evaluating our phase contrast images, we also noticed that the large and extra-large EB formed using the cone-shaped bottomed AggreWell plate (Figures 2E and H respectively) did not generate the same distinct primitive endoderm layer on the exterior surface seen in the small EB (Figures 2A to C) or the larger EB formed in the Microwell (Figure 2D) and the Hanging Drop (Figure 2F). Therefore, we stained the large EB for the presence

of primitive endoderm marker, GATA-4, on 2 day EB and found that the EB formed by AggreWell (Figure 6B) expressed endoderm marker GATA-4 only at the cone-shaped tips while the EB formed by Microwell and Hanging Drop expressed the primitive endoderm maker evenly throughout the surface of the EB (Figures 6A and C).

We proceeded to quantitatively examine the development all three germ layers at day 2 and 7, using nestin to identify the primitive ectoderm and brachyury for mesoderm differentiation. The percentage of cells expressing the early mesoderm marker, brachyury, ranged from 42-47%, but did not differ significantly between EB formed from Microwell, AggreWell, or Hanging Drop (Figures 6D and E). The percentage of cells expressing, nestin, the marker for primitive ectoderm, was less than 50% in the AggreWell on day 2, compared with the Microwell and Hanging Drop

F	Microwell	AggreWell	Hanging Drop
500 cells per EB	505 ± 16	51 2± 9	502 ± 33
Seeding Concentration (cells/ml)	1.75×10^5	3×10^5	0.33×10^5
Circularity	0.82 ± 0.03	0.80 ± 0.06	0.77 ± 0.011
1000 cells per EB	969 ± 2	1200 ± 28	967 ± 47
Seeding Concentration (Cells/ml)	3×10^5	6×10^5	0.67×10^5
Circularity	0.83 ± 0.04	0.73 ± 0.05	0.76 ± 0.10

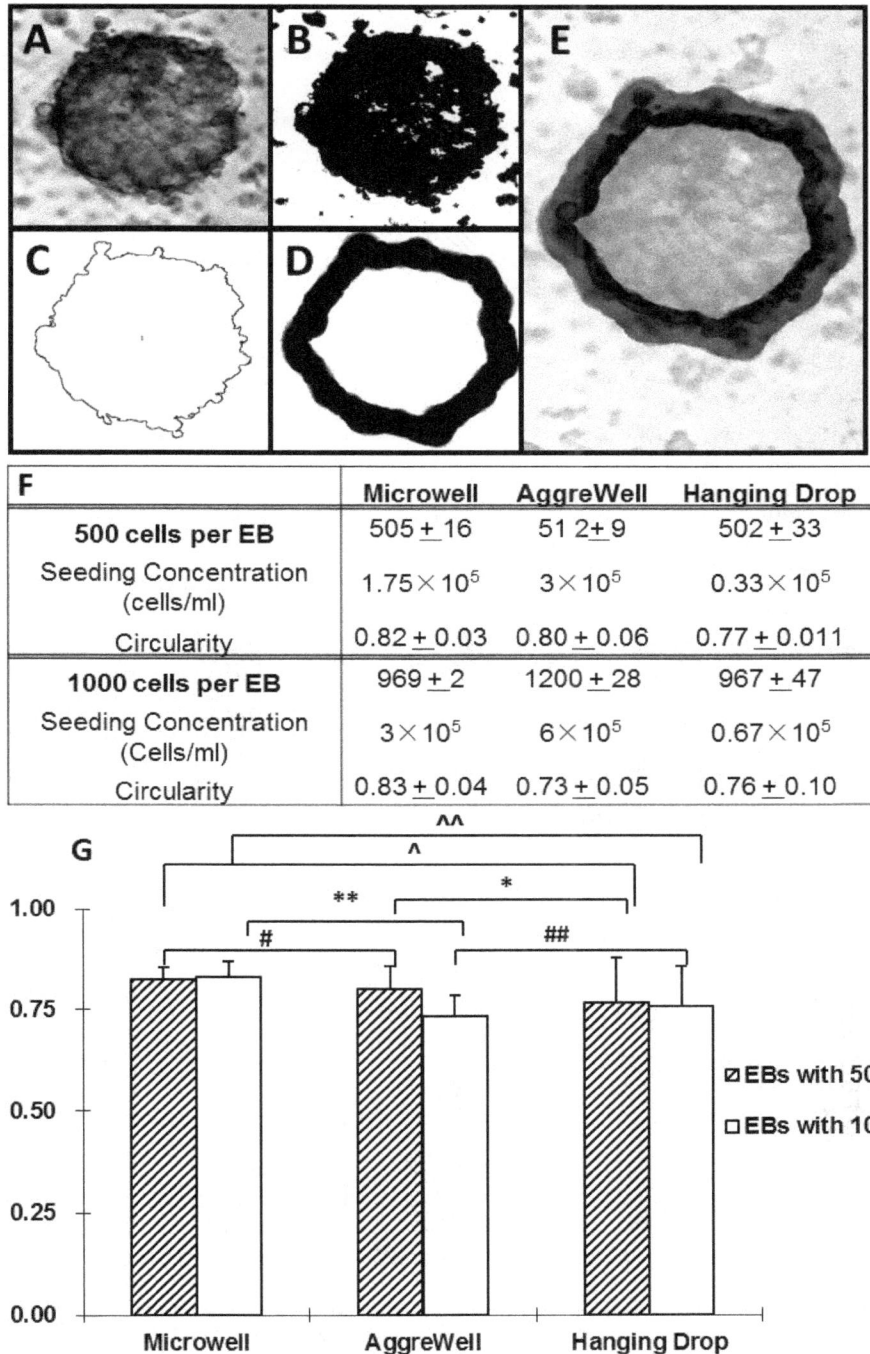

Figure 3. The EB formed in the Honeycomb Microwell are more circular compared with EB formed in AggreWell and Hanging Drop. Images of EB were further processed for determining the circularity, a more sensitive measurement of roundness, using ImageJ software. Images include: A) bright field image, B) high contrast image, C) rendered outline of cell aggregate, D) outline modified with mean filter, and E) composite image of original bright field image with modified outline overlay. Scale bar is 50 µm. F) Tables indicating cell numbers (with standard deviations in cell numbers), seeding concentration and circularity values for the targeted 500 cells/EB and 1000 cells/EB. G) Graph depicting the circularity values calculated from the area and perimeter measurements of the inside of the modified outline. The data indicates that at 500-cell EB, the Microwell (comparison noted by ^, $P<0.0001$) and AggreWell (*, $P< 0.0001$) are both superior to the Hanging Drop, and the Microwell is superior to the AggreWell (#, $P<0.01$) while at 1000 cells per EB, the Microwell forms more circular EB compared with both the AggreWell (**, $P<0.0001$) and Hanging Drop (^^, $P < 0.0001$), but the Hanging Drop is superior to the AggreWell (##, $P <0.01$).

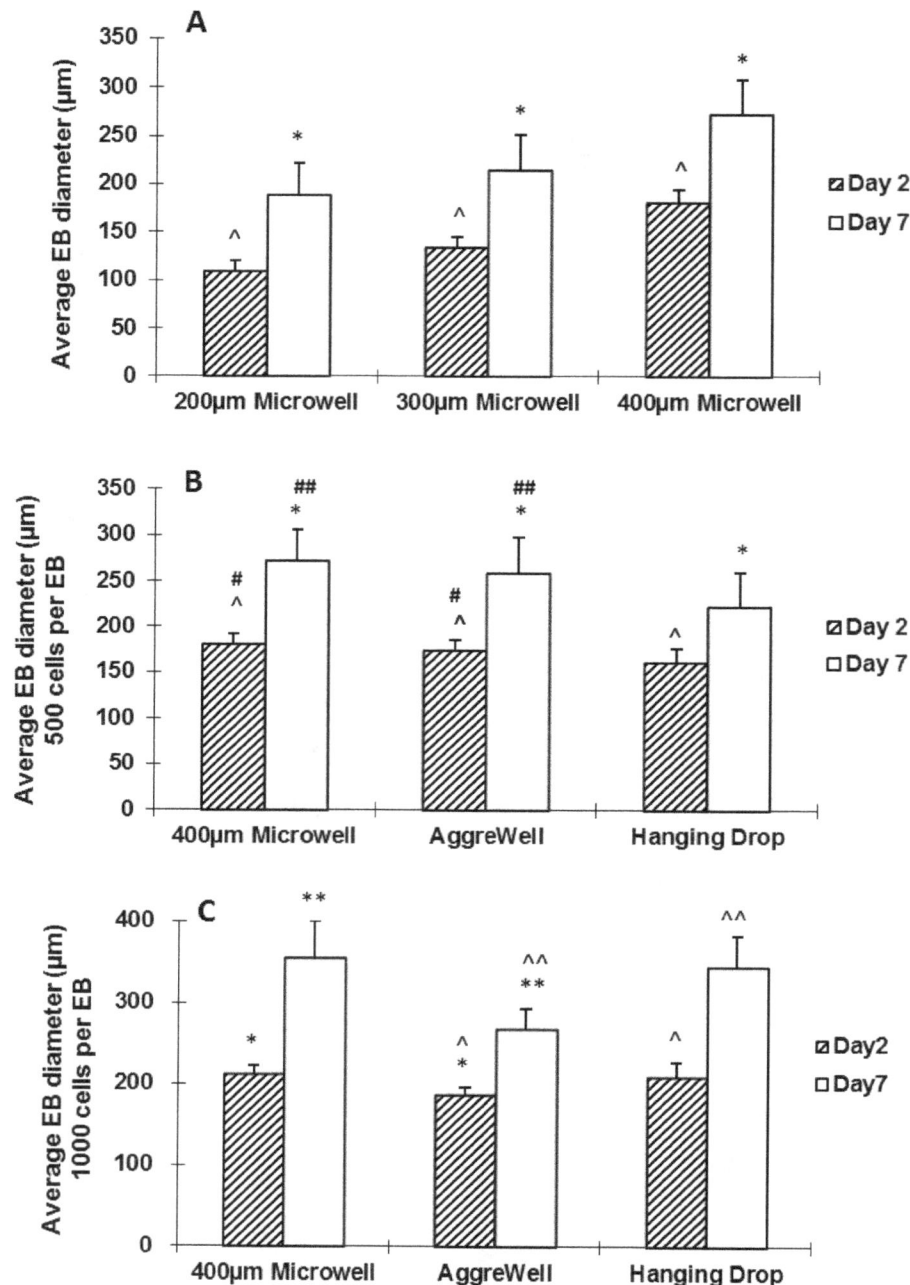

Figure 4. Honeycomb Microwell, AggreWell and Hanging Drop all generate well-controlled sized EB. The diameters of the EB generated in the **A)** 200-, 300- and 400-µm sized Honeycomb Microwells indicate show that the size of the Microwells dictate the size of EBs, **B)** small 500-cell EB and large 1000-cell EB were then measured following generation in the 400 µm sized Honeycomb Microwells, AggreWells, and Hanging Drops. The EB were moved to suspension cultures after 2 days and allowed to grow further. The diameters of the EB were measured again after 7 days. The differences between all diameters were statistically significant (Figure A and B; ^, P<0.0001, except differences between the 400µm-sized Microwell and AggreWell (noted by # were significant at P<0.001) on day 2. At day 7, all differences were statistic (Figure A and B, *, P<0.0001, except differences between the 400µm-sized Microwell and AggreWell were significant at ##, P<0.1). More importantly, the small standard deviations in the sizes indicate the ability for all methods to control the EB sizes by adjusting cell numbers. **C)** We also examined the EB sizes for the larger EB using only the largest 400 µm-sized Microwell. We noted that the diameter of the EB generated in the AggreWell plate were smaller than the EB in the Microwell at both day 2 and 7 (* and **, P<0.0001) or Hanging Drop (^ and ^^, P<0.0001).

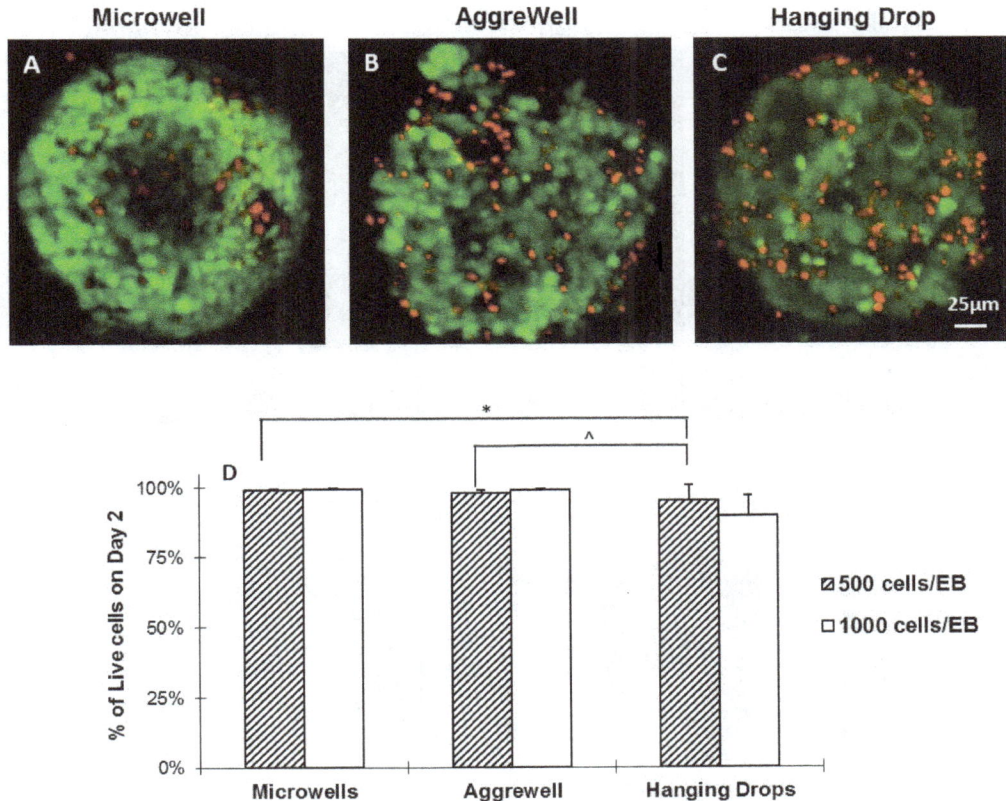

Figure 5. The EB formed in the Honeycomb Microwell and AggreWell, and Hanging Drops contain viable cells. The larger 1000-cell EB generated using A) 400-μm sized Honeycomb Microwells, B) AggreWell plates, and C) Hanging Drops were stained with calcein-acetoxymethyl ester (green) for live cells and ethidium homodimer (red) for dead cells on day 2 and imaged using z-stacked confocal microscopy. In order to quantify the numbers of live and dead cells, the small 500-cell EB and large 1000-cell EB were dissociated into single cells and analyzed using a flow cytometer. D) The viability (% of live cells) in EB formed via AggreWell, Honeycomb Microwell and Hanging Drops was consistently greater than 90% for all methods. The differences in viability of cells in EB between formation methods was not generally significant, but cells in the EB formed in the Microwell (*, $P<0.1$) and AggreWell (^, $P<0.1$) both contained more viable cells compared with the cells in the EB formed via Hanging Drop.

expressing over 55% nestin positive cells (Figure 6F). However, the nestin expression by day 7 was comparable in the EB from Microwells and AggreWells at 50%, and slightly lower in the Hanging Drop (Figure 6G). Conversely, the expression of primitive endoderm marker, GATA-4, was significantly lower in the EB formed in AggreWell plates on day 2 (Figure 6G). Although the cone-shaped morphology of the EB in the AggreWell do recover after transfer to suspension cultures (not shown), the GATA-4 development of the primitive endoderm in these EB does not recover and remains below 1% compared with 6-8% expression in EB from Microwell and Hanging Drop (Figure 6I).

EB from human ESC

Perhaps most importantly, the Microwell and AggreWell plates are able to generate EB from single human ESC with greater efficiency than conventional Hanging Drop Methods (Figure 7). This is a significant advancement in the field, as the generation of human EB from single cells is notoriously challenging. We expect the success of these well plates in generating EB from single cells is due to the physical aggregation of individual cells that bring the cells into closer contact with one another compared with a Hanging Drop.

DISCUSSION AND CONCLUSIONS

The formation of the EB is a principal step in the differentiation of ESC. When factors that maintain the undifferentiated state of an ESC are removed from the culture medium and placed in suspension culture, the ESC will spontaneously self-assemble into an EB

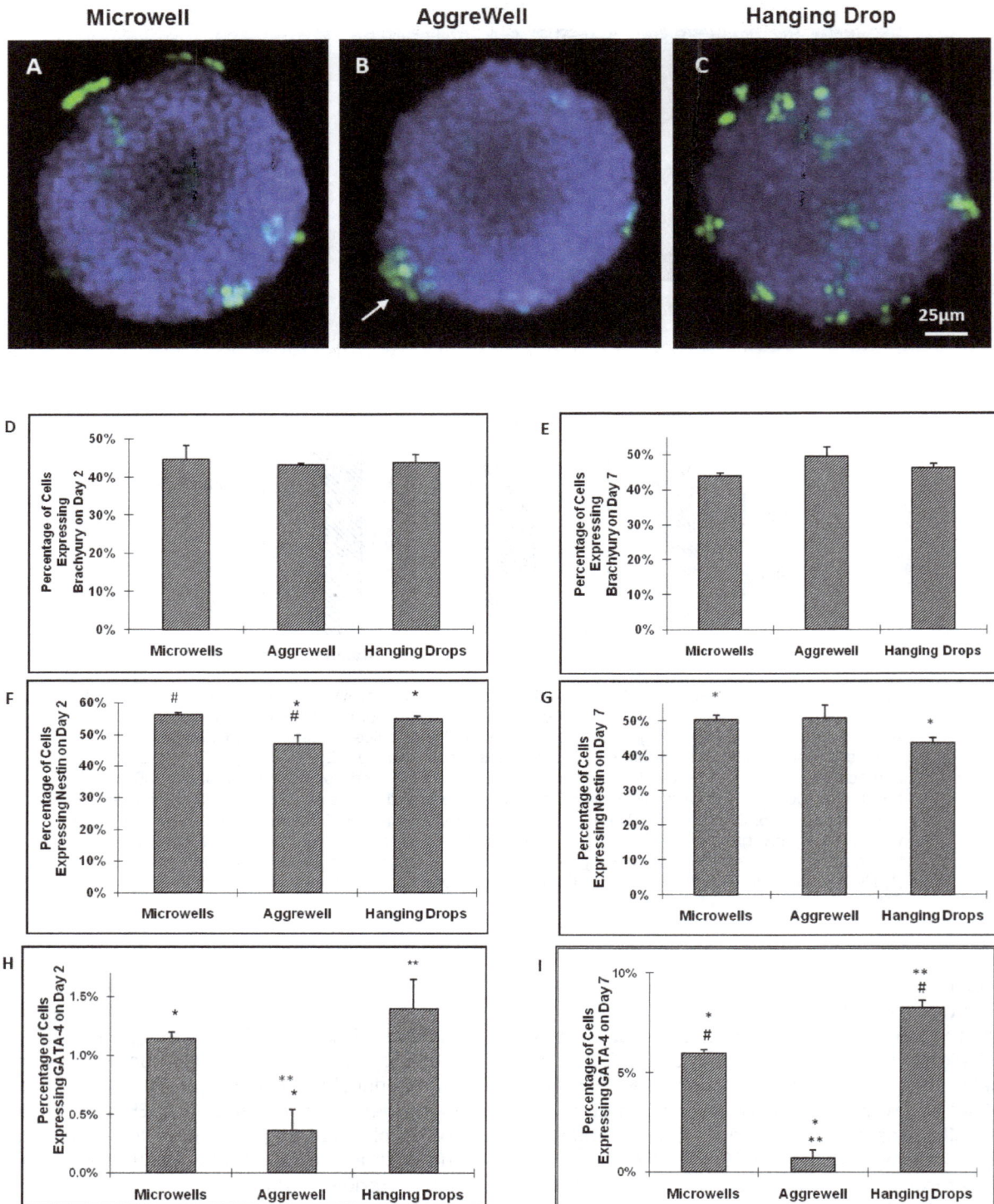

Figure 6. Primitive Endoderm, Ectoderm and Mesoderm in EB. The larger EB generated using A) 400-μm sized Honeycomb Microwells, B) AggreWell plates, and C) Hanging Drops were stained on day 2 with primitive endoderm marker, GATA-4 (green), and cell nucleus, Draq5 (blue) and imaged using z-stacked confocal microscopy. Note that the GATA-4 expression in the EB formed using the AggreWell plate is only present on the cone-shaped tip (indicated by arrow) rather than throughout the surface layer of the EB. These were then quantitatively analyzed by staining germ layers for FACS analysis on both days 2 and 7 with D and E) mesoderm marker, brachyury, (F and G) primitive ectoderm marker, nestin, and (H and I) primitive endoderm marker, GATA-4 . Note that the GATA-4 expression in the EB formed using the AggreWell plate is most significantly reduced compared with the GATA-4 expression in EB formed using the Microwells and Hanging Drops (* and **, P<0.01) and (#, P<0.05).

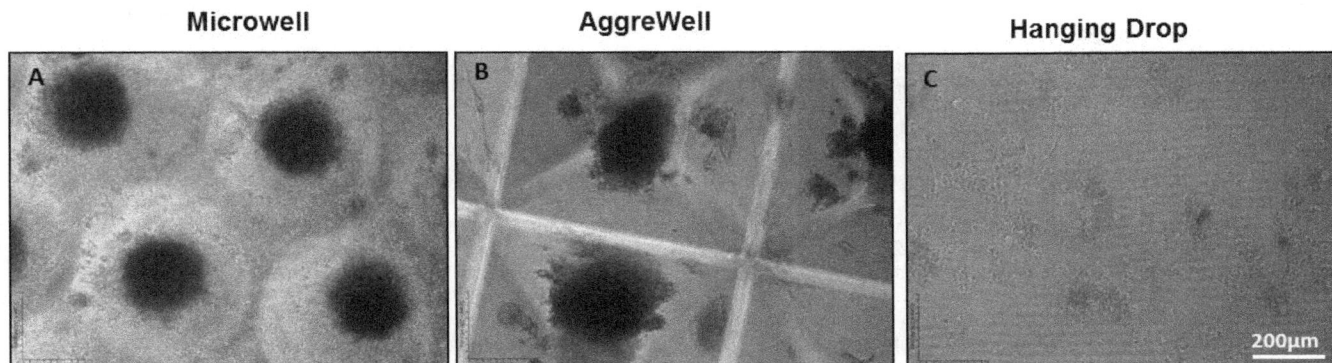

Figure 7. EB generation from human ESC. We were also able to successfully generate EB from single human ESC using both the (A) Microwell and (B) AggreWell plates more readily compared with conventional the (C) Hanging Drop method.

(Kurosawa, 2007). The initial size of the EB is dependent on the number of cells which initially self-assemble via cell–to-cell adhesion receptors (Bratt-Leal et al., 2009). Thereafter, the differentiation of the cells within the EB proceeds. After 2–4 days of suspension culture, the primitive endoderm forms on the surface of the EB, giving rise to a structure called the "simple EB". After 4 days, a "cystic" EB develops, characterized by the formation of a central cavity and columnar epithelium with a basal lamina (Khoo et al., 2005). Upon continued culture, the EB will give rise to all three germ layers and differentiate into a large variety of cell types.

Because the differentiation of stem cells within an EB is thought to recapitulate, at least in part, the developing embryo, the non-uniform shape of EB in the larger EB from the AggreWell plate is assumed to be an undesirable factor. This manuscript has shown that the endoderm marker, GATA-4, does not develop normally with a cone-shaped EB. On day 2 of development, the GATA-4 is concentrated at the cone-shaped tips of the AggreWell EB and is reduced throughout the remaining surface of the EB. Although the cone-shape of these EB do recover into a spherical shape after transfer from the AggreWell plates into suspension culture, the GATA-4 expression remains significantly reduced in these EB compared with EB formed using the Microwell and Hanging Drop. Most importantly, use of the AggreWell plate for initial EB formation would most likely limit the derivation of endodermal tissue products, like cells in the gastrointestinal and respiratory tracts, and should be avoided in these protocols.

This study is the first indicating that the shape of an EB can influence the fate of the differentiating ESC, however; the initial size of an EB is already known to be an important physical parameter influencing the proportion of cells differentiating towards some specific lineages (Ng et al., 2005; Park et al., 2007). In addition, the shape of individual human mesenchymal stem cells affects their differentiation efficiency towards adipocyte or osteoblast lineages (McBeath et al., 2004). Therefore, it is not surprising that the shape of an EB might play a role in the fate of the differentiating cells within the EB.

ACKNOWLEDGMENT

Funding was provided from the California Institute of Regenerative Medicine (RN2-00921-1) and Shrink Nanotechnologies, Inc. Although not evaluated in our study, it should be noted that StemCell Technologies very recently released a larger well plate, the AggreWell™800, and claims that this plate can generate EBs from 1000 up to 20,000 cells per EB.

REFERENCES

Amit M, Carpenter MK, Inokuma MS, Chiu CP, Harris CP, Waknitz MA, Itskovitz-Eldor J, Thomson JA (2000). Clonally derived human embryonic stem cell lines maintain pluripotency and proliferative potential for prolonged periods of culture. Dev. Biol., 227: 271-8.

Bratt-Leal AM, Carpenedo RL, McDevitt TC (2009). Engineering the embryoid body microenvironment to direct embryonic stem cell differentiation. Biotechnol. Prog., 25: 43-51.

Chen CS, Pegan J, Luna J, Xia B, McCloskey K, Chin WC, Khine M (2008). Shrinky-dink hanging drops: a simple way to form and culture embryoid bodies. J. Vis. Exp., (13).

Choi YY, Chung BG, Lee DH, Khademhosseini A, Kim JH, Lee SH (2010). Controlled-size embryoid body formation in concave Microwell arrays. Biomater., 31: 4296-303.

Dang SM, Gerecht-Nir S, Chen J, Itskovitz-Eldor J, Zandstra PW (2004). Controlled, scalable embryonic stem cell differentiation culture. Stem Cells 22: 275-82.

Dang SM, Zandstra PW (2005). Scalable production of embryonic stem cell-derived cells. Methods Mol. Biol., 290: 353-64.

Desbaillets I, Ziegler U, Groscurth P, Gassmann M (2000). Embryoid bodies: an in vitro model of mouse embryogenesis. Exp. Physiol., 85: 645-51.

Hwang YS, Cho J, Tay F, Heng JY, Ho R, Kazarian SG, Williams DR, Boccaccini AR, Polak JM, Mantalaris A (2009a). The use of murine embryonic stem cells, alginate encapsulation, and rotary microgravity bioreactor in bone tissue engineering. Biomater., 30: 499-507.

Hwang YS, Chung BG, Ortmann D, Hattori N, Moeller HC, Khademhosseini A (2009b). Microwell-mediated control of embryoid body size regulates embryonic stem cell fate via differential expression of WNT5a and WNT11. Proc. Natl. Acad. Sci. USA., 106,

16978-83.

Itskovitz-Eldor J, Schuldiner M, Karsenti D, Eden A, Yanuka O, Amit M, Soreq H, Benvenisty N (2000). Differentiation of human embryonic stem cells into embryoid bodies compromising the three embryonic germ layers. Mol. Med., 6: 88-95.

Khoo ML, McQuade LR, Smith MS, Lees JG, Sidhu KS, Tuch BE (2005). Growth and differentiation of embryoid bodies derived from human embryonic stem cells: effect of glucose and basic fibroblast growth factor. Biol. Reprod., 73: 1147-56.

Koike M, Kurosawa H, Amano Y (2005). A Round-bottom 96-well Polystyrene Plate Coated with 2-methacryloyloxyethyl Phosphorylcholine as an Effective Tool for Embryoid Body Formation. Cytotechnol., 47: 3-10.

Kurosawa H (2007). Methods for inducing embryoid body formation: in vitro differentiation system of embryonic stem cells. J. Biosci. Bioeng., 103, 389-98.

Lock LT, Tzanakakis ES (2009). Expansion and differentiation of human embryonic stem cells to endoderm progeny in a microcarrier stirred-suspension culture. Tissue Eng. Part A, 15: 2051-63.

McBeath R, Pirone DM, Nelson CM, Bhadriraju K, Chen CS (2004). Cell shape, cytoskeletal tension, and RhoA regulate stem cell lineage commitment. Dev. Cell, 6: 483-95.

Ng ES, Davis RP, Azzola L, Stanley EG, Elefanty AG (2005). Forced aggregation of defined numbers of human embryonic stem cells into embryoid bodies fosters robust, reproducible hematopoietic differentiation. Blood 106, 1601-3.

Nguyen D, Sa S, Pegan JD, Rich B, Xiang G, McCloskey KE, Manilay JO, Khine M (2009). Tunable shrink-induced Honeycomb Microwell arrays for uniform embryoid bodies. Lab Chip. 9, 3338-44.

Odorico JS, Kaufman DS, Thomson JA (2001). Multilineage differentiation from human embryonic stem cell lines. Stem Cells, 19: 193-204.

Park J, Cho CH, Parashurama N, Li Y, Berthiaume F, Toner M, Tilles AW, Yarmush ML (2007). Microfabrication-based modulation of embryonic stem cell differentiation. Lab Chip, 7: 1018-28.

Schroeder M, Niebruegge S, Werner A, Willbold E, Burg M, Ruediger M, Field LJ, Lehmann J, Zweigerdt R (2005). Differentiation and lineage selection of mouse embryonic stem cells in a stirred bench scale bioreactor with automated process control. Biotechnol. Bioeng., 92: 920-33.

Valamehr B, Jonas SJ, Polleux J, Qiao R, Guo S, Gschweng EH, Stiles B, Kam K, Luo TJ, Witte ON, Liu X, Dunn B, Wu H (2008). Hydrophobic surfaces for enhanced differentiation of embryonic stem cell-derived embryoid bodies. Proc. Natl. Acad. Sci. USA., 105: 14459-64.

27

An investigation on cephalometric parameters in Iranian population

Esmaeilzadeh Mahdi*, Nejadali Abolfazl, Kazemzadeh Fariba and Borhani Mohammad

Basic Sciences Department, Nikshahr Branch, Islamic Azad University, Nikshahr, Iran.

Anthropometry is the biological science of human body measurement. Anthropometry is applied in medical profession such as maxillofacial surgery, growth and development studies, plastic surgery, bioengineering and non-medical branches such as like shoe-making and eye-glasses industries. The aim of this study was to determine Iranian cephalometric parameters and cranial and facial anthropometric ratios. This cross sectional analytical study was done randomly on 137 people from Nikshahr (Iran) with normal face patterns. Facial and cranial ratios was estimated and compared. Data was analyzed by SPSS software. The regression line and the growth coefficient were determined for each Parameter. Finally, the mean values of these parameters were determined. At birth, Iranian population have hypereuryprosopic face and hypercephalic cranium form. While getting older, the midface height increased, face became more prominent, chin became shorter and face and cranium changed to eurycephalic and hyperleptoprosopic forms, respectively. Due to the wide racial combinations in Iran, more studies, with wider sample size, should be conducted among different Iranian race.

Key words: Anthropometry, craniofacial, Nikshahr.

INTRODUCTION

Over the last century wars, poverty, and political turmoil in Europe, Asia and Africa have led to sharply increased migration of numerous peoples to North America. These newcomers represent a much broader spectrum of ethnic groups than were seen in earlier waves of immigration.This influx of diverse people has important implications for craniofacial surgeons and other medical professionals whose work involves analysis and correction of morphological disfigurements and anomalies of the head and face (Farkas, 1994).

Although physical anthropologists have long been aware of differences in facial measurements among ethnic groups (Muzj, 1979; Topinard, 1885), for centuries the neoclassical facial canons established during the Renaissance (Beall, 1984) went unchallenged .

Only in recent years has the validity of these canons

been systematically investigated (Farkas et al., 1985; Le et al., 2002). These initial studies, based on anthropometric techniques, led to comparisons of facial anthropometric differences among Iranian and Canadian populations that were not presented in earlier investigations. To our knowledge, the present study is the broadest yet conducted in terms of geographical reach and diversity of subjects.

As a part of physical anthropology, anthropometry measures and examines linear and angular skeletal dimensions on living individuals (Mariclode, 1997). Understanding anthropometric parameters of face and cranium gives researchers and clinicians considerable insight into craniofacial growth and development which, in turn, has many practical applications including classification, diagnosis and treatment of craniofacial anomalies (Ainsowrth, 1979; Ramanathan and Chellappa, 2006), correction of craniofacial deformities using maxillofacial and plastic surgical methods and forensic medicine. By finding the mean value of anthropometric parameters in normal samples of a population, it is possible to create a

---

*Corresponding author. E-mail: mehdi_dna@yahoo.com.

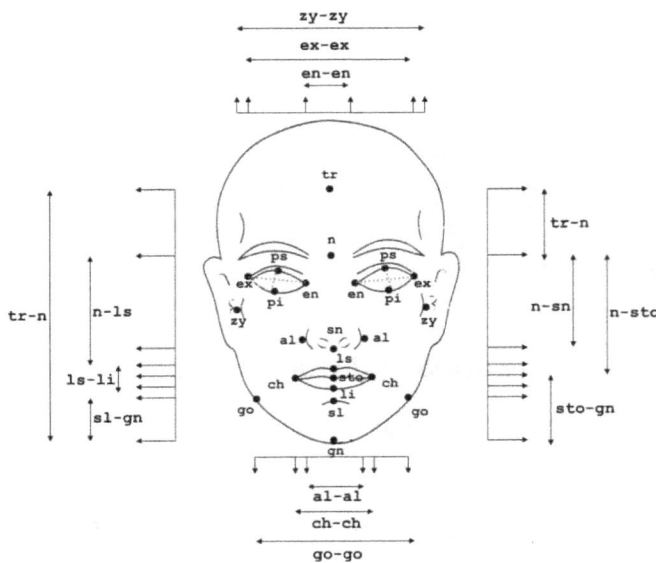

Figure 1. Face Anthropometry: of the 57 facial landmarks, 13 were chosen for the study.

template for facial analysis of this specific population (Mariclode, 1997).

As anthropometric and cephalometric parameters vary considerably depending on age, sex, geographical habitat and ethnic backgrounds of human beings (Mariclode, 1997; Williams et al., 1995), each anthropometric study should be conducted on a particular and predetermined age range, sex or ethnic group (Afak and Turgut, 1998). In his study, Porter compared anthropometric parameters of African-American males with those of North American whites and found significant differences (Porter, 2004). In a similar study, Choeks et al. (2004) showed significant differences between facial anthropometric measurements of Korean-American women and those of North American women. Farkas et al. (2005) depicted that the breadth of nose in Asians and Africans is larger than in North American whites. However, Middle Eastern people have nasal width similar to that of North American whites. By analyzing the anthropometric features of a group of 18-21-year-old Croatians, Buretic-Toljanovic showed that cranial measurements are influenced by geographical conditions (Buretic-Toljanovic et al., 2007).

The aim of this study was to determine Iranian cephalometric parameters and cranial and facial anthropometric ratios.

MATERIALS AND METHODS

This study was conducted on a sample of 137 four-to-eleven-year-old Iranian people. This study was performed as a research project with the permission of Ethics Board of Shirvan Islamic Azad University. Before the study Informed Consent was obtained from all human adult participants and from the parents or legal

guardians of minors for all participants. The informed consent clearly indicated: title of research project, researcher name and colleagues, research goals and Organization name of research executive.

All the participants had Angle Class I dental occlusion and no history of orthodontic treatment, tooth extraction, maxillofacial surgery, cleft lip and palate or other facial anomalies. A D40 Nikon digital camera with 18/135 lens (Nikon inc., Japan, 2007) was used to take frontal full-face photographs of each child while his/her head was in natural head position (NHP). The samples were relaxed during imaging and no special facial expressions such as smiling, laughing or frowning were detectable in their faces. A 10-mm wide sticker on each sample forehead was employed to calculate the image magnification.

Images were transferred to a computer and classified according to the age of samples. Using Adobe Photoshop software (Adobe Inc., USA), the points indicating the desired anthropometric landmarks were put on each image. The newly developed software "Smile Analyzer", by the Orthodontic Department of Mashhad Dental School, was used to measure the anthropometric parameters on each image (Jahanbin et al., 2010). This software has specifically been designed for precise measuring of desired distances or angles on images and radiographs.

The following measurements were analysed in this study (Figure 1):

1. The width of the nose or Alare width (al-al).
2. The width of the mouth or the distance between Cheilion points (ch-ch).
3. Intercanthal width or the distance between left and right Endocanthion points (en-en).
4. Binocular width or the distance between left and right Exocanthion points (ex-ex).
5. Forehead width or the distance between soft tissue Frontotemporale points (ft'-ft').
6. Intergonial width or the distance between left and right soft tissue Gonion points (go'-go').
7. Facial height or the distance between soft tissue Nasion and Gnathion (n'- gn').
8. The height of the nose or the distance between soft tissue Nasion and Subansal points (n'-sn).
9. The depth of the upper third of face; distance between Tragion and soft tissue Glabella (t-g').
10. The depth of the lower third of face; distance between Tragion and soft tissue Gnathion (t-gn').
11. The depth of the middle third of face; distance between Tragion and Subnasal points (t-sn).
12. Cranial base width or the distance between Tragion points (t-t).
13. Facial width or the distance between soft tissue Zygion points (zy'-zy').

Data was analyzed using t-test, ANOVA and linear regression models of the SPSS software (SPSS Inc., Chicago, Il, USA). Furthermore, the mean anthropometric measurement of Iranians was compared with Canadians.

RESULTS

Anthropometric parameter study in Iranian face indicates a gradual increase in mean alare width by age although it suddenly drops at age 8. A sharp increase in mouth width between 5 and 6 years of age followed by a steady growth. Ch-ch / age equation shows more growth in mouth compared to nose width by age (Table 1). Intercanthal width showed a sharp drop between 4 and 5

Table 1. Linear regression equation between anthropometric ratio and age in Iranian face.

Anthropometric ratio	Regression equation	Anthropometric ratio	Regression equation
al-al	0.5 × age + 26.8	n'-sn	1.4 × age + 29
ch-ch	1.25 × age + 30	t-g'	1.2 × age + 57
en-en	0.35 × age + 25	t-gn'	2.2 × age + 76.6
ex-ex	1.35 × age + 68	t-sn	1.28 × age + 53.4
ft'-ft'	1.6 × age + 89	t-t	2.1 × age + 99
go'-go'	1.7 × age + 77.3	zy'-zy'	2.17 × age + 90.98
n'-gn'	2.6 × age + 71.5		

years if age and then a dramatic increase between 5 and 6 followed by a gradual rise after 7. Also, the binocular width followed a noticeable increase between 5 and 6 years of age to reach a plateau between 6 and 7 and then, it raised gradually. The binocular width / age equation revealed a binocular growth rate of about four times as growth rate as the intercanthal width (Table 1). Forehead width increased more steadily, compared to the parameters above mentioned. Furthermore, data analysis demonstrated two growth acceleration periods between 5 to 6 and 9 to 11 years of age in intergonial width separated by an almost steady state. According to these findings, facial height increased gradually by age, although it accelerated at 5-6 and 9-10 intervals. Based on the n-gn / age equation, facial height has the largest growth rate among anthropometric measurements of the face (Table 1). The height of nose accelerated in three age ranges: 4-6, 7-8 and 9-10. The equation showed changes is nasal height as age increased. The depth of the upper third of face increased dramatically between 5-6 and 10-11 years. This study illustrated that the depth of the lower third of face increased steadily except for two plateaus between 6-7 and 10-11 years of age. The t-gn equation suggested rapid growth in this part of face. A growth-related change in the depth of the middle third of the face resembles those of the lower third. However, the growth rate was slower according to t-sn equation. Results showed that the growth curve of the cranial base width followed a sharp rise between 5 and 6 years to reach a plateau and then increase gradually after 7. The equation indicated a relatively fast growth in cranial base width. Facial width growth rate increased almost gradually except for a sharp rise between 5 and 6 years. The zy-zy equation shows that facial width has a rapid growth rate compared to most other parts of the face (Table 1).

DISCUSSION

As each ethnic group possesses its own specific facial and cranial form which changes with age as well, it is essential to specify the ethnic group and the age range to determine the anthropometric standards. The aim of our

study was to measure 13 anthropometric parameters on facial frontal images of 137 people from Nikshar (Iran) to assess facial and cranial and anthropometric ratios in Iranian population.

Our findings showed that craniofacial dimensions change at different rates at each age range, as did other investigators (Enlow and Hans, 1996; Proffit et al., 2007). The changes may be faster at an age but insignificant at another age. Interestingly, in almost all measured dimensions we found significant growth acceleration between 5 and 6 years of age. Another growth spurt was also seen between 9 and 11, although it was less significant.

Comparisons of the linear regression equations suggest that different craniofacial dimensions do not grow similarly: some parts grow at much slower pace compared to others. The intercanthal width had the least growth rate followed by the alar width while facial height and then facial width showed fastest growth.

The intercanthal width growth curve displayed a dramatic rise before 7 years of age. The growth of this dimension is related to the growth of brain and cranial base which is essentially complete by this age (Proffit et al., 2007). The orbital dimensions also reach the adult size at about 7, the reason why intercanthal growth continues much slowler after this age.

As body grows by age, facial height increases more than facial width. Thus, nasal cavities length progressively increases to facilitate air flow to expanding lungs. We found faster growth rate in facial height compared to width.

Being able to predict an individual's facial form at different ages has many practical applications. For instance, in forensic medicine, by analyzing a picture of a kidnapped child, the experts can guess how his /her facial form is after many years. Based on our findings and of other researchers and using artificial intelligence technology, computer programs can be designed to reconstruct facial forms of the individuals from a specific ethnicity at different ages.

Considering the differences in the facial and cranial anthropologic ratios and size among Iranian population (Resident in Nikshahr) and due to the wide ethnic combination in Iran, studies should be conducted covering wider scope.

REFERENCES

Afak SY, Turgut HB (1998). Weight, head and face measurements in Turkish newborns of central Anatolia. Gazi Med. J., 9: 116-120.

Ainsowrth H (1979). Numerical evaluation of facial pattern in children with isolated pulmonary stenosis. Arch. Dis. Child., 54(9): 62-69.

Beall CM (1984). Origins of the study of human growth. By Edith Boyd. Edited by B. S. Savara and J. F. Schilke. Portland: University of Oregon Health Sciences Center Foundation. 1980. xxviii + 676 pp., figures, tables, bibliography. $65.00 (cloth). Am. J. Phys. Anthropol., 64(1): 93-94.

Buretic-Toljanovic A, Giacometti J, Ostojic S, Kapovic M (2007). Sex-specific differences of craniofacial traits in Croatia: the impact of environment in a small geographic area. Ann. Hum. Biol., 34(3): 296-314.

Choeks K, Sclafani A, Litner J (2004). The Korean American woman's face: anthropometric measurements and quantitative analysis of facial aesthetics Arch Facial Plats Surg., 6: 244-252.

Enlow DH, Hans MG (1996). Essentials of facial growth. 2nd ed. Philadelphia WB Saunders Company. 79-98, 206.

Farkas LG (1994). Anthropometry of the head and face. 2nd ed: New York: Raven Press.

Farkas LG, Hreczko T, Kolar JC, Munro IR (1985). Vertical and horizontal proportions of the face in young adult North American Caucasians: Revision of neoclassical canons. Plast. Reconstr. Surg., 75: 328–337.

Farkas LG, Katic MJ, Katic MJ (2005). International anthropometric study of facial morphometry in various ethnic groups/ races. Craniofac Surg., 16(4): 615-646.

Jahanbin A, Mahdavi Shahri N, Baghayeripour M (2010). Anthropometric measurements of lip-nose complex in 11-17 years old males of Mashhad using photographic analysis. The Iranian J. Otorhinolaryngol., 59(1): 25-30.

Le T, Farkas LG, Ngim RCK, Levin LS, Forrest CR (2002). Proportionality in Asian and North American Caucasian faces using neoclassical facial canons as criteria. Aesth. Plast. Surg., 26: 64–69.

Mariclode C (1997). Biological anthropology. Naderi A, translator. 1st Ed ed: Tehran Gostar Publisher.

Muzj E (1979). The human face. A casual evaluation or a genetic program? Responsabilita` del sapere. Rome: Centro di comparazione e sintesi.

Porter J (2004). The average African American male face: an anthropometric analysis. Arch Facial Surg., 6: 78-81.

Proffit WR, Fields HW, Sarver DM (2007). Contemporary Orthodontics. 4th ed ed. St. Louis: Mosby Elsevier.

Ramanathan N, Chellappa R (2006). Modeling age progression in young faces. University of Maryland West Point, 31(3): 387-394.

Topinard C (1885). Ele´ments d'anthropologie ge´ne´rale. Paris: A Delahaye et E Legrosivier.

Williams P, Dyson M, Dussak JE, Bannister LH, Berry MM, Collins P (1995). Skeletal system in Gray's anatomy. 38th Ed ed. London: Elbs with Churchill Livingston.

28

Study of cephalometric parameters among 4 to 11 years old Persian girls resident in Iran

Esmaeilzadeh Mahdi*, Kazemzadeh Fariba and Adibfar Nader

Department of Basic Sciences, Bojnourd Branch, Islamic Azad University, Bojnourd, Iran.

Anthropometry is applied in medical profession such as maxillofacial surgery, growth and development studies, plastic surgery, bioengineering and non-medical branches such as like shoe-making and eye-glasses industries. The aim of the present study was to determine facial and cranial parameters among 4 to 11 years old Mashhadian young girls in order to assess growth pattern in this age range. 564 Persian girls aged between 4 to 11 years old resident in Mashhad and had normal facial patterns were selected from schools and kindergartens of the town. At first, frontal photographs in natural head position were taken by an expert person and after scanning all the photographs. They were transmitted to smile analyzer software to measure 13 anthropometric parameters. Pared t-test was used for statistic analysis. Anthropometric results obtained from 4 to 11 years old Fars girls residing in the city of Mashhad show that there is a special discipline in growth of different parts of face and skull. We reached formulas that may have a wide range of applications such as prediction of facial situation of an individual before or after his/her present situation. Such predictions can be helpful in forensic medicine, for instance in finding the lost kids. Reaching a normal range for face dimensions through studies about growth will play an important role in maxillofacial surgery and plastic surgery as well as in study of growth disorders. There are different factors such as ecologic, racial, age, and sex besides all genetic factors that influence the dimensions of human body and the way of its growth and development. Since the Iranian race is a compound one and consists of different races, and there are a wide variety of body dimensions in different parts of Iran, it is necessary to conduct more studies according to the geographical factors and the facial and skull anthropometric data shall be studied in different tribes of Iran. Obtaining an average for each of the variables in Iranian race, the anthropometry studies shall be conducted for different tribes.

Key words: Anthropometry, parameters, girls, growth pattern.

INTRODUCTION

As a part of physical anthropology, anthropometry measures and examines linear and angular skeletal dimensions on living individuals (Chamella, 1997). Understanding anthropometric parameters of face and cranium gives researchers and clinicians considerable insight into craniofacial growth and development which, in turn, has many practical applications including classification, diagnosis and treatment of craniofacial anomalies (Ainsowrth, 1979; Narayanan and Rama, 2006), correction of craniofacial deformities using maxillofacial and plastic surgical methods and forensic medicine. By finding the mean value of anthropometric parameters in normal samples of a population, it is possible to create a template for facial analysis of this specific population (Chamella, 1997).

As anthropometric and cephalometric parameters vary considerably depending on age, sex, geographical habitat and racial and ethnic backgrounds of human beings (Chamella, 1997; Williams et al., 1995) each anthropometric study should be conducted on a particular and predetermined age range, sex or racial group (Afak

*Corresponding author. E-mail: mehdi_dna@yahoo.com.

and Turgut, 1998). In his study, Porter compared anthropometric parameters of African-American males with those of North American whites and found significant differences (Porter, 2004). In a similar study, Choeks et al. (2004) showed significant differences between facial anthropometric measurements of Korean-American women and those of North American women (Choeks et al., 2004). Farkas et al. (2005) depicted that the breadth of nose in Asians and Africans is larger than in North American whites. However, Middle Eastern people have nasal width similar to that of North American whites (Farkas et al., 2005). By analyzing the anthropometric features of a group of 18 to 21 year old Croatians, Buretic-Toljanovic et al. (2007) showed that cranial measure-ments are influenced by geographical conditions Buretic-Toljanovic et al. (2007). The aim of our study was to measure the anthropometric parameters of 4 to 11 year old girls of Fars ethnic origin.

MATERIALS AND METHODS

This study was conducted on 564 (4 to 11 year old girls) of Fars ethnic origin. All the participants had Angle Class I dental occlusion and no history of orthodontic treatment, tooth extraction, maxillofacial surgery, cleft lip and palate or other facial anomalies.

A D40 Nikon digital camera with 18/135 lens (Nikon inc., Japan, 2007) was used to take frontal full-face photographs of each child while his/her head was in natural head position (NHP). The samples were relaxed during imaging and no special facial expressions such as smiling, laughing or frowning were detectable in their faces. A 10 mm wide sticker on each sample forehead was employed to calculate the image magnification.

The images transferred to a computer and classified according to the age of samples. Using Adobe Photoshop software (Adobe Inc., USA), the points indicating the desired anthropometric landmarks were put on each image. Newly developed software by the Orthodontic Department of Mashhad Dental School which is called "Smile Analyzer" was used to measure the anthropometric parameters on each image. This software has specifically been designed for precise measuring of desired distances or angles on images and radiographs.

Thirteen following measurements were taken in this study:

1. The width of the nose or Alare width (al-al),
2. The width of the mouth or the distance between Cheilion points (ch-ch),
3. Intercanthal width or the distance between left and right Endocanthion points (en-en),
4. Binocular width or the distance between left and right Exocanthion points (ex-ex),
5. Forehead width or the distance between soft tissue Frontotemporale points (ft'-ft'),
6. Intergonial width or the distance between left and right soft tissue Gonion points (go'-go'),
7. Facial height or the distance between soft tissue Nasion and Gnathion (n'-gn'),
8. The height of the nose or the distance between soft tissue Nasion and Subnasal points (n'-sn),
9. The depth of the upper third of face or the distance between Tragion and soft tissue Glabella (t-g'),
10. The depth of the lower third of face or the distance betwee Tragion and soft tissue Gnathion (t-gn'),
11. The depth of the middle third of face or the distance between Tragion and Subnasal points (t-sn),

12. Cranial base width or the distance between Tragion points (t-t),
13. Facial width or the distance between soft tissue Zygion points (zy'-zy'),

Data was analyzed using t-test, ANOVA and linear regression models of the SPSS software (SPSS Inc., Chicago, II, USA).

RESULTS

As Table 1 shows 564 (4 to 11 year old girls) of Fars ethnic origin participated in the study. The table also includes the mean and standard deviation of 13 anthropometric measurements.

Anthropometric parameter al-al

Figure 1 shows a gradual increase in mean alare width by age although it suddenly drops at 8. The linear regression model shows the following equation between alare width and age.

al-al = 0.5 × age + 26.8

Anthropometric parameter ch-ch

Figure 2 depicts a sharp increase in mouse width between 5 and 6 years of age followed by a steady growth. The following ch-ch / age equation shows more growth in mouse compared to nose width by age.

ch-ch = 1.25 × age + 30

Anthropometric parameter en-en

According to Figure 3, intercanthal width shows a sharp drop between 4 and 5 and then a dramatic increase between 5 and 6 followed by a gradual rise after 7. Based the linear regression equation, intercanthal width grows much more slowly compared to mouse and nose widths.

en-en = 0.35 × age + 25

Anthropometric parameter ex-ex

Figure 4 indicate that the binocular width follows a noticeable increase between 5 and 6 years of age to reach a plateau between 6 and 7 and then, it rises gradually. The binocular width / age equation reveals that the binocular growth rate is about four times as growth rate as the intercanthal width.

ex-ex = 1.35 × age + 68

Anthropometric parameter ft'-ft'

Forehead width increases more steadily, compared to

Table 1. Mean and SD of measured parameters by age.

Age	4	5	6	7	8	9	10	11
No	74	66	62	87	62	64	86	63
al-al (mm)	28.95	29.14	30.12	30.43	30.22	31.08	31.58	32.90
(Mean±SD)	± 1.61	± 1.74	± 2.10	± 1.96	± 1.85	± 2.22	± 2.15	± 2.12
ch-ch (mm)	35.14	35.43	38.36	38.86	39.54	40.92	42.39	44.10
(Mean±SD)	±2.59	± 2.49	±2.78	± 2.49	± 2.77	± 2.98	± 3.37	± 3.37
en-en (mm)	26.38	25.87	27.74	27.68	27.93	28.10	28.35	28.68
(Mean±SD)	± 1.89	± 2.27	± 2.30	± 1.99	± 1.92	± 2.33	± 2.28	± 2.62
ex-ex (mm)	73.12	73.57	77.59	77.75	79.12	79.66	81.21	82.84
(Mean±SD)	± 3.49	± 4.25	± 4.11	± 3.50	± 3.43	± 3.64	± 4.00	± 3.74
ft'-ft' (mm)	95.39	96.44	99.07	100.34	102.32	102.86	104.87	106.54
(mean±SD)	± 4.19	±4.81	± 9.33	± 4.54	± 5.60	± 4.63	± 4.86	± 4.52
go'-go' (mm)	83.57	84.71	89.21	90.15	90.55	90.64	93.69	96.79
(Mean±SD)	± 5.55	± 5.66	± 5.14	± 5.29	± 5.11	± 5.75	± 6.47	± 6.11
n'-gn' (mm)	82.26	84.03	88.45	89.54	92.11	93.39	98.63	100.31
(Mean±SD)	± 5.43	± 5.31	± 5.24	± 4.51	± 4.50	± 5.38	± 5.99	± 5.67
n'-sn (mm)	35.23	36.27	37.60	37.87	40.11	40.85	43.92	44.63
(Mean±SD)	± 6.26	± 2.62	± 2.49	± 2.26	± 2.96	± 3.38	± 2.94	± 2.87
t-g' (mm)	61.71	62.04	65.88	65.89	66.58	67.22	68.30	71.27
(Mean±SD)	± 4.74	± 4.50	± 4.69	± 3.85	± 4.64	± 5.16	± 4.83	± 5.09
t-gn' (mm)	84.88	88.15	90.82	91.20	94.09	96.53	99.62	99.92
(Mean±SD)	± 4.80	± 5.90	± 4.55	± 4.70	± 4.97	± 5.46	± 6.08	± 5.37
t- sn (mm)	58.04	60.01	61.97	61.88	63.72	65.09	66.54	67.01
(Mean±SD)	± 3.58	± 3.96	± 3.76	± 4.13	± 4.44	± 4.35	± 4.98	± 4.72
t-t (mm)	106.93	109.42	114.45	114.50	115.70	117.76	120.22	123.60
(Mean±SD)	±5.94	± 6.01	± 5.64	± 5.23	± 5.64	± 6.08	± 6.52	± 6.30
zy'-zy' (mm)	99.35	101.10	105.70	106.76	107.65	109.73	112.68	115.25
(Mean±SD)	± 5.56	± 5.68	± 5.23	± 5.30	± 5.38	± 6.29	± 6.39	± 5.90

other above mentioned parameters (Figure 5). The linear regression equation for ft'-ft' / age is as follows:

$$ft'\text{-}ft' = 1.6 \times age + 89$$

Anthropometric parameter go'-go'

Figure 6 demonstrates two growth acceleration periods between 5 to 6 and 9 to 11 years in intergonial width separated by an almost steady state. The intergonial width at any age is predictable using the following equation:

$$go'\text{-}go' = 1.7 \times age + 77.3$$

Anthropometric parameter n'-gn'

According to Figure 7, facial height increases gradually

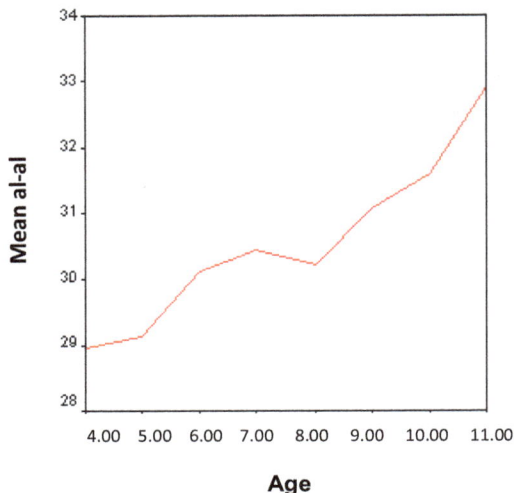

Figure 1. Mean al-al by age.

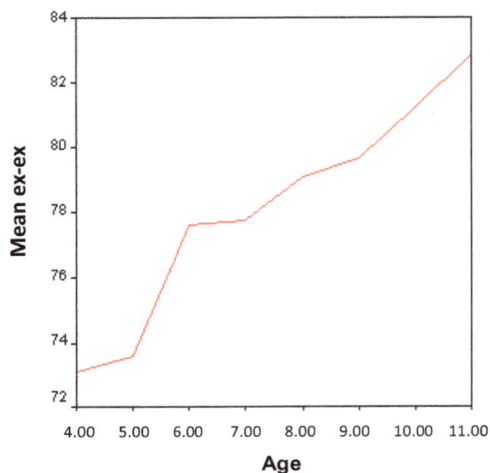

Figure 2. Mean ch-ch by age.

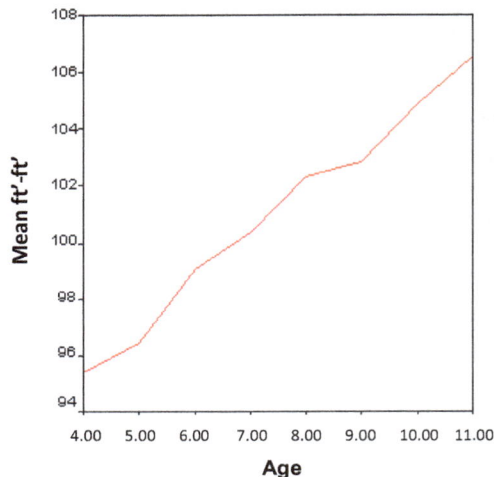

Figure 3. Mean en-en by age.

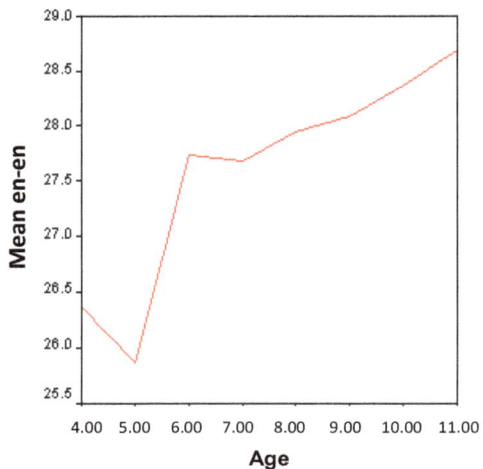

Figure 4. Mean ex-ex by age.

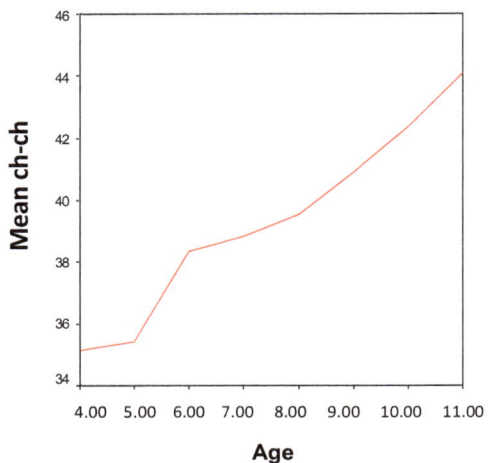

Figure 5. Mean ft'-ft' by age.

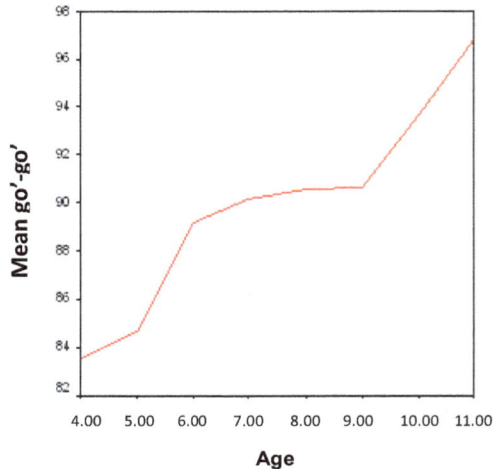

Figure 6. Mean go'-go' by age.

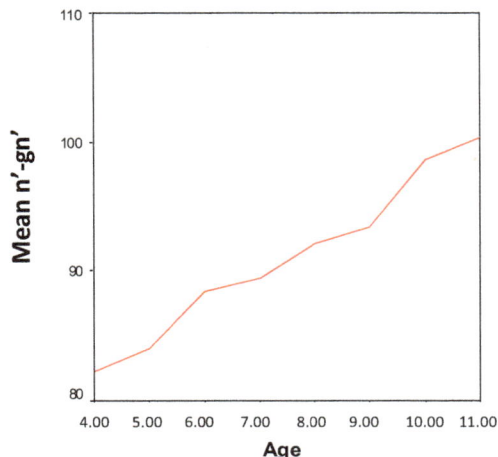

Figure 7. Mean n'-gn' by age.

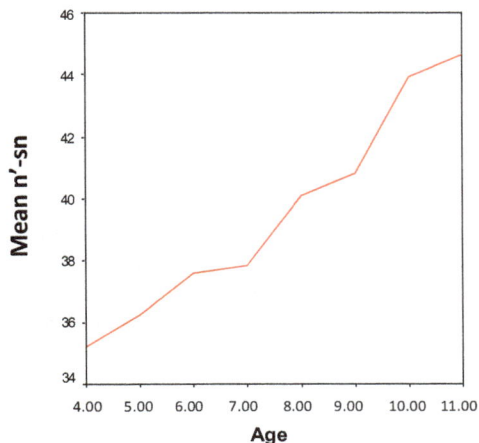

Figure 9. Mean t-g' by age.

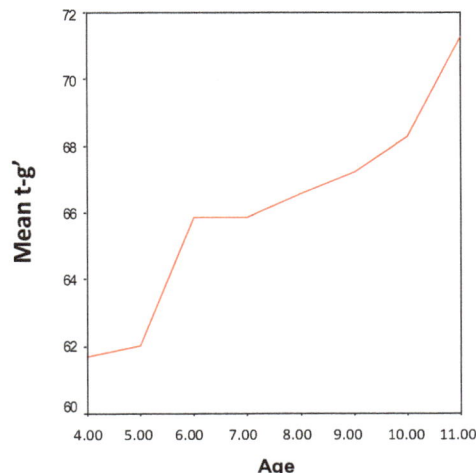

Figure 8. Mean n'-sn by age.

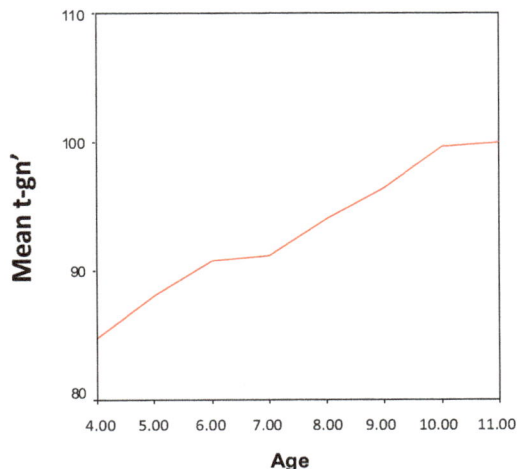

Figure 10. Mean t-gn' by age.

by age, although it accelerates at 5-6 and 9-10 intervals. Based on the following equation, facial height has the largest growth rate among anthropometric measurements of the face.

$$n'\text{-}gn' = 2.6 \times age + 71.5$$

Anthropometric parameter n'-sn

The height of nose accelerates in three age ranges: 4 to 6, 7 to 8 and 9 to 10 (Figure 8). The equation shows the changes is nasal height as age increases:

$$n'\text{-}sn = 1.4 \times age + 29$$

Anthropometric parameter t-g'

The depth of the upper third of face increases

dramatically between 5 to 6 and 10 to 11 years (Figure 9). The linear regression equation for t-g' / age is as follows:

$$t\text{-}g' = 1.2 \times age + 57$$

Anthropometric parameter t-gn'

Figure 10 illustrates that the depth of the lower third of face increases steadily except for two plateaus between 6 to 7 and 10 to 11 years of age. The following equation suggests rapid growth in this part of face:

$$t\text{-}gn' = 2.2 \times age + 76.6$$

Anthropometric parameter t-sn

Growth-related changes in the depth of the middle third of

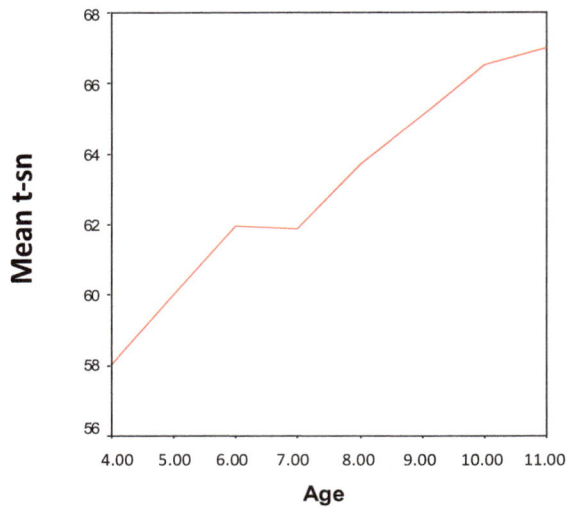

Figure 11. Mean t-sn by age.

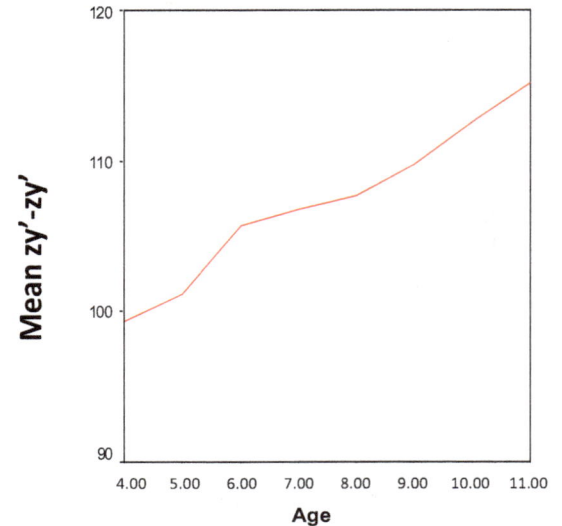

Figure 12. Mean t-t by age.

the face resemble those of the lower third (Figure 11). However, the growth rate is slower according to the equation.

t-sn = 1.28 × age + 53.4

Anthropometric parameter t-t

As Figure 12 represents, the growth curve of the cranial base width follows a sharp rise between 5 and 6 years to reach a plateau and then increase gradually after 7. The equation indicates a relatively fast growth in cranial base width.

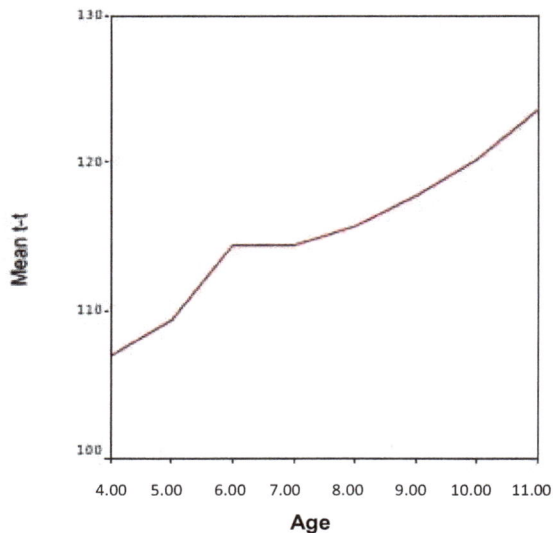

Figure 13. Mean zy'-zy' by age.

t-t = 2.1 × age + 99

Anthropometric parameter zy'-zy'

Facial width growth rate increases almost gradually except for a sharp rise between 5 and 6 years (Figure 13). The following equation shows that facial width has a rapid growth rate compared to most other parts of the face:

zy'-zy' = 2.17 × age + 90.98

DISCUSSION

As each racial and ethnic group possesses its own specific facial and cranial form which changes by age as well, to determine the anthropometric standards, it is essential to specify that anthropometric parameters are determined among which ethnic group and in which age range. The aim of our study was to measure 13 anthropometric parameters on facial frontal images of 564 (4 to 11 year old girls) of Fars ethnic origin to determine and predict growth-related changes in this age range.

The photo-anthropometric method Ferrario et al. (2003) was employed to measure the desired anthropometric dimensions. The method is relatively simple and fast, the required equipments are easily accessible and the findings are reliable. During imaging, the subject should stand still with no detectable facial expression. The photographs should be taken with a high resolution camera to ensure that all angles and lines are easily identifiable. To reduce the measuring error, all measurements were performed by the same operator.

However, some landmarks such as Cheilion or Endocanthion are more precisely identifiable compared to landmarks like Gonion.

Our findings show that craniofacial dimensions change at different rates at each age range, as other investigators mentioned as well (Enlow and Hans, 1996; Proffit et al., 2007). In other words, the changes may be faster at an age but insignificant at another age. Interestingly, in almost all measured dimensions we found significant growth acceleration between 5 and 6 years of age. Another growth spurt was also seen between 9 and 11, although it was less significant.

Comparing the linear regression equations suggests that different craniofacial dimensions do not grow similarly. Some parts grow at much slower pace compared to others. The intercanthal width has the least growth rate followed by the alar width while facial height and then facial width show fastest growth.

The intercanthal width growth curve displays a dramatic rise before 7 years of age. The growth of this dimension is related to the growth of brain and cranial base which is essentially complete by this age (Proffit et al., 2007). The orbital dimensions also reach the adult size at about 7. This is the reason why intercanthal growth continues much slowly after age 7.

As body grows by age, facial height increases more than facial width. Thus, nasal cavities length progressively increases to facilitate air flow to expanding lungs. We found faster growth rate in facial height compared to width.

Being able to predict an individual's facial form at different ages has many practical applications. For instance, in forensic medicine, by analyzing a picture of a kidnapped child, the experts can guess how his /her facial form is after many years. Based on our and other researchers' findings and by using artificial intelligence technology, computer programs can be designed to reconstruct facial forms of the individuals from a specific ethnicity at different ages.

Conclusions

Based on our findings in Fars 4 to 11 year old girls, we concluded that:

1. By age, craniofacial dimensions change at different rates.
2. Different craniofacial dimensions do not grow at consistent rates. Some parts grow slower compared to others.
3. The intercanthal width has the slowest growth.
4. Facial height shows the fastest growth.
5. Using linear regression equations, at any age, each craniofacial dimension can easily and precisely be determined.

REFERENCES

Chamella M (1997). Biological anthropology. Translated to Persian by Naderi A. 1st Ed. Tehran Gostar Publisher, p: 57-114.

Ainsowrth H (1979). Numerical evaluation of facial pattern in children with isolated pulmonary stenosis. Arch. Dis. Child, 54(9):62-9.

Narayanan R, Rama C (2006). Modeling age progression in young faces. University of Maryland West Point, 31(3):387-394.

Williams PL, Dyson M, Dussak JE, Bannister LH, Berry MM, Collins P (1995). Skeletal system in Gray's anatomy. 38th Ed. London. Elbs with Churchill Livingston, p: 607-612.

Afak SY, Turgut HB (1998). Weight, head and face measurements in Turkish newborns of central Anatolia. Gazi Med. J., 9:116-20.

Porter J (2004). The average African American male face: an anthropometric analysis. Arch Facial Surg., 6:78-81.

Choeks K, Sclafani A, Litner J (2004). The Korean American woman's face: anthropometric measurements and quantitative analysis of facial aesthetics. Arch Facial Plats Surg., 6:244-52.

Farkas LG, Katic MJ, Katic MJ (2005). International anthropometric study of facial morphometry in various ethnic groups/ races. Craniofac Surg., 16(4):615-46

Buretic-Toljanovic A, Giacometti J, Ostojic S, Kapovic M (2007). Sex-specific differences of craniofacial traits in Croatia: the impact of environment in a small geographic area. Ann. Hum. Biol., 34(3):296-314.

Ferrario V, Sforza C, Dellavia C (2003). A quantitative three-dimensional assessment of abnormal variations in facial soft tissues of adult patients with cleft lip and palate. Cleft Palate Craniofac J., 40:544-9.

Enlow DH, Hans MG (1996). Essentials of facial growth. 2nd ed. Philadelphia .WB Saunders Company, p: 122-165.

Proffit WR, Fields HW, Sarver DM (2007). Contemporary Orthodontics. 4th ed. St. Louis: Mosby Elsevier.

29

Regeneration of germlings and seedlings development from cauline leaves of *Sargassum thunbergii*

Feng Li[1], Shoutuan Yu[2], Yuze Mao[1] and Naihao Ye[1]*

[1]Yellow Sea Fisheries Research Institute, Chinese Academy of Fishery Sciences, Qingdao 266071, China.
[2]Hongdao Street Office of QingdaoChengyang District, China.

A method of producing artificial *Sargassum thunbergii* seedlings is urgently desired to meet the increasing demand of raw materials used for aquaculture and for the sake of environmental protection. This is the first report of attempt to study the vegetative propagation of this species using cauline leaves under laboratory conditions. On average, 45.75% of the excised leaves survived and one leaf could produce several new individuals. Adventitious burgeons grew into branches about 2 mm in length after 3 months of culture. The new individuals were cutoff and could be used as seedlings for raft culture and seabed restoration after a further culture to elongate enough for hand-planting.

Key words: *Sargassum thunbergii,* artificial seedling, brown seaweed, cauline leaves.

INTRODUCTION

Sargassum spp. have drawn lot of attentions for its active substances, such as antitumor polysaccharide (Fujihara et al., 1984; Zhuang et al., 1995; Jo et al., 2005), anticoagulative polysaccharide (Athukorala et al., 2006), anthelmintic components (Lee and Min, 1970), as well as peroxynitrite-scavenging and reactive oxygen scavenging constituents (Seo et al., 2004; Park et al., 2005). *Sargassum thunbergii* (Mert.) O. Kuntze, named rat-tail alga in China after its morphological characteristics, is becoming increasingly valuable as a resource of diet for aquaculture, especially with regards to feeding sea cucumber (Liu et al., 2004). During the past ten years, wild stocks of the species have been subject to intense harvesting by local inhabitants. With increasingly high levels of harvesting, some places of natural stands of *S. thunbergii* have collapsed (Zou et al., 2005). Recently, sexual breeding of *S. thunbergii* have been achieved under laboratory conditions (Wang et al., 2006). However, availability of a production technique for artificial seedling breeding is needed and it would be advantageous over the current method for large-scale breeding.

Knowledge regarding the biology of *S. thunbergii* is exceedingly limited and this precludes the development of seedling cultivation techniques aimed at alleviating the negative impact of indiscriminate harvesting currently affecting the wild stocks. Under natural conditions, vegetative propagation from holdfasts is an important method sustaining the population (Yoshida et al., 1999). Practices have proved that seedlings with holdfasts grow faster and have lower morality than those without, and this is the key reason makes *S. thunbergii* cultivation different from other commercial algae, such as *Gracilaria lemaneiformis*, which can be successfully cultured using any parts of its vegetative filaments (Hurtado-Ponce et al., 1992; Nelson et al., 2001; Ryder et al., 2004). Studies reveal that zygotic embryos of *Sargassum* for making seedlings are only formed seasonally (Tokuda et al, 1987; Nanba and Okuda, 1992; Sun et al., 2007). Tissue culture of several other species of *Sargassum* has been conducted. Spontaneous formation and development of adventitious embryos from the cauline leaves of *S. macrocarpum* are observed in laboratory culture. Swellings from the single leaf of *S. macrocarpum* became cylindrical protuberances and grew into 'daughter' thalli, which detach from their 'mother' thalli and develop into individual thalli exhibiting the same morphological processes as zygotic embryos (Yoshida et al., 1999) However, factors induce adventitious embryos in *S. macrocarpum* are still uncertain and the process is complicated and time consuming (Yoshida et al., 1999). In the present study, we report the regeneration of germlings developed from cauline leaves of *S. thunbergii* and their

*Corresponding author. E-mail: yenh@ysfri.ac.cn.

Figure 1. Seedling regeneration of *S. thunbergii from the* cauline leaves under laboratory conditions: (A) cauline branches obtained from the wild, (B) leaves used for the regeneration, (C) a newly developed germlings from the cauline leaves, 3 months after the culture, (D) the developing germlings on the leave, 4 months after the culture, (E) an individual attaching to the stone after 1 month of field culture; (F) a *S. thunbergii* plant with several newly-formed cauline branches after 2 months of field culture, (G) a mature *S. thunbergii* plant after 3 months of field culture (Bar, 1 cm).

availability as artificial seedlings.

weekly changed.

MATERIALS AND METHODS

Samples were collected between August and November 2005 from the intertidal zone (35.35°N, 119.30°E) of the Second Bathing Beach, Qingdao, China using a pump placed 2 m depth under the water surface, and filtered with plankton nets (200 μm net and 20 μm net nested inside), then, autoclaved and prepared into enriched seawater with ES (enriched seawater) (Mclachlan, 1979). In the laboratory, intact plants were isolated, washed several times with sterile seawater, sterilized with 1% sodium hypochlorite for 2 min, then rinsed with autoclaved seawater and placed into a sterile aquarium (d=40 cm, h=30 cm) containing enriched seawater, and maintained at 15°C, with illumination of 50 photons m-2s-1 provided by cool-white fluorescent tubes and with a photonperiod of 12:12 h. The branches of *S. thunbergii* fronds were cut (2 to 3 cm in length) and rinsed with antibiotic enriched seawater solution made of ampicillin, penicillin, rifampicin, nystatin (0.2mg ml^{-1} each) andof GeO (0.1g ml^{-1}) for about 6 h. The leaves were cutoff using a surgeon knife (sigma) and cultured in flasks (5 L) containing autoclaved enriched seawater (1:80, w/w), which were bubbled 24 h in the incubator (Ningbo Jiangnan, GXZ-430ABC, Ningbo, China). Ten flasks were included in the treatment and culture media were

RESULTS AND DISCUSSION

The percentage of the survived excised leaves from *S. thunbergii* plant was 45.75% (n=201) and regeneration of adventitious embryos was observed (Figure 1). Longer cauline branches (Figure 1A) were selected and cut into segments for sterilization, then glossy and intact leaves were cutoff using a surgeon knife (Figure 1B). After 3 months of culture, semispherical swellings, about 2 mm in diameter, arose from the petioles end of the cauline leaves and became cylindrical protuberances (Figure 1C). Each protuberance developed into a new germling on the mother plant leaf (Figure 1D). New germlings were easily detached from the mother plant by scraping with forceps. The morphological characteristics of these germlings were quite similar to those developed from zygotic embryos with a few small cauline leaves (Figure 1D). Within several weeks of detachment, rhizoids emerged at the basal part of the germlings and they attached to the culture substances (Figure 1E). The cauline leaves, which began to produce adventitious embryos, were gradually covered with one or several of newly developed germlings. The growth of each germling was very slow. It took 3 to 5 months to grow big enough for the convenience of manual operation. The germlings could be stored in culture vessels for about half a year under the culture conditions described previously with the medium being renewed weekly. After several months in laboratory culture, germlings were taken out of the vessels and cultured outdoor with the substances, such as stones (Figure 1E). Rhizoids developed fast at the basal part of the germlings and they attached to the culture substances and leaves were produced consecutively at the top of the axis of each plant (Figure 1F). After another 3 months of culture, the length of the largest cauline branch was more than 10 cm (Figure1G). The primary branches showed rapid elongation in the field.

Their daily growth rate was about 0.5 cm/day from early March to late June. At that time, the germlings had three to thirteen cauline branches (Figure 1G).

All plants began to mature after 4 months of field culture. They were all, males or females, the same as their matrixes indicating that all new germlings were clones. Vegetative propagation of the genus *Chondrus* (Chen and Taylor, 1978), *Gracilaria* (Goldstein, 1973; Santelices and Varela, 1995) and *Eucheuma* (Doty, 1987) through thallus fragmentation is well known and it has allowed a rapid expansion in farming of these species all over the world, suggesting that vegetative propagation is a good method of generating propagules for large scale cultivation. However, there is an exceedingly limited amount of information regarding the regulatory effect of physical factors such as temperature, light, etc on the regeneration processes of *S. thunbergii*. In this study, incubation at 15°C under 50 photons·m^{-2} · s^{-1} and 12:12 h LD illumination were suitable conditions used in the seedling breeding as they showed high burgeon ratio and fast growth rate of the germlings. Considering the sustenance of the resource, it is becoming increasingly apparent that the present status imposes a severe pressure on the wild stocks of *S. thunbergii*, and the available information indicates that some stands have already been abandoned by fishermen as a result of over-exploitation. From the current observations, it is clear that the leaf-origin germlings of *S. thunbergii* can grow and mature the same as the cauline branches in the wild and, therefore, can be useful for artificial seedling production. Also, the results indicate that the produced seedlings could be stored for a long term in the laboratory without loss of growth activity and vegetative regeneration capability.

ACKNOWLEDGEMENTS

This work was supported by the National Natural Science Foundation of China (40706050, 40706048 and 30700619), the National Science and Technology Pillar Program (2008BAC49B04), National special fund for transgenic project (2009ZX08009-019B), Natural Science Foundation of Shandong Province (2009ZRA02075), Qingdao Municipal Science and Technology plan project (09-2-5-8-hy) and the Hi-TechResearch and Development Program(863) of China (2009AA10Z106).

REFERENCES

Athukorala Y, Jung WK, Vasanthan T, Jeon YJ (2006). An anticoagulative polysaccharide from anenzymatic hydrolysate of *Ecklonia cava*. Carbohydr. Polym., 66: 184-191.
Chen LC, Taylor AR (1978). Medullary tissue culture of the red alga *Chondrus crispus*. Can. J. Bot., 56: 883-886.
Doty MS (1987). The production and use of Eucheuma. In: Doty MS, Caddy JF and Santelices B (eds), Case studies of seven commercial seaweed resources. FAO Fisheries Tech. Pap. 281. FAO, Rome, 123-161.
Fujihara M, Iizima N, Yamamoto I, Nagumo T (1984). Purification and chemical and physical characterization of an antitumour polysaccharide from the brown seaweed *Sargassum fulvellum*. Carbohydr. Res., 125: 97-106.
Goldstein M (1973). Regeneration and vegetative propagation of the agarophyte *Gracilaria debilis* (Forsskal) Boerg. (Rhodophyceae). Bot. Mar., 16: 226-228.
Hurtado-Ponce AQ, Samonte GPB, Luhan MR, Guanzon JN (1992). *Gracilaria* (Rhodophyta) farming in Panay, Western Visayas, Philippines. Aquaculture, 105: 233-240.
Jo EH, Cho SD, Ahn NS, Jung JW, Yang SR, Park JS, Hwang JW, Lee SH, Park JR, Kim SJ, Park HK, Lee YS, Kang KS (2005). Inhibition of human breast carcinoma by BLC (*Sargassum fulvellum*) and BLC/HEN egg *in vitro* and *in vivo*. Korean J. Vet. Res., 45: 85-91.
Lee WH, Min KN (1970). Detection of anthelmintic components of *thunbergii Kuntze*. Korean J. Pharmacognol., 1: 19-22.
Liu X, Zhu G, Zhao Q, Wang L, Gu B (2004). Studies on hatchery techniques of the sea cucumber, *Apostichopus japonicas*. In: advance in sea cucumber aquaculture and management, FAO Fisheries Technical Paper 463, pp. 287-288.
McLachlan J (1979). Growth media marine. In: Stein JR (ed) Handbook of phycological methods. Culture methods and growth measurements. Cambridge University Press, Cambridge, pp. 25-51.
Nanba N, Okuda T (1992). Egg release of five Fucalean species in Tsuyazaki, Japan. Nippon Suisan Gakkaishi, 58: 659-663.
Nelson SG, Glenn EP, Conn J, Moore, D, Walsh T, Akutagawa M (2001). Cultivation of *Gracilaria parvispora* (Rhodophyta) in shrimp-farm effluent ditches and floating cages in Hawaii: a two phase polyculture system. Aquaculture, 193: 239-248.
Park PJ, Heo SJ, Park EJ, Kim SK, Byun HG, Jeon BT, Jeon YJ (2005). Reactive oxygen scavenging effect of enzymatic extracts from *Sargassum thunbergii*. J. Agric. Food Chem., 53: 6666-6672.
Ryder E, Nelson SG, McKeon C, Glenn EP, Fitzsimmons K, Napolean S (2004). Effect of water motion on the cultivation of the economic seaweed *Gracilaria parvispora* (Rhodophyta) on Molokai, Hawaii. Aquaculture, 238: 207-219.
Santelices B, Varela D (1995). Regenerative capacity of *Gracilaria* fragments: effects of size, reproductive state and position along the axis. J. appl. Phycol., 7: 501-506.
Seo Y, Lee HJ, Park KE, Kim YA, Ahn JW, Yoo JS, Lee BJ (2004). Peroxynitrite-scavenging constituents from the brown alga *Sargassum thunbergii*. Biotechnol. Bioprocess Eng., 9: 212-216.
Sun X, Wang F, Zhang L, Wang X, Li F, Liu G, Liu Y (2007). Observations on morphology and structure of receptacles and pneumathode of *Sargassum thunbergii*. Mar. Fish., Res., 28: 125-131.
Tokuda H, Ohno M, Ogawa H (1987). The Resources and cultivation of seaweeds. Midori-shobou, Tokyo, pp. 354.
Wang F, Sun X, Li F (2006). Studies on sexual reproduction and seedling-rearing of *Sargassum thunbergii*. Mar. Fish. Res., 27:1-6.
Yoshida G, Uchida T, Arai S, Terawaki T (1999). Development of

adventive embryos in cauline leaves of *Sargassum macrocarpum* (Fucales, Phaeophyta). Phycol. Res., 47: 61-64.

Zhuang C, Itoh H, Mizuno T, Ito H (1995). Antitumor active fucoidan from the brown seaweed umitoranoo (*Sargassum thunbergii*). Biosci. Biotechnol. Biochem., 59: 563-567.

Zou J, Li Y, Liu Y, Zhang T, Wang Y (2005). Studies on the biological characters and technology of raft culture of *Sargassum thunbergii*. Shandong Fish., 22: 22-28.

30

Antibacterial activity of crude seed extracts of *Buchholzia coriacea* E. on some pathogenic bacteria

T. I. Mbata[1]*, C. M. Duru[2] and H. A. Onwumelu[3]

[1]Department of Microbiology, Federal Polytechnic Nekede, Owerri, Nigeria.
[2]Department of Biotechnology, Federal University of Technology, Owerri, Nigeria.
[3]Department of Chemistry, Nnamdi Azikiwe University, Awka, Nigeria.

The antibacterial efficacy of hot water and methanol extracts of dried seeds of *Buchholzia coriacea* against *Escherichia coli, Staphylococcus aureus, Salmonella typhimurium, Bacillus cereus,* and *Vibrio cholerae* were determined using the Agar-gel diffusion method. The minimum inhibitory concentration (MIC), minimum bactericidal concentration (MBC) and phytochemistry of the extracts were also evaluated. Results obtained showed that the methanol extracts of the dried seed was potent, inhibiting the isolates with diameter zone of inhibition ranging from 7.0 - 35.0 mm. The extracts inhibited the growth of the bacterial isolates in a concentration dependant manner with MICs ranging between 9.3 - 50 mg/ml, and MBCs of 4.1 - 17.4 mg/ml. Phytochemical analysis of dried seed extracts revealed the presence of alkaloids, anthraquinones, carbohydrates, cardiac glycosides, flavonoids, glycosides, resins, saponin, steroidal rings, steroidal terpenes and tannins. The findings from this study could be of interest and suggests the need for further investigations in terms of toxicological studies and purification of active components with the view to using the plant in novel drug development.

Key words: Antibacterial activity, phytochemical analysis, *Buchholzia coriacea*, bacterial isolates.

INTRODUCTION

Traditional medicine is widespread throughout the world and it can be described as the total combination of knowledge and practices, whether explicable or not, used in diagnosing, preventing or eliminating a physical, mental or social disease and which may rely exclusively on past experience and observation handed down from generation, verbally or written (Sofowora, 1984). Medicinal plants has been defined by WHO consultative group as any plant which in one or more of its organs contains substances that can be used for therapeutic purposes or which are precursors for the synthesis of useful drugs (Andrews, 1982).

For many years medicine depended exclusively on leaves, flowers and barks of plants; only recently have synthetic drugs came into use and in many instances, there are carbon copies of chemicals identified in plants (Conway, 1973). In orthodox medicine, a plant may be subjected to several chemical processes before its active ingredient is extracted, refined and made ready for consumption while in traditional medicine a plant is simply eaten raw, cooked or infused in water or native wine or even prepared as food (Conway, 1973).

Buchholzia coriacea E. (Capparidaceae) is a forest tree with large, glossy, leathery leaves and conspicuous cream white flowers in racemes at the end of the branches. The plant is easily recognized by the compound pinnate leaves and the long narrow angular fruits containing large, usually aligned seeds. In Nigeria the plant has various common names including; 'Ovu' (Bini), and 'Aponmu' (Akure). *B. coriacea* is found widely distributed in other African countries such as Ivory Coast and Gabon (Keay et al., 1964; Koudogbo et al., 1972).

The plant's fruit is about 5 inches long and 2 - 3 inches in diameter and resembles avocado pear, yellowish when ripe with a yellow flesh containing a few large, blackish seeds about 1 inch long. They are edible and taste peppery.

It has been used for years to meet a variety of illnesses; since it has been used continually over many generations it is likely that the kola (seed) actually has an effect against illnesses.

The leaves and stem bark of Buchholzia in various for-

*Corresponding author. E-mail: theoiyke@yahoo.com.

mulations, decoctions and concoction exhibit antihelmintic, antimicrobial and cytotoxicity effects on microorganisms (Ajaiyeoba et al., 2001; Ajaiyeoba et al., 2003; Nweze and Asuzu, 2006; Ezekiel and Onyeoziri, 2009). In Ghana fresh bark of the plants were used for earache (Irvine, 1961).

Despite the various reports, information on the antibacterial properties of seeds of the plants on gastrointestinal pathogens is scare. The study was therefore undertaken in order to evaluate the antibacterial activities and phytochemical profile of the crude extracts of seeds of *B. coriacea* on some gastrointestinal bacterial pathogens.

MATERIALS AND METHODS

Plants collection

The seeds of *B. coriacea* were collected from Awka, Nigeria and were authenticated by a Taxonomist at the International Institute of Tropical Agriculture (IITA) Ibadan, Nigeria. The seeds were air-dried for 5 days to constant weight, cut into pieces and grinded into powder using a sterile electric blender. The powder was then used for extraction of bioactive components.

Extraction of plant material

Aqueous (water) and organic (methanol) solvents were used for extraction of the active components of the plant part. For aqueous extraction, hot water extraction method as described by Asuzu (1986) was used. 20 g of each of the grounded seeds were extracted by successive soaking for 2 days using 40 ml of hot distilled water in a 250 ml sterile conical flask. The extracts were filtered using Whatman filter paper and the filtrates concentrated in vacuum at 60 °C. The concentrated filtrate, now the extracts were then stored in universal bottles in the refrigerator at 4 °C prior to use. For organic extraction, 25 g of the powdered plant part was extracted in 250 ml of 95% methanol for 6 h using the soxhlet apparatus as described by Harbone (1993). The volatile oil obtained was concentrated by evaporation using water bath at 100 °C for 1 h.

Preparation of crude extract

Each of the extracts were reconstituted by dilution (methanol crude extract in 50% Dimethylsulphoxide (DMSO) and aqueous extracts in sterile distilled water) to various concentrations of 250, 200, 150, 100 and 50 mg/ml) as described by Akujobi et al. (2004) and used for antibacterial susceptibility testing.

Photochemical screening

This was carried out according to the methods described by Trease and Evans (1989).

Test bacteria

Clinical isolates of *Bacillus cereus, Escherichia coli, Salmonella typhimurium, Staphylococcus aureus* and *Vibrio cholerae* used for this work were collected from the Bacteriology Laboratory Center, University of Nigeria, Teaching Hospital, Enugu. The bacterial isolates were further purified by subculturing each isolate onto fresh plates of Nutrient Agar (NA). The pure isolates were identified using standard biochemical methods (Holt et al., 1994) and then maintained as described by Cruickshank et al. (1980).

Determination of antibacterial susceptibility of extracts

This was carried out using the agar-gel diffusion method as described by Osadebe and Ukwueze (2004). In this method, broth culture of the test isolates (0.1 ml) containing 1×10^5 cells/ml of organism was aseptically inoculated by spreading evenly onto the surface of NA plates using a bent sterile glass rod. Six wells (5.0 mm diameter) were then made in the plates using a sterile cork borer. The fifth and sixth wells served as negative and positive control. The sterile distilled water served as the negative control, ciprofloxacin used as the positive control. The bottom of the wells 1 - 4 was sealed with one drop of the sterile nutrient agar to prevent diffusion of the extract under the agar. Fixed volumes (0.1 ml) of the extracts were transferred into the wells 1 - 4 using a sterile Pasteur pipette. The control wells were filled with 0.1 ml of distilled water and ciprofloxacin. The plates were allowed on the bench for 40 min for pre-diffusion of the extract (Esimone et al., 1998) and then incubated at 37 °C for 24 h. Antibacterial activity of the extracts were determined by measurement of the resulting zone diameters of inhibition (mm) against each test bacteria using a ruler. The experiment was carried out in triplicates and the mean values of the results were taken as antibacterial activity (Abayomi, 1982; Junaid et al., 2006).

Determination of minimum inhibitory concentration (MIC) and minimum bactericidal concentration (MBC)

The MIC and MBC of the potent extracts was determined according to the macro broth dilution technique (Boron and Fingold, 1990). Standardized suspensions of the test organism was inoculated into a series of sterile tubes of nutrient broth containing dilutions (250, 200, 150, 100 and 50 mg/ml) of leaf extracts and incubated at 37 °C for 24 h. The MICs were read as the least concentration that inhibited any visible growth (absence of turbidity) of the test organisms. For MBC determination, a loopful of broth from each of the tubes that did not show any visible growth (no turbidity) during MIC determination was subcultured onto extract fresh free NA plates, and further incubated for 24 h at 37 °C. The least concentration, at which no visible growth was observed, was noted as the MBC.

RESULTS AND DISCUSSION

Results of preliminary phytochemical screening of the seed extracts of *B. coriacea* are shown in Table 1. Results showed the presence of alkaloids, anthraquinones, carbohydrates, cardiac glycosides, flavonoids, glycosides, resins, saponin, steroidal rings, steroidal terpenes and tannin. The presence of phytochemicals in the seed extracts (Table 1) showed that the extracts possess antibacterial properties. These results are in agreement with similar study by Ajaiyeoba et al. (2003).

Table 2 shows the results of antibacterial effects of seed extracts of the plant against the test bacteria. Results showed that the activity of the extracts against the test bacteria decreased with decrease in the concentration with the methanol extracts demonstrating higher activity (35 mm, 250 mg/ml,) than the hot water extracts (3 mm, 50 mg/ml). This could be because the

Table 1. Phytochemical analysis of dried seed extracts of *Buchholzia coriacae*.

Phytochemicals	Hot water seed extract	Methanol seed extract
Alkaloids	+	++
Anthraquinone	+	+
Carbohydrates	++	++
Cardiac glycosides	+	+
Flavonoids	+	++
Glycosides	++	++
Resins	++	++
Saponin	+	++
Steroidal ring	+	+
Steroidal terpenes	++	++
Tannin	+	++

+ = Present.

Table 2. Antibacterial activity of dried seed extracts of *Buchholzia coriacea* on test Isolates.

Isolates	Mean zone diameter of inhibition (mm)						Extracts	
Staph. aureus	30	25	18	16	12	17	0	METH
	22	18	18	10	8	15	0	HH$_2$OD
S. typhimurium	25	20	16	12	10	20	0	METH
	18	17	10	6	5	18	0	HH$_2$OD
B. cereus	35	25	20	17	12	16	0	METH
	24	18	16	11	10	15	0	HH$_2$OD
V. cholerae	24	19	15	11	7	17	0	METH
	8	13	9	7	3	15	0	HH$_2$OD
E. coli	18	16	12	10	7	21	0	METH
	13	11	8	6	3	18	0	HH$_2$OD
Conc. of extracts (mg/ml)	250	200	150	100	50	+ve control	-ve control	

HH$_2$OD = Hot water (Dried seed), METD = Methanol (Dried seed).

active component must be a highly poplar compound. It has been observed that the more polar the solvent the higher the yield of extraction (Chang et al., 1977), although the inhibitory effects of aqueous extract of medicinal plants has been reported (Tignokpa et al., 1986; Olayinka et al., 1992; Omer et al., 1998).

Figure 1 and Table 3 showed the minimum inhibitory concentration (MIC) and minimum bactericidal concentration (MBC) of the extracts on the test isolates respectively. Results showed that the values obtained are quite higher for the aqueous (hot) extracts than that of methanol extracts suggesting that extraction with methanol could produce better active antimicrobial phytochemicals which are contained in the seed. The presence of phytochemicals in the seed extracts (Table 1) showed

that the extracts possess antibacterial properties. These results are in agreement with similar study by Ajaiyeoba et al. (2001). The observed antibacterial effects of the seeds on the bacterial isolates though *in-vitro* is an indication that the seed extracts could be effective in the management of infections cause by these organisms (Tables 1 and 2).

Conclusion

Results of the study showed that seed extracts of *B. coriacea* possessed phytochemical substances that can be used as components of new antimicrobial agents. Therefore there is need for further investigations in terms

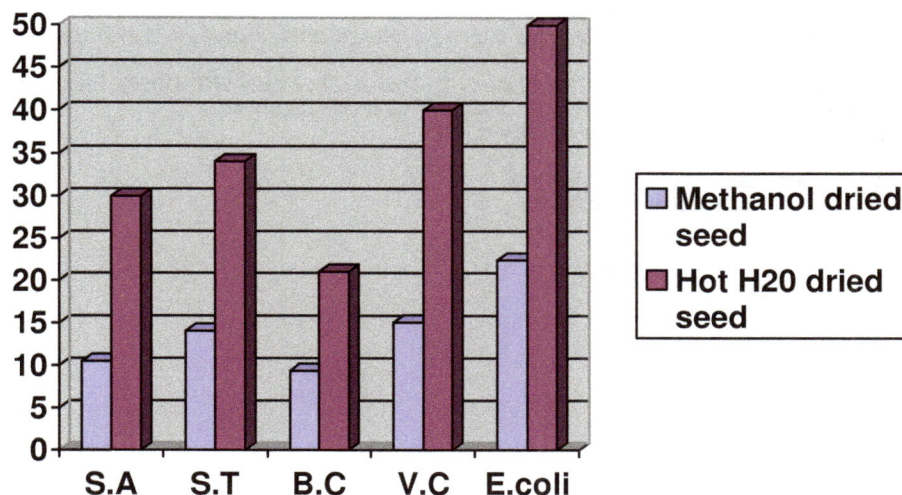

Figure 1. Minimum Inhibitory Concentration of seed extracts of *Buchholzia coriacea*. HH₂OD= Hot water (Dried seed), METH= Methanol (Dried seed), S.A = *Staphylococcus aureus* S.T= *Salmonella typhimurium*, B.C = *Bacillus cereus*, V.C =. *Vibrio cholerae*, E. coil = *Escherichia coli*

Table 3. Minimum bactericidal concentration (MBC) of seed extracts of *Buchholzia coriacea*.

Isolates	Extracts	MBC (mg/ml)
S. aureus	METD	4.1
	HH₂OD	9.3
S. typhimurium	METD	5.2
	HH₂OD	11.2
B. cereus	METD	3.9
	HH₂OD	8.1
V. cholerae	METD	6.1
	HH₂OD	12.4
E. coli	METD	10.5
	HH₂0D	17.4

HH₂OD= Hot water (Dried seed), METH= methanol (Dried seed).

of toxicological studies and purification of active components with the view to using the plant in novel drug development. The study has also justified the traditional usage of this plant as health remedy.

REFERENCES

Abayomi S (1982). The State of Medicinal Plants Research in Nigeria. University of Ife Press p. 200

Ajaiyeoba EO, Onocha PA, Olarenwaju OT (2001). *In-vitro* antihelmintic properties of *Buchholzia coriacea* and *Gynandropsis gynandra* extracts. Pharm. Biol. 39 (3): 217-22.

Ajaiyeoba EO, Onocha PA, Nwozo SO, Sama W (2003). Antimicrobial and cytotoxicity evaluation of *Buchholzia coriacea* stem bark. Fitoterapia 74 (7-8): 706-709.

Akujobi C, Anyanwu BN, Onyeze C, Ibekwe VI (2004). Antibacterial ac-

tivities and preliminary phytochemical screening of four medicinal plants. J. Appl. Sci. 7 (3): 4328-4338

Andrews JA (1982). Bibliography on Herbs, Herbal Medicine, Natural Foods and Unconventional Medical Treatment, Libraries Unlimited Inc USA.

Asuzu IU (1986). Pharmacological evaluation of folklore of *Sphenostylis slenocarpa*. J. Ethanopharmacol 16: 236-267

Boron JE, Fingold SM (1990). Method for testing antimicrobial effectiveness. In: Bailey Scotts Diagnostic Microbiology Mosby, CV (8th edition), Missouri.

Chang SS, Ostric-Matis JB, Hsieh OA, Hung CL (1977). Natural antioxidants from rosemary and sage. J. Food Sci. 42: 1102-1106

Conway D (1973). The magic of herbs. Jonathan cape, London.

Cruickshank R, Duguid JP, Marmion BP, Swain RHP (1980). Medical Microbiology. 12th edition, Church Living, Edinburgh.

Esimone CO, Adikwu MU, Okonta JM (1998). Preliminary antimicrobial screening of the ethanolic extract from the lichen *Usnea subfloridans* L. J. Pharm. Res. Dev. 3(2): 99-101.

Ezekiel OO, Onyeoziri NF (2009). Preliminary studies on the antimicrobial properties of Buchholzia coriacea (wonderful kola). Afr. J. Biotechnol. 8 (3): 472-474

Harbone NV (1993). Phytochemical method. A guide to modern technique of plant analysis, 2nd edition, Fakenhamn Press Ltd London.

Keay RWJ, Onochie CFA, Standfield DP (1964). Nigerian trees Federal Department of Forest Research, Ibadan, Nigeria Vol. 1.

Holt JG, Krieg NR, Sneath PHA, Stanley JT, Williams ST (1994). Bergeys Manual of Determinative Bacteriology 9th edition. The Williams and Wilkins Company, Baltimore USA.

Irvine FR (1961). Woody plant of Ghana with special reference to their uses. London Oxford University Press,

Junaid SA, Olabode AO, Onwuliri FC, Okowsi AEJ, Agina SE (2006). The antimicrobial properties of *Ocimum gratisssimum* extracts on some elected bacterial gastrointestinal isolate. Afr. J. Biotechnol. 5 (22): 2315-2321

Koudogbo B, Delaveau P, Adjanohoun E (1972). Study of an African Cepparidaceae, *Buchholzia coriacea* Engler. Ann Pharm Fr. 30 (2): 93-98.

Nweze NE, Asuzu IU (2006). The antihelmintic effect of Buchholzia coriacea seed. Niger. Vet. J. 27 (2): 60-65

Olayinka DO, Onoruiwe O, Lot TY (1992). Cardiovascular effects in rodents of methanolic extract of *Khaya senegalensis* Phytother. Res. 6: 282-284

Omer MEA, Aimagboul AZ, Egami AA (1998). Sudanese plants used in folkloric medicine screening for antimicrobial activity. Fitoterapia 69: 542-545

Osadebe PO, Ukwueze SE (2004). Comparative study of the phyto-chemical and antimicrobial properties of the Eastern Nigerian species of African Mistletoe (*Loranthus micranthus*) sourced from different host trees. J. Biol. Res. Biotech 2 (1): 18-23

Sofowora EA (1984). Medicinal Plants and Traditional Medicine in Africa. 4th edition, John Wiley and sons Ltd, Chichester pp.96-105.

Tignokpa M, Laurens A, Mboup S, Sylla O (1986). Popular Medicinal Plants of the markets in Darkar (Senegal). Inter. J. Crude Drug Res. 24: 75-80

Trease GE, Evans WC. (1989). Pharmacognosy 13th edition. Bailliere. Tindall Ltd London.

Permissions

All chapters in this book were first published in JDBTE, by Academic Journals; hereby published with permission under the Creative Commons Attribution License or equivalent. Every chapter published in this book has been scrutinized by our experts. Their significance has been extensively debated. The topics covered herein carry significant findings which will fuel the growth of the discipline. They may even be implemented as practical applications or may be referred to as a beginning point for another development.

The contributors of this book come from diverse backgrounds, making this book a truly international effort. This book will bring forth new frontiers with its revolutionizing research information and detailed analysis of the nascent developments around the world.

We would like to thank all the contributing authors for lending their expertise to make the book truly unique. They have played a crucial role in the development of this book. Without their invaluable contributions this book wouldn't have been possible. They have made vital efforts to compile up to date information on the varied aspects of this subject to make this book a valuable addition to the collection of many professionals and students.

This book was conceptualized with the vision of imparting up-to-date information and advanced data in this field. To ensure the same, a matchless editorial board was set up. Every individual on the board went through rigorous rounds of assessment to prove their worth. After which they invested a large part of their time researching and compiling the most relevant data for our readers.

The editorial board has been involved in producing this book since its inception. They have spent rigorous hours researching and exploring the diverse topics which have resulted in the successful publishing of this book. They have passed on their knowledge of decades through this book. To expedite this challenging task, the publisher supported the team at every step. A small team of assistant editors was also appointed to further simplify the editing procedure and attain best results for the readers.

Apart from the editorial board, the designing team has also invested a significant amount of their time in understanding the subject and creating the most relevant covers. They scrutinized every image to scout for the most suitable representation of the subject and create an appropriate cover for the book.

The publishing team has been an ardent support to the editorial, designing and production team. Their endless efforts to recruit the best for this project, has resulted in the accomplishment of this book. They are a veteran in the field of academics and their pool of knowledge is as vast as their experience in printing. Their expertise and guidance has proved useful at every step. Their uncompromising quality standards have made this book an exceptional effort. Their encouragement from time to time has been an inspiration for everyone.

The publisher and the editorial board hope that this book will prove to be a valuable piece of knowledge for researchers, students, practitioners and scholars across the globe.

List of Contributors

Bui Dinh Thach
Institute of Tropical Biology, Vietnam Academy of Science and Technology, Vietnam

Lenguyen Tu Linh
Institute of Tropical Biology, Vietnam Academy of Science and Technology, Vietnam

Nguyen ThiThuy Van
Institute of Tropical Biology, Vietnam Academy of Science and Technology, Vietnam

Trinh Thi Ben
Institute of Tropical Biology, Vietnam Academy of Science and Technology, Vietnam

Mai Truong
Institute of Tropical Biology, Vietnam Academy of Science and Technology, Vietnam

Nguyen Huu Ho
Institute of Tropical Biology, Vietnam Academy of Science and Technology, Vietnam

Anju Kurup
Division of Developmental Biology, Department of Zoology, Faculty of Science, The M S University of Baroda, Vadodara, Gujarat-390002, India

A.V. Ramachandran
Division of Developmental Biology, Department of Zoology, Faculty of Science, The M S University of Baroda, Vadodara, Gujarat-390002, India

Anant Prakash
Inorganic Bioinorganic Research Laboratory, Department of Chemistry, Government Post Graduate College, Ranikhet-263645 (Uttarakhand) India

Mukesh Pal Gangwar
Department of chemistry, Mahatama Jyotiba Phule Rohilkhand University, Bareilly-243001(Uttar Pradesh) India

K. K. Singh
Shri Ram Murti Smarak College of Engineering and Technology (S. R. M. S. CET) Bareilly-243001 (Uttar Pradesh) India

Nai Hao Ye
Yellow Sea Fisheries Research Institute, Chinese Academy of Fishery Sciences, Qingdao, 266071, China

Hong Xia Wang
Key Laboratory of Experimental Marine Biology, Institute of Oceanology, Chinese Academy of Sciences, Qingdao 266071, China

Zheng Quan Gao
School of Life Sciences, Shandong University of Technology, Zibo, 255049, China

Guangce Wang
Key Laboratory of Experimental Marine Biology, Institute of Oceanology, Chinese Academy of Sciences, Qingdao 266071, China

Christiane Heinemann
Max Bergmann Center of Biomaterials and Institute of Materials Science, Technische Universität Dresden, Budapester Str. 27, D-01069 Dresden, Germany

Sascha Heinemann
Max Bergmann Center of Biomaterials and Institute of Materials Science, Technische Universität Dresden, Budapester Str. 27, D-01069 Dresden, Germany

Corina Vater
Department of Orthopedic Surgery, University Hospital Carl Gustav Carus, Fetscherstr 74, D-01307 Dresden, Germany

Hartmut Worch
Max Bergmann Center of Biomaterials and Institute of Materials Science, Technische Universität Dresden, Budapester Str. 27, D-01069 Dresden, Germany

Thomas Hanke
Max Bergmann Center of Biomaterials and Institute of Materials Science, Technische Universität Dresden, Budapester Str. 27, D-01069 Dresden, Germany

Waguih Mohamed Abouzeid
Research Department, Alexandria Dental Research Center, Alexandria University, Egypt

Samiha Ahmed Mokhtar
High Institute of Public Health, Alexandria University, Egypt

Nehad Hassan Mahdy
High Institute of Public Health, Alexandria University, Egypt

Mohamed Sherif Ahmed
Faculty of Medicine, Alexandria University, Egypt

Fayek Salah El Kwsky
Medical Research Institute, Alexandria University, Egypt

Feng Li
Yellow Sea Fisheries Research Institute, Chinese Academy of Fishery Sciences, Qingdao 266071, China

Shoutuan Yu
Hongdao Street Office of QingdaoChengyang District, China

Yuze Mao
Yellow Sea Fisheries Research Institute, Chinese Academy of Fishery Sciences, Qingdao 266071, China

Naihao Ye
Yellow Sea Fisheries Research Institute, Chinese Academy of Fishery Sciences, Qingdao 266071, China

M. Guru Prasad
Institute of Frontier Technology, Regional Agricultural Research Station, Acharya N G Ranga Agricultural University, Tirupati,-517 502, A. P., India

P. Sudhakar
Institute of Frontier Technology, Regional Agricultural Research Station, Acharya N G Ranga Agricultural University, Tirupati,-517 502, A. P., India

T. N. V. K. V. Prasad
Institute of Frontier Technology, Regional Agricultural Research Station, Acharya N G Ranga Agricultural University, Tirupati,-517 502, A. P., India

M. Guru Prasad
Institute of Frontier Technology, Regional Agricultural Research Station, Acharya N G Ranga Agricultural University, Tirupati-517 502, A. P. India

T. N. V. K. V. Prasad
Institute of Frontier Technology, Regional Agricultural Research Station, Acharya N G Ranga Agricultural University, Tirupati-517 502, A. P. India

P. Sudhakar
Institute of Frontier Technology, Regional Agricultural Research Station, Acharya N G Ranga Agricultural University, Tirupati-517 502, A. P. India

Mohammed Shahril Azwan
School of Mechanical Engineering, USM, Engineering Campus, 14300 Nibong Tebal, Pulau Pinang, Malaysia

Inzarulfaisham Abd Rahim
School of Mechanical Engineering, USM, Engineering Campus, 14300 Nibong Tebal, Pulau Pinang, Malaysia

A. Rajasekaran
Department of Pharmaceutical Chemistry, KMCH College of Pharmacy, Coimbatore, Tamilnadu, India

M. Periasamy
Department of Pharmaceutical Chemistry, KMCH College of Pharmacy, Coimbatore, Tamilnadu, India

S. Venkatesan
Department of Pharmaceutical Chemistry, KMCH College of Pharmacy, Coimbatore, Tamilnadu, India

Mahender Singh Rathore
Biotechnology Unit, Department of Botany, Jai Narain Vyas University, Jodhpur, Rajasthan-342033 India

Narpat Singh Shekhawat
Biotechnology Unit, Department of Botany, Jai Narain Vyas University, Jodhpur, Rajasthan-342033 India

Z. Jamalpoor
Trauma Research Center, Baqiyatallah University of Medical Sciences, Tehran, Iran
Department of Physiology, Baqiyatallah University of Medical Sciences, Tehran, Iran

M. Ebrahimi
Chemical Injury Research Center, Baqiyatallah University of Medical Sciences, Tehran, Iran
Department of Organ Anatomy, Yamaguchi University Graduate School of Medicine, Ube, Japan

N. Amirizadeh
Iranian Blood Transfusion Organization, Tehran, Iran

K. Mansoori
Department of Physical Medicine and Rehabilitation, Shafayahyaian Rehabilitation Hospital, Iran University of Medical Sciences, Tehran, Iran

A. Asgari
Department of Physiology, Baqiyatallah University of Medical Sciences, Tehran, Iran

M. R. Nourani
Trauma Research Center, Baqiyatallah University of Medical Sciences, Tehran, Iran
Chemical Injury Research Center, Baqiyatallah University of Medical Sciences, Tehran, Iran

Frank Winterroth
Department of Biomedical Engineering, University of Michigan, Ann Arbor, MI, USA

Junho Lee
School of Dentistry, University of Michigan, Ann Arbor, MI, USA

Shiuhyang Kuo
Department of Oral and Maxillofacial Surgery, University of Michigan, Ann Arbor, MI, USA

J. Brian Fowlkes
Department of Biomedical Engineering, University of Michigan, Ann Arbor, MI, USA
Department of Radiology, University of Michigan, Ann Arbor, MI, USA

Stephen E. Feinberg
Department of Biomedical Engineering, University of Michigan, Ann Arbor, MI, USA
Department of Oral and Maxillofacial Surgery, University of Michigan, Ann Arbor, MI, USA

Scott J. Hollister
Department of Biomedical Engineering, University of Michigan, Ann Arbor, MI, USA

Kyle W. Hollman
Department of Biomedical Engineering, University of Michigan, Ann Arbor, MI, USA
Sound Sight Research, Livonia, MI, USA

Anil Dhingra
Department of Conservative Dentistry and Endodontics, Subharti Dental College, Meerut (U.P.), Meerut, Uttar Pradesh India

Viresh Chopra
Department of Conservative Dentistry and Endodontics, Subharti Dental College, Meerut (U.P.), Meerut, Uttar Pradesh India

Shalya Raj
Department of Conservative Dentistry and Endodontics, Subharti Dental College, Meerut (U.P.), Meerut, Uttar Pradesh India

Rehab H. Abd-Allah
Department of Pharmacognosy, Faculty of Pharmacy, University of Zagazig, Zagazig 44519, Egypt

Ehsan M. Abo Zeid
Department of Pharmacognosy, Faculty of Pharmacy, University of Zagazig, Zagazig 44519, Egypt

Mahmoud M. Zakaria
Department of Pharmacognosy, Faculty of Pharmacy, University of Zagazig, Zagazig 44519, Egypt

Samih I. Eldahmy
Department of Pharmacognosy, Faculty of Pharmacy, University of Zagazig, Zagazig 44519, Egypt

Salah A. Latief Mohammed
Textile Engineering Department, College of Engineering, Sudan University of Science and Technology, Khartoum North, Sudan

Denis E. Solomon
96 Standishgate, Wigan WN1 1XA, England

Omotoso Olumuyiwa Temitope
Department of Zoology, Faculty of Science, Ekiti State University, P.M.B. 5363, Ado-Ekiti, Ekiti, Nigeria

Iskra Ventseslavova Sainova
Department of Experimental Morphology, Institute of Experimental Morphology, Pathology and Anthropology with Museum, Bulgarian Academy of Sciences, "Acad. G. Bonchev" Str., 1113 Sofia, Bulgaria

Ilina Vavrek
Department of Experimental Morphology, Institute of Experimental Morphology, Pathology and Anthropology with Museum, Bulgarian Academy of Sciences, "Acad. G. Bonchev" Str., 1113 Sofia, Bulgaria

Velichka Pavlova
Department of Experimental Morphology, Institute of Experimental Morphology, Pathology and Anthropology with Museum, Bulgarian Academy of Sciences, "Acad. G. Bonchev" Str., 1113 Sofia, Bulgaria

Teodora Daneva
Institute of Biology and Immunology of Reproduction, Bulgarian Academy of Sciences, Sofia, Bulgaria

Ivan Iliev
Department of Experimental Morphology, Institute of Experimental Morphology, Pathology and Anthropology with Museum, Bulgarian Academy of Sciences, "Acad. G. Bonchev" Str., 1113 Sofia, Bulgaria

Lilija Yossifova
Department of Experimental Morphology, Institute of Experimental Morphology, Pathology and Anthropology with Museum, Bulgarian Academy of Sciences, "Acad. G. Bonchev" Str., 1113 Sofia, Bulgaria

Elena Gardeva
Department of Experimental Morphology, Institute of Experimental Morphology, Pathology and Anthropology with Museum, Bulgarian Academy of Sciences, "Acad. G. Bonchev" Str., 1113 Sofia, Bulgaria

Elena Nikolova
Department of Experimental Morphology, Institute of Experimental Morphology, Pathology and Anthropology with Museum, Bulgarian Academy of Sciences, "Acad. G. Bonchev" Str., 1113 Sofia, Bulgaria

N. Anand Laxmi
DCP, NDRI, Karnal, India

G. Gunjan
DCP, NDRI, Karnal, India

M. Prateesh
IVRI, Barrielly, India

Jörg Neunzehn
Max Bergmann Center of Biomaterials and Institute of Material Science, Technische Universität Dresden, Budapester Str. 27, D-01069 Dresden, Germany

Sascha Heinemann
Max Bergmann Center of Biomaterials and Institute of Material Science, Technische Universität Dresden, Budapester Str. 27, D-01069 Dresden, Germany

Hans Peter Wiesmann
Max Bergmann Center of Biomaterials and Institute of Material Science, Technische Universität Dresden, Budapester Str. 27, D-01069 Dresden, Germany

Maryam Moezizadeh
Department of Operative Dentistry, School of Dentistry, Shahid Beheshti University of Medical Sciences, Tehran, Iran

Melanie Köllmer
Department of Biopharmaceutical Sciences, University of Illinois, Chicago, IL 60612-7231, USA

Jason S. Buhrman
Department of Biopharmaceutical Sciences, University of Illinois, Chicago, IL 60612-7231, USA

Yu Zhang
Department of Biopharmaceutical Sciences, University of Illinois, Chicago, IL 60612-7231, USA

Richard A. Gemeinhart
Department of Biopharmaceutical Sciences, University of Illinois, Chicago, IL 60612-7231, USA
Department of Bioengineering, University of Illinois, Chicago, IL 60607-7052, USA
Department of Ophthalmology and Visual Sciences, University of Illinois, Chicago, IL 60612-4319, USA

Priscilla C. Aveline
Inserm Unit U658, Hospital Porte Madeleine, Orleans, France

Julie I. Bourseguin
Inserm Unit U658, Hospital Porte Madeleine, Orleans, France

Gaël Y. Rochefort
Inserm Unit U658, Hospital Porte Madeleine, Orleans, France

Silin Sa
Graduate Group in Biological Engineering and Small-scale Technologies, University of California, Merced, USA

Diep T. Nguyen
Department of Biomedical Engineering, University of California, Irvine, USA

Jonathan D. Pegan
School of Engineering, University of California, Merced, USA

Michelle Khine
Department of Biomedical Engineering, University of California, Irvine, USA

Kara E. McCloskey
Graduate Group in Biological Engineering and Small-scale Technologies, University of California, Merced, USA
School of Engineering, University of California, Merced, USA

Esmaeilzadeh Mahdi
Basic Sciences Department, Nikshahr Branch, Islamic Azad University, Nikshahr, Iran

Nejadali Abolfazl
Basic Sciences Department, Nikshahr Branch, Islamic Azad University, Nikshahr, Iran

Kazemzadeh Fariba
Basic Sciences Department, Nikshahr Branch, Islamic Azad University, Nikshahr, Iran

Borhani Mohammad
Basic Sciences Department, Nikshahr Branch, Islamic Azad University, Nikshahr, Iran

Esmaeilzadeh Mahdi
Department of Basic Sciences, Bojnourd Branch, Islamic Azad University, Bojnourd, Iran

Kazemzadeh Fariba
Department of Basic Sciences, Bojnourd Branch, Islamic Azad University, Bojnourd, Iran

Adibfar Nader
Department of Basic Sciences, Bojnourd Branch, Islamic Azad University, Bojnourd, Iran

Feng Li
Yellow Sea Fisheries Research Institute, Chinese Academy of Fishery Sciences, Qingdao 266071, China

Shoutuan Yu
Hongdao Street Office of QingdaoChengyang District, China

Yuze Mao
Yellow Sea Fisheries Research Institute, Chinese Academy of Fishery Sciences, Qingdao 266071, China

Naihao Ye
Yellow Sea Fisheries Research Institute, Chinese Academy of Fishery Sciences, Qingdao 266071, China

T. I. Mbata
Department of Microbiology, Federal Polytechnic Nekede, Owerri, Nigeria

C. M. Duru
Department of Biotechnology, Federal University of Technology, Owerri, Nigeria

H. A. Onwumelu
Department of Chemistry, Nnamdi Azikiwe University, Awka, Nigeria